COMPUTATIONAL METHODS FOR NEXT GENERATION SEQUENCING DATA ANALYSIS

Wiley Series on

Bioinformatics: Computational Techniques and Engineering

A complete list of the titles in this series appears at the end of this volume.

COMPUTATIONAL METHODS FOR NEXT GENERATION SEQUENCING DATA ANALYSIS

Edited by

ION I. MĂNDOIU
ALEXANDER ZELIKOVSKY

Published by John Wiley & Sons, Inc., Hoboken, New Jersey
Published simultaneously in Canada

For general information on our other products and services or for technical support, please contact our Customer Care Department within the United States at (800) 762-2974, outside the United States at (317) 572-3993 or fax (317) 572-4002.

Wiley also publishes its books in a variety of electronic formats. Some content that appears in print may not be available in electronic formats. For more information about Wiley products, visit our web site at www.wiley.com.

Library of Congress Cataloging-in-Publication Data:

Names: Măndoiu, I. Ion, editor of compilation. | Zelikovsky, Alexander, editor
  of compilation.
Title: Computational methods for next generation sequencing data analysis /
  edited by Ion I. Măndoiu, Alexander Zelikovsky.
Description: Hoboken, New Jersey : John Wiley & Sons, 2016. | Includes
  bibliographical references and index.
Identifiers: LCCN 2016010861 (print) | LCCN 2016014704 (ebook) | ISBN
  9781118169483 (cloth) | ISBN 9781119272168 (pdf) | ISBN 9781119272175
  (epub)
Subjects: LCSH: Nucleotide sequence–Methodology. | Nucleotide sequence–Data
  processing.
Classification: LCC QP620 .C648 2016 (print) | LCC QP620 (ebook) | DDC
  611/.0181663–dc23
LC record available at http://lccn.loc.gov/2016010861

Cover image courtesy of Gettyimages/Andrew Brookes

Printed in the United States of America

10 9 8 7 6 5 4 3 2 1

CONTENTS IN BRIEF

CONTENTS

CONTRIBUTORS

Vanessa Aguiar-Pulido, Bioinformatics Research Group (BioRG), School of Computing and Information Sciences, Florida International University, Miami, FL, USA

Sahar Al Seesi, Department of Computer Science and Engineering, University of Connecticut, Storrs, CT, USA

Alexander Artyomenko, Department of Computer Science, Georgia State University, Atlanta, GA, USA

Niko Beerenwinkel, Department of Biosystems Science and Engineering, ETH Zurich, Basel, Switzerland

Adrian Caciula, Department of Computer Science, Georgia State University, Atlanta, GA, USA

David S. Campo, Division of Viral Hepatitis, Centers of Disease Control and Prevention, Atlanta, GA, USA

Michael Campos, Miller School of Medicine, University of Miami, Miami, FL, USA

Stefan Canzar, Center for Computational Biology, McKusick-Nathans Institute of Genetic Medicine, Johns Hopkins University School of Medicine, Baltimore, MD, and Toyota Technological Institute at Chicago, Chicago, IL, USA

Jeong-Hyeon Choi, Cancer Center, Medical College of Georgia, Georgia Regents University, Augusta, GA, USA; Department of Biostatistics and Epidemiology, Medical College of Georgia, Georgia Regents University, Augusta, GA, USA

Chong Chu, Department of Computer Science and Engineering, University of Connecticut, Storrs, CT, USA

Zoya Dimitrova, Division of Viral Hepatitis, Centers of Disease Control and Prevention, Atlanta, GA, USA

Jorge Duitama, Agrobiodiversity Research Area, International Center for Tropical Agriculture (CIAT), Cali, Colombia

Eleazar Eskin, Department of Computer Science, University of California, Los Angeles, CA, USA

Mitch Fernandez, Bioinformatics Research Group (BioRG), School of Computing and Information Sciences, Florida International University, Miami, FL, USA

Liliana Florea, Center for Computational Biology, McKusick-Nathans Institute of Genetic Medicine, Johns Hopkins University School of Medicine, Baltimore, MD, USA

Olga Glebova, Department of Computer Science, Georgia State University, Atlanta, GA, USA

Xuan Guo, Department of Computer Science, Department of Biology, Georgia State University, Atlanta, GA, USA

Steven J. Hallam, Graduate Program in Bioinformatics and Department of Microbiology and Immunology, University of British Columbia, Vancouver, BC, Canada

Niels W. Hanson, Graduate Program in Bioinformatics, University of British Columbia, Vancouver, BC, Canada

Elena Harris, Department of Computer Science, California State University, Chico, CA

Wenrui Huang, Bioinformatics Research Group (BioRG), School of Computing and Information Sciences, Florida International University, Miami, FL, USA

Mazhar I. Khan, Department of Pathobiology and Veterinary Science, University of Connecticut, Storrs, CT, USA

Yury Khudyakov, Division of Viral Hepatitis, Centers of Disease Control and Prevention, Atlanta, GA, USA

Kishori M. Konwar, Department of Microbiology and Immunology, University of British Columbia, Vancouver, BC, Canada

Bing Li, Department of Computer Science, Department of Biology, Georgia State University, Atlanta, GA, USA

James Lindsay, Department of Computer Science and Engineering, University of Connecticut, Storrs, CT, USA

Rasiah Loganantharaj, Bioinformatics Research Lab, The Center for Advanced Computer Studies, University of Louisiana, Lafayette, LA, USA

Stefano Lonardi, Department of Computer Science and Engineering, University of California, Riverside, CA, USA

Nicholas Mancuso, Department of Computer Science, Georgia State University, Atlanta, GA, USA

Ion I. Măndoiu, Department of Computer Science and Engineering, University of Connecticut, Storrs, CT, USA

Igor Mandric, Department of Computer Science, Georgia State University, Atlanta, GA, USA

Serghei Mangul, Department of Computer Science, University of California, Los Angeles, CA, USA

Tobias Marschall, Centrum Wiskunde & Informatica, Amsterdam, Netherlands

Kalai Mathee, Herbert Wertheim College of Medicine, Florida International University, Miami, FL, USA

Giri Narasimhan, Bioinformatics Research Group (BioRG), School of Computing and Information Sciences, Florida International University, Miami, FL, USA

Ekaterina Nenastyeva, Department of Computer Science, Georgia State University, Atlanta, GA, USA

Rachel O'neill, Department of Molecular and Cell Biology, University of Connecticut, Storrs, CT, USA

Yi Pan, Department of Computer Science, Department of Biology, Georgia State University, Atlanta, GA, USA

Sumathi Ramachandran, Division of Viral Hepatitis, Centers of Disease Control and Prevention, Atlanta, GA, USA

Thomas A. Randall, Integrative Bioinformatics, National Institute of Environmental Health Sciences, Research Triangle Park, NC, USA

Juan Riveros, Bioinformatics Research Group (BioRG), School of Computing and Information Sciences, Florida International University, Miami, FL, USA

Alexander Schönhuth, Centrum Wiskunde & Informatica, Amsterdam, Netherlands

Jonathan Segal, Herbert Wertheim College of Medicine, Florida International University, Miami, FL, USA

Huidong Shi, Cancer Center, Medical College of Georgia, Georgia Regents University, Augusta, GA, USA Department of Biochemistry, Medical College of Georgia, Georgia Regents University, Augusta, GA, USA

Pavel Skums, Division of Viral Hepatitis, Centers of Disease Control and Prevention, Atlanta, GA, USA

Ren Sun, Department of Molecular and Medical Pharmacology, University of California, Los Angeles, CA, USA

Sing-hoi Sze, Department of Computer Science and Engineering and Department of Biochemistry and Biophysics, Texas A&M University, College Station, TX, USA

Yvette Temate-tiagueu, Department of Computer Science, Georgia State University, Atlanta, GA, USA

Armin Töpfer, Department of Biosystems Science and Engineering, ETH Zurich, Basel, Switzerland

Bassam Tork, Department of Computer Science, Georgia State University, Atlanta, GA, USA

Nicholas C. Wu, Department of Integrative Structural and Computational Biology, The Scripps Research Institute, La Jolla, CA, USA

Shang-ju Wu, Department of Computer Science, University of British Columbia, Vancouver, BC, Canada

Yufeng Wu, Department of Computer Science and Engineering, University of Connecticut, Storrs, CT, USA

Ning Yu, Department of Computer Science, Department of Biology, Georgia State University, Atlanta, GA, USA

Alexander Zelikovsky, Department of Computer Science, Georgia State University, Atlanta, GA, USA

Erliang Zeng, Department of Computer Science and Engineering, University of Notre Dame, Notre Dame, IN, USA

Jin Zhang, McDonnell Genome Institute, Washington University in St. Luis, MO, USA

PREFACE

Massively parallel DNA sequencing and RNA sequencing have become widely available, reducing the cost by several orders of magnitude and placing the capacity to generate gigabases to terabases of sequence data into the hands of individual investigators. These so-called *next-generation sequencing (NGS)* technologies have dramatically accelerated biological and biomedical research by enabling the comprehensive analysis of genomes and transcriptomes to become inexpensive, routine, and widespread. The ensuing explosion in the volume of data has spurred numerous advances in computational methods for NGS data analysis.

This book aims to provide an in-depth survey of some of the most important recent developments in this area. It is neither intended as an introductory text nor as a comprehensive review of existing bioinformatics tools and active research areas in NGS data analysis. Rather, our intention is to make a carefully selected set of advanced computational techniques accessible to a broad readership, including graduate students in bioinformatics and related areas and biomedical professionals who want to expand their repertoire of computational techniques for NGS data analysis. We hope that our emphasis on in-depth presentation of both algorithms and software for computational data analysis of current high-throughput sequencing technologies will best prepare the readers for developing their own algorithmic techniques and for successfully implementing them in existing and novel NGS applications.

The book features 18 chapters authored by bioinformatics experts who are active contributors to the respective subjects. The chapters are intended to be largely independent, so that readers do not have to read every chapter nor have to read them in a particular order. The chapters are grouped into the following four parts:

- Part I focuses on computing and experimental infrastructure for NGS data analysis, including chapters on cloud computing, a modular pipeline for metabolic pathway reconstruction, pooling strategies for massive viral sequencing, and high-fidelity sequencing protocols.

- Part II concentrates on analyses of DNA sequencing data and includes chapters on the classic scaffolding problem, detection of genomic variants, two chapters on finding insertions and deletions, and two chapters on the analysis of DNA methylation sequencing data.
- Part III is devoted to analyses of RNA-seq data. Two chapters describe algorithms and compare software tools for transcriptome assembly: one chapter focuses on methods for alternative splicing analysis and the other chapter focuses on tools for transcriptome quantification and differential expression analysis.
- Part IV explores computational tools for NGS applications in microbiomics. The first chapter concentrates on error correction of NGS reads from viral populations, then two chapters describe methods for viral quasispecies reconstruction, and the last chapter surveys the state of the art and future trends in microbiome analysis.

We are grateful to all the authors for their excellent contributions, without which this book would not have been possible. We hope that their deep insights and fresh enthusiasm will help in attracting new generations of researchers to this dynamic field. We would also like to thank Yi Pan and Albert Y. Zomaya for nurturing this project since its inception, and the editorial staff at Wiley Interscience for their patience and assistance throughout the project. Finally, we wish to thank our friends and families for their continuous support.

ION I. MĂNDOIU

Storrs, Connecticut

ALEXANDER ZELIKOVSKY

Atlanta, Georgia

ABOUT THE COMPANION WEBSITE

This book is accompanied by a companion website:

www.wiley.com/go/Mandoiu/NextGenerationSequencing

The book companion website contains the color version of a few selected figures

Figure 2.3, Figure 2.5, Figure 2.6, Figure 2.13, Figure 3.1, Figure 3.9,
Figure 7.5, Figure 8.3, Figure 8.4, Figure 9.4, Figure 9.8, Figure 9.9,
Figure 9.12, Figure 9.14, Figure 12.3, Figure 12.4, Figure 12.5, Figure 15.3,
Figure 16.1, Figure 16.6, Figure 16.7, Figure 16.11, Figure 16.12, Figure 16.13,
Figure 18.1, Figure 18.2, Figure 18.3, Figure 18.4, Figure 18.5, Figure 18.7.

PART I

COMPUTING AND EXPERIMENTAL INFRASTRUCTURE FOR NGS

1

CLOUD COMPUTING FOR NEXT-GENERATION SEQUENCING DATA ANALYSIS

Xuan Guo, Ning Yu, Bing Li, and Yi Pan

Department of Computer Science, Department of Biology, Georgia State University, Atlanta, GA, USA

1.1 INTRODUCTION

Since the automated Sanger sequencing method dominated in the 1980s (1), considered as the first-generation sequencing technology, researchers first have the opportunity to construct steadily an effective ecosystem for the production and consumption of genomic information. A large number of computational tools have been developed to decode the biological information from the sequence databases in the ecosystem. Due to the expensive cost of using the first-generation sequencing technology, only a few bacteria, whose organisms possess relatively small and simple genomes, were sequenced to publish. However, along with the completion of the Human Genome Project in the beginning of the 21st century, studies on large-scale genome analysis became feasible depending on an unprecedented proliferation of genomic sequence data, which was unimaginable only a few years ago. The advent of newer methods of sequencing, known as next-generation sequencing (NGS) technologies (2), threatens the conventional genome informatics ecosystem in terms of the storage space, as well as the efficiencies of transitional tools when analyzing such huge amounts of data. The medical discoveries of the future will largely rely on our ability to dig out the "treasure" from the massive biological data. Thus,

Computational Methods for Next Generation Sequencing Data Analysis, First Edition.
Edited by Ion I. Măndoiu and Alexander Zelikovsky.
© 2016 John Wiley & Sons, Inc. Published 2016 by John Wiley & Sons, Inc.
Companion website: www.wiley.com/go/Mandoiu/NextGenerationSequencing

unprecedented demands are placed on the storage and analysis approaches for big data. Moreover, voluminous data may consume all network bandwidth available to the organization and cause traffic trouble in the network because of the uploading and downloading for large data sets. In addition, local data centers will constantly suffer other issues, including control of data access, sufficient input/output, data backup, power supply, and cooling of computing resources. All of these obstacles have led to the solution in the form of cloud computing, which has become a significant technology in big data era and exerted revolutionary influences on both academy and industry.

1.2 CHALLENGES FOR NGS DATA ANALYSIS

Since the 1980s, the genomic ecosystem (Figure 1.1 (3)) for production and consumption of genomic information consists of sequencing lab, archives, power users, and casual users. The sequencing labs submitted their data to big archival databases, such as GenBank of National Center for Biotechnology Information (NCBI) (4), European Bioinformatics Institute EMBL database (5), and Sequence Read Archive (SRA, previously known as Short Read Archive) (6). Most of these databases maintain, organize, and distribute sequencing data, and also provide data access and associated tools to both power users and casual users freely. Most users obtain information either via websites created by archival databases or by value-added integrators.

The basis for the above ecosystem is Moore's law (7), which describes a long-term trend first introduced in 1965 by Intel co-founder Gordon Moore. Moore's law stated that "the number of transistors that can be placed on an integrated circuit board is increasing exponentially, with a rate of doubling in roughly 18 months" (8). The trend has remained true for approximately 40 years across multiple changes in semiconductor and manufacturing techniques. Similar phenomena have been noted for disk storage: hard drive capacity doubles roughly annually (Kryder's law) (9); and network capacity that the cost of sending a bit of information over optical networks halves every 9 months (Nielsen's law and Butter's law) (10). Along with the improvement of genome sequencing technology, the increasing rate of time for DNA sequencing was approximating the growth of computing and storage capacity at the beginning. The archival databases and computational biologists did not need to worry about running out of disk storage space or not having access to sufficiently powerful

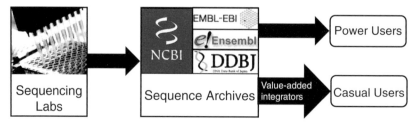

Figure 1.1 The old genome informatics ecosystem prior to the advent of next-generation sequencing technologies (3).

networks because the slight difference between two rates allowed them to upgrade their capacity ahead of the curve.

However, a deluge of biological sequence data has been generated since the Human Genome Project was completed in 2003. The advent of NGS technologies in the mid-2000s increases the slope of the DNA sequencing curve abruptly and now threatens the conventional genome informatics ecosystem. The commercially available NGS technologies, including 454 Sequencer (11), Solexa/Illumina (12), and ABI SOLiD (13), generated a tsunami of petabyte-scale genomic data, which flooded biological databases as never before. In terms of the prices of hard disk and DNA sequencing, we illustrate this by using a long-term trend (Figure 1.2) (7) plotted by Stein (14). Note that exponential curves are drawn as straight lines in the logarithmic scale. According to the figure, it is clear that the cost of storing a byte of data was halved every 14 months during 1990–2010. On the contrary, the cost of sequencing a base was halved every 19 months during 1990–2004, which is more slow than the unit cost of storage did. After the widespread use of NGS technologies, the cost of sequencing a base was halved down to every 5 months, which leads to the drop in the cost of genome sequencing several times faster than the cost of storage. It is not difficult to predict that it will cost us less to sequence a base of DNA than to store it on a hard disk sometime shortly. There is no guarantee to accelerate the trends all the time, but recently announced results by Illumina (15), Pacific Biosystems (16), Helicos (17), and Ion Torrent (18) ensure the continuing of the trends for at least another half-century. The development of NGS makes the current ecosystem face four challenges from the perspectives of storage, transportation, analysis, and economy.

- **Storage**. The tsunami of genomic data from NGS projects threats public biological databases in terms of space and cost. For example, just after the first 6 months of the 1000 Genomes Project, the raw sequencing data deposited in GenBank's Sequence Read Archive (SRA) division (19) were two times

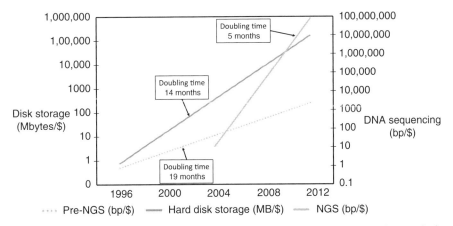

Figure 1.2 Historical trends in storage prices versus DNA sequencing costs (7). Source: Stein et al. 2010. Creative Commons Attribution License 4.0.

larger than all of the data deposited into GenBank in last 30 years (7). Another instance involved NCBI that it announced to discontinue the access service to the high-throughput sequence data due to the unaffordable cost for SRA service (20).

- **Transportation**. The uploading and downloading of huge amounts of data can easily exhaust all the network capacity available to researchers. It is reported that annual worldwide sequencing capacity is currently beyond 13 Pbp (21). Both power users and value-added genome integrators must directly or indirectly download the data from archival databases via the Internet and store copies in local storage systems to analyze them to provide web service. The mirroring of data sets across the network in multiple local storage systems are increasingly cumbersome, error-prone, expensive, and even getting worse when updates are made to databases and all mirrors are needed to be refreshed.

- **Analysis**. The massive amounts of sequence data generated by NGS put the computational burden on traditional analysis significantly. Take sequence assembly of the human genome, for example. Velvet (22), a popular sequential assembly program, needs at least 2 TB memory and several weeks to fully assemble the human genome based on the data from Illumina platform. The single desktop computer is not powerful enough to give us the results in an acceptable time. On the other hand, if we try to cast traditional programs on computing clusters, the coding experience for traditional high-performance computing is not easy to be acquired.

- **Economy**. The load of servers for accessing genome databases and web services usually fluctuates hourly, daily, and seasonally, so large data centers, such as NCBI, UCSC, and other genome data providers, are forced to choose either a cluster to meet average daily requirements or a powerful one to handle peak usage. No matter choosing which option, a large portion of computing resources will stay idle waiting for activities, such as a new large genome data set is submitted, or a major scientific conference is getting close. In addition, as long as the services are online, all the computers require electricity and maintenance, which is not a small amount of the cost.

1.3 BACKGROUND FOR CLOUD COMPUTING AND ITS PROGRAMMING MODELS

A promising solution to address these four challenges mentioned earlier hides in cloud computing, which has been an emerging trend in the scientific community (23). The cloud symbol is often employed to depict the term of "cloud computing" in Internet flowcharts. Based on virtualization technologies, cloud computing provides a variety of services from the hardware level to the application level, and all the services are charged on a pay-per-use basis. Therefore, scientists can have immediate access to needed resources, such as computation power and storage space of large distributed infrastructures, without planning, and release them to save cost as soon as experiments finish.

1.3.1   Overview of Cloud Computing

The general notions in cloud computing can be categorized into two broad types: cloud and cloud technologies. The cloud offers a large pool of easily usable and accessible resources that are scalable to allow optimum utilization (24). A fundamental basis of cloud technologies is virtualization, that is, a single physical machine can host multiple virtual machines (VMs). A VM is a software application that can load a single digital image of resources, often known as a whole system snapshot, and emulate a physical computing environment. In addition, a VM image can be duplicated entirely, including operating system (OS) and its associated applications. Taking the advent of virtualization, the components of infrastructure in the cloud are reusable. At any time point, a particular element in the cloud can be used by a certain user, while at other time points the same element can be employed by other subscribed users. There is no fixed one-to-one relationship between the data or software or physical computing resources. The distinction between traditional computing and virtualization is shown in Figure 1.3. In comparison to the tradition computing, an extra virtualization management layer, Hypervisor, is placed between a physical machine layer and resource images layer. Hypervisor acts as a bridge to translate and transport requests from applications running on VMs to manage physical hardware, such as CPU, memory, hard disks, and network connectivity (25).

Cloud resources for NGS data contain various services, including data storage, data transportation, parallelization of transitional tools, and web services of the data analysis. Basically, cloud services for NGS data can be classified into four categories: Hardware as a Service (HaaS), Platform as a Service (PaaS), Software as a Service (SaaS), and Data as a Service (DaaS). More details of these services, including the definitions, cloud-based methods, and NGS applications, will be covered in the next section. With the contributions from open source communities, such as Hadoop, cloud computing becomes more and more popular and practicable in both fields of industry and academy. In brief, users, especially software developers, can pay more attentions on design and arrangement of distributed subtasks for large data sets than for the program deployment on the cloud.

1.3.2   Cloud Service Providers

In this section, three representative cloud service providers will be introduced, that is, Amazon Elastic Compute Cloud, Google App Engine, and Microsoft Azure.

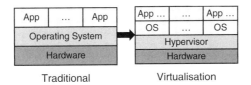

Figure 1.3   Traditional computing versus physical machine with virtualization. Source: O'Driscolla 2013 (25). Reproduced with permission of Elsevier.

Amazon Elastic Compute Cloud (EC2) (26) provides Linux-based or Windows-based virtual computing environments. For users working with NGS data, EC2 offers an integration of public databases embedded in Amazon Web Services (AWS). The integrated databases include the archives from GenBank, Ensembl, 1000 Genomes, Model Organism Encyclopedia of DNA Elements, UniGene, Influenza Virus, and so on. Users can create their Windows-based or Linux-based VMs or load pre-built specific images on the servers they rent. Users just need to upload the VM images to the storage service of EC2, that is, Amazon Simple Storage Service (S3). EC2 only charges users when the allocated VM is alive. As far as we know, the cost for S3 starts at 15 cents per GB per month, and it is 2 cents per hour for using EC2 (27).

Google App Engine (GAE) (28) allows users to build and run web apps on the same systems that are powering Google applications. Various developing and maintenance services are provided by GAE, including fast development and deployment, effortless scalability, and easy administration. Additionally, GAE supports Application Programming Interfaces (APIs) for data management, verification of Google Accounts, image processing, URL fetching, email services, the web-based administration console, and so on. Currently, all applications are allowed to use up to 1 GB storage and other resources free for a month (28).

Microsoft Azure (29) provides users on-demand compute and storage to host, scale, and manage web applications on Microsoft data centers through the Internet. The administrations of applications, such as uploading and updating data, starting and stopping applications, and so on are accessed by a web-based live desktop application. The transfers of every file are protected using Secure Socket Layers combining with users' live ID. Microsoft Azure aims to be a platform to facilitate implementation of SaaS applications.

1.3.3   Programming Models

Cloud programming is about what and how to program on cloud platforms. Although there are many choices of service providers, platforms, and software available in cloud computing, the key point to take advantage of cloud computing technology hinges on the programming model. We take two popular programming models, MapReduce and task programming model, as examples to illustrate some basic principles of how to design cloud applications.

1.3.3.1   MapReduce Programming Model   MapReduce is a popular programming model introduced firstly by Google for dealing with data-intensive tasks (30). It allows programmers to apply transformations to each data record and to think in a data-centric manner. In the framework of MapReduce programming model, a job includes three continuous phases: Mapping, Sorting & Merging, and Reducing (31). In mapping, the mapper is the primary unit function, and multiple mappers can be created to fetch and process data records without repeat. The outputs from mapper are all in the form of key/value pair. The sorting & merging phase sorts and groups the outputs according to their keys. In reducing, the reducer is the primary unit function, and multiple reducers can be created to apply further calculations on the grouped outputs. Programs following the MapReduce manner can be automatically executed in parallel on the platforms supporting MapReduce. Developers only need to focus on how to fit their sequential methods into the MapReduce model. The general formation of the mapper is described as follows:

$$map() :: (key_1, value_1) \rightarrow arraylist(key_2, value_2) \tag{1.1}$$

The mapper is an initial ingestion and transformation to process input records in parallel. The mapper takes each data record in the form of key/value pair as input and outputs a collection of key/value. The design of key/value can be customized based on users' demand.

The sorting & merging phase sorts the collection of key/value pairs from all mappers in order of the keys. The pairs with the same key are combined and passed to the same reducer.

$$reduce() :: (key_2, value_2) \rightarrow list(value_3) \tag{1.2}$$

The reducer acts as the aggregation and summarization that all associated records are processed together if necessary as a single entity.

In MapReduce programming model, a complete round of three phases is often considered as a job. Figure 1.4 illustrates the basic workflow of MapReduce programming model. There are several extensions of the basic MapReduce programming model. Three of them are shown in Figure 1.5. As indicated by their names, "map-only" means there is no sorting & merging phase and reducing phases; "map-reduce" is the standard version of MapReduce framework; and "iterative map-reduce" stands for a

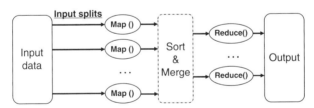

Figure 1.4 The workflow of MapReduce programming model (32). Source: http://hadoop .apache.org. The Apache Software Foundation.

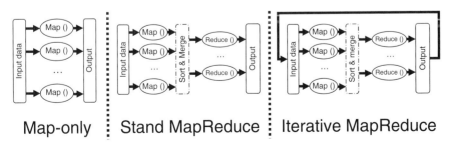

Figure 1.5 Three MapReduce programming models.

multiple rounds of standard map-reduce. An implementation of MapReduce model, Hadoop, will be used as an example to illustrate the MapReduce programming model.

Hadoop   Hadoop (32) is an open source software framework for developing data-intensive applications in the cloud. Currently, it can only run on the Linux-based cluster. Hadoop natively supports Java, and it can also be extended to support other languages, such as Python, C, and C++. Hadoop has implemented the model of MapReduce: inputs are partitioned into logical records and processed independently by mappers; results from multiple mappers are sorted and merged into distinct groups; and groups are passed to separate reducers for more calculation. The architecture of Hadoop is shown in Figure 1.6 and the workflow is *Input()* → *Map()* → *Sort()/Merge()* → *Reduce()* → *Output*. The components of Hadoop are organized as a *master-slave* structure. When developing Hadoop applications, we only need to specify three items: Java classes defining key/value pairs, mapper, and reducer (32).

The foundation of Hadoop to support the MapReduce model is the Hadoop Distributed File System (HDFS). HDFS integrates the shared file system as one file system logically. It also provides a Java-based API to handle file operations. The service of HDFS is functionally based on two processes, NameNode and DataNode. NameNode is in charge of control services, and DataNode is in charge of block storage and retrieval services (32). For the scheduling of jobs on each VMs or operating system, Hadoop uses another two processes, TaskTracker and JobTracker. TaskTracker schedules the execution order of each mapper and reducer on slave computing nodes. JobTracker is in charge of job submissions, job monitoring, and distribution of tasks to TaskTrackers (32). In order to obtain high reliability, input data are mirrored into multiple copies in HDFS, referred as "replica." As long as at least one replica is still alive, TaskTracker is able to continue the job without

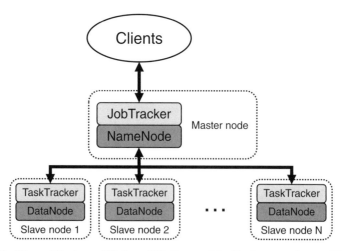

Figure 1.6   The architecture of Hadoop (32). Source: http://hadoop.apache.org. The Apache Software Foundation.

reporting storage failure. Note that master node holds the NameNode and JobTracker services, and slave nodes hold the TaskTracker and DataNode services.

1.3.3.2   Task Programming Model   Some scientific problems can be easily split into multiple independent subtasks. Taking BLAST search, for example, a bunch of query sequences can be searched independently if a set of duplicate databases is available and equipped with separate network accessibility. This is the basis for task programming model. The typical framework of task programming model is shown in Figure 1.7.

In the task programming model, subtasks are initially configured by developers and inserted into a task queue. Each entity in this queue contains scheduling information encoded as text messages. A task pool, held by a master node, is used to distribute task entities and coordinate other computing nodes. Task programming model provides a simple way to guarantee fault tolerance: one task can be processed by multiple computing nodes if the task is reported failed, and task pool only deletes the task in the queue when it has been completed. In the following, Microsoft Azure is used as an implementation to illustrate the task programming model.

Microsoft Azure   is a cloud service platform provided by Microsoft (29). The architecture of Microsoft Azure is shown in Figure 1.8. It virtualizes hardware resources and abstracts them as Virtual Machines. Any number of conceptually identical VMs

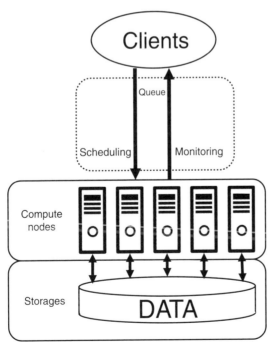

Figure 1.7   The task programming model (33). Source: Gunarathne 2011 (33). Reproduced with permission of John Wiley and Sons.

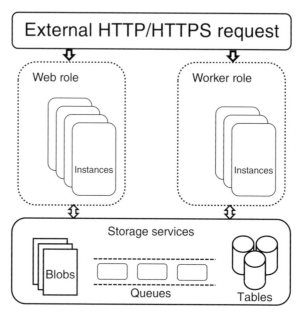

Figure 1.8 The architecture of Microsoft Azure (33). Source: Gunarathne 2011 (33). Reproduced with permission of John Wiley and Sons.

can be readily added to or removed from an application in the abstraction layer above the physical hardware resources, which enhances the administration, availability, and scalability.

A Microsoft Azure application can be divided into several logical components, called "roles", with distinguishing functions. A role contains a particular set of codes, such as a .NET assembly, and an environment where the codes can be executed. Developers can customize the number and scale of instances (VMs) for their applications. There are three types of roles: Web role, Worker role, and VM role. The web role is in charge of front-end web communications, and it is based on Internet Information Services (IIS) compatible technologies, such as ASP.NET, PHP, and Node.js. A worker role, similar to traditional windows desktop environment, performs tasks in the background. The duties include data process and communicate with other role instances. A VM role is used to store the image of Windows Server operating system and can be configured to meet necessary physical environment for running the image. Each role can have multiple VM instances. Unlike Hadoop, instances of worker role can communicate internally or externally in Microsoft Azure.

Most commonly used storage/communication structures in Microsoft Azure are BLOB, Queue, and Azure Table. BLOB, which stands for Binary Large OBject, works as containers, similar to directories in the Unix-based system. Users can set their BLOBs as either public or private and access them by the URLs with account names and access keys. The Queue is a basic structure to support message passing function. The message in the Queue will not disappear permanently until a computing node explicitly deletes it. This feature ensures the fault tolerance as discussed before (34). Azure Table storage is a key/attribute store with a schema-less design where each entity is indexed with row and column, and it can also support query operations, such as traditional database operations.

1.4 CLOUD COMPUTING SERVICES FOR NGS DATA ANALYSIS

In this section, we use case studies to illustrate how the cloud computing services provide support for NGS data analysis. Currently, four main cloud services are available for NGS data, that is, Hardware as a service (HaaS), Platform as a Service (PaaS), Software as a service (SaaS), and Data as a service (DaaS). A summarization of these cloud services is shown in Table 1.1. For SaaS, six methods are used as examples with elaboration on their descriptions, algorithms, and parallel solutions. Four typical biological problems are covered by these six methods, that is, BLAST, comparative genomic, sequence mapping, and SNP detection.

1.4.1 Hardware as a Service (HaaS)

Hardware as a Service, also known as Infrastructure as a service (IaaS), provides users with computing resources, such as storage service and virtualized OS image, through the Internet. Based on the demand from users, HaaS vendors dynamically resize the computing resources and deploy necessary software to build virtual machines. Different users often have different resource requirements, so scalability and customization are two essential features for HaaS. And users only pay for the cloud resources that they use. We briefly introduce several popular HaaS platforms.

AWS is estimated to take 70% of the total HaaS market share. AWS has some offerings, including the Elastic Compute Cloud (EC2) and Simple Storage Service (S3). EC2 provides servers on which users can build VM images. S3 is an online storage service. The HaaS market is changing rapidly with significant fluctuations because HP, Microsoft, Google, and other large companies are all competing for market supremacy. HP releases its cloud platform solution, HPCloud, which integrates servers, storage, networking, and security into an automated system. HPCloud is built on OpenStack, a cloud HaaS software initially developed by Rackspace and NASA. The management of cloud resources in HPCloud is a hybrid service, which combines security and convenience in a private cloud with cost-effectiveness. There are some other HaaS providers, such as Microsoft Azure, Google Compute Engine, and Rackspace. Rackspace is also a hybrid cloud and is able to combine two or more types of cloud, such as private and public, through Virtual Private Networking (VPN) technology typically. The above-mentioned HaaS platforms do share some common features like access to providers' data center authorized by paying nominal fees and the charge depending on the alive CPU usage, the storage space for the data, and the amount of data transferred.

1.4.2 Platform as a Service (PaaS)

PaaS offers users a platform with necessary software and hardware to develop, test, and deploy cloud applications. In PaaS, VMs can be scaled automatically

TABLE 1.1 Cloud Resources for NGS Data Analysis

Applications	Functions & Descriptions & Availabilities
Hardware as a Service (HaaS):	
Amazon Web Services	Provided by Amazon; http://aws.amazon.com/
HPCloud	Provided by HP; http://www.hpcloud.com/?
Microsoft Azure	Provided by Microsoft; http://www.windowsazure.com/
Google Compute Engine	Provided by Google; http://cloud.google.com/products/ compute-engine?
Rackspace	Hybrid Cloud; http://www.rackspace.com/
Platform as a Service (PaaS):	
Google App Engine	http://developers.google.com/appengine/?
Microsoft Azure	http://www.windowsazure.com/?
MapReduce/Hadoop	http://hadoop.apache.org/?
Software as a Service (SaaS):	
AzureBlast (35)	BLAST (a parallel BLAST running on the cloud computing platform of Microsoft Azure)
CloudBLAST (36)	BLAST (a cloud-based implementation of BLAST) http://ammatsun.acis.ufl.edu/amwiki/index.php/ CloudBLAST_Project
BlastReduce (37)	BLAST (Hadoop-based BLAST) http://www.cbcb.umd.edu/ software/blastreduce/
RSD (38)	Comparative genomics (reciprocal smallest distance algorithm for ortholog detection on EC2) http://roundup .hms.harvard.edu
CloudBurst (39)	Genomic sequence mapping (highly sensitive short read mapping with MapReduce) http://cloudburst-bio .sourceforge.net
SeqMapReduce (40)	Genomic sequence mapping (parallelizing sequence mapping by using Hadoop) http://www.seqmapreduce.org
CloudAligner (41)	Genomic sequence mapping (fast and full-featured MapReduce-based tool for sequence mapping) http://cloudaligner.sourceforge.net
SEAL (42)	Genomic sequence mapping (short read pair mapping and duplicate removal tool by using Hadoop) http://biodoop-seal.sourceforge.net/.
Crossbow (43)	Genomic sequence analysis (read mapping and SNP calling using cloud computing)
CrossTSS (44)	Genomic sequence analysis (a parallel haplotype block partition and SNPs selection method by using the Hadoop)
Contrail (45)	Genomic sequence analysis (cloud-based *de novo* assembly of large genomes) http://contrail-bio.sourceforge.net
CloudBrush (46)	Genomic sequence analysis (a genome assembler based on string graphs and MapReduce) https://github.com/ice91/ CloudBrush
Myrna (47)	RNA sequencing analysis (differential gene expression tool for RNA-Seq) http://bowtie-bio.sourceforge.net/myrna

TABLE 1.1   *(Continued)*

Applications	Functions & Descriptions & Availabilities
Eoulsan (48)	RNA sequencing analysis (a modular and scalable framework based on the Hadoop) http://transcriptome.ens.fr/eoulsan/
FX (49)	RNA sequencing analysis(RNA-Seq analysis tool) http://fx.gmi.ac.kr
HadoopBAM (50)	Sequence file management (an integration layer between analysis applications and BAMfiles) http://sourceforge.net/projects/hadoop-bam
SeqWare (51)	Sequence file management (query engine supporting databasing information for large genomes) http://seqware.sourceforge.net
GATK (52)	Sequence file management (a gene analysis toolkit for next-generation resequencing data) http://www.broadinstitute.org/gatk/
Data as a Service (DaaS):	
AWS Public Data Sets	(Cloud-based archives of GenBank, 1000 Genomes, and so on) http://aws.amazon.com/publicdatasets

and dynamically to meet applications' demands. Because the deployment and assignment of hardware are in a transparent manner, users can pay more attentions on the development of cloud-based programs. Typically, the environment delivered by PaaS comes with programming language execution environments, web servers, and databases. Some popular platforms have been introduced in the beginning, such as Google App Engine, Microsoft Azure, and MapReduce/Hadoop. When considering cloud-based databases, DaaS can be also treated as an instance of PaaS. Here, we separate DaaS from PaaS and will discuss DaaS later.

1.4.3   Software as a Service (SaaS)

SaaS provides on-demand software as web services and facilitates remote access to data analyses in various types. The analysis of NGS data involves many biological issues, such as sequence mapping, sequence alignment, sequence assembly, expression analysis, sequence analysis, orthology detection, functional annotation of personal genomes, detection of epistatic interactions, and so on (25). SaaS eliminates the necessity of complicated local deployment, simplifies software maintenances, and ensures up-to-date cloud-based services for all possible users with access to the Internet. Since it is impractical to cover all cloud-based NGS data analysis tools available nowadays, four representative categories are carefully selected to elaborate the thinking in the cloud for solving problems that arise from NGS data.

1.4.3.1   BLAST   Basic Local Alignment Search Tool (BLAST) (35) is one of the most widely used sequence analysis programs provided by NCBI. Meaningful information from the query sequence can be extracted by comparing it to the NCBI databases using BLAST. The pairwise comparison is trivial when only limited number of sequences is needed to be compared, but the number of sequences in NCBI's databases are extremely large. For instance, 361 billion nucleotide bases were reported in Reference Sequence (RefSeq) Database up to November 10, 2013. Without doubt, it is computational intensive even when one query is submitted to a huge database by using pairwise comparison. Several cloud-based applications have been proposed to parallel BLAST on commercial cloud platforms. The basic strategies of them are very similar. Because the queries of sequences are independent, they can be executed simultaneously on a set of separate computers with a partial or complete database. In the following, two cloud applications, AzureBlast (53) and CloudBLAST (36), are discussed to illustrate the idea.

AzureBlast Lu et al. (53) proposed a parallel BLAST, named AzureBlast, running on the Microsoft Azure. The workflow of AzureBlast is shown in Figure 1.9. Instead of partitioning the database into segmentations, they use a query-segmentation data-parallel pattern to split the query sequences into several disjoint sets. The reason for this is that the queries on segmentations need less communication among instances than the query on several parts of the database. Given some sequences as the input, AzureBlast partitions the input sequences into multiple files and allocates them to worker instances to start the comparisons. The results are merged from all worker instances. The experiments of AzureBlast demonstrate that Microsoft Azure can very well support the BLAST based on its scalable and fault-tolerant computation and storage services (53).

CloudBLAST Matsunaga et al. (36) proposed a WAN-based implementation of BLAST, called CloudBLAST. In CloudBLAST, the parallelization, deployment, and management of applications are built and evaluated on Hadoop platform. Similar to

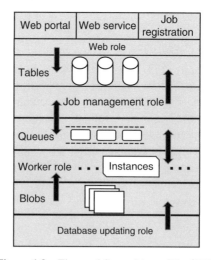

Figure 1.9   The workflow of AzureBlast (53).

AzureBlast, input query sequences are split at first and then the grouped sequences are passed to mappers to run BLAST program separately. The results from mappers are stored on a local disk and combined as the final results. Demonstrated by the experiments of CloudBLAST, cloud-based applications built on Internet-connected resources for bioinformatics issues can be considerably efficient. CloudBLAST's performance was experimentally contrasted against a publicly available tool, mpiBLAST (54), on the same cloud configuration. mpiBLAST is a free parallel implementation of NCBI BLAST running on clusters with job-scheduling software such as PBS (Portable Batch System). By using 64 processors, both tools gained nearly equivalent performance with speedups (31) of 57 of CloudBLAST against 52.4 of mpiBLAST (36), respectively.

1.4.3.2 Comparative Genomics Comparative genomics is a study of understanding functional similarities and differences as well as evolutionary relationships between genomes by comparing genomic features, such as DNA sequence, genes, and regulatory sequences, across different biological species or strains. One computationally intensive application in comparative genomics is the Reciprocal Smallest Distance algorithm (RSD) (38). RSD is used to detect orthologous sequences between multiple pairs of genomes. It has three steps: (i) employ BLAST to generate a set of hits between query sequences and references, (ii) use alignment tools on each protein sequence and take PAML (38) to obtain the maximum likelihood estimation of the number of amino acid substitutions, and (iii) call BLAST again to re-calculate the maximum likelihood distance to determine whether the pair of sequences is correct orthologous pair or not. Wall et al. (38) proposed a cloud-based tool, named RSD-cloud, by fitting legacy RSD into MapReduce model on EC2. RSD-cloud has two primary phases, that is, BLAST and estimation of evolutionary distance. In the first phase, mappers use BLAST to generate hits for all genomes. In the second phase, mappers conduct ortholog computation to estimate orthologs and evolutionary distances for all genomes. As shown in Figure 1.10, two blocks in step 2 illustrate the above two paralleled phases. All results from RSD-cloud directly go into Amazon S3. Experiments showed that it is able to run more than 300,000 RSD-cloud processes within the EC2 to compute the orthologs for all pairs of 55 genomes by using 100 high-capacity computing nodes (38). The total computation time was less than 70 hours and the cost was $6,302 USD.

1.4.3.3 Genomic Sequence Mapping Genomic sequence mapping aims to locate the relative positions of genes or DNA fragments on the reference chromosomes. Generally, there are two types of genomic sequence mapping: (i) genetic mapping, using classical genetic techniques, such as pedigree analysis, to depict the features of genome, and (ii) physical mapping, using modern molecular biology techniques for the same goal. Current cloud-based solutions for genomic sequence mapping belong to the second type. Similar to BLAST, genomic sequence mapping can also be paralleled in terms of independence of the sequence queries although extra processes may be needed. In the following, two cloud tools, CloudBurst (39) and CloudAligner (41), are used to illustrate the cloud-based solutions for genomic sequence mapping.

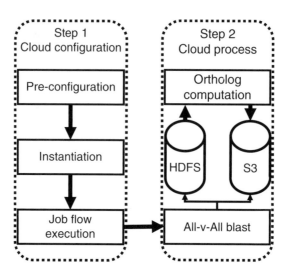

Figure 1.10   Workflow of RSD using the MapReduce framework on the EC2 (38). Source: Wall, http://bmcbioinformatics.biomedcentral.com/articles/10.1186/1471-2105-11-259. Used under CC BY 2.0 http://creativecommons.org/licenses/by/2.0/.

CloudBurst   Schatz (39) designed a parallel algorithm, named CloudBurst, which is a seed-and-extend read mapping algorithm on Hadoop platform based on a popular read mapping program, RMAP (55). According to MapReduce model, CloudBurst modifies RMAP to run on multiple machines in parallel. The workflow of CloudBurst with two phases, Map phase and Reduce phase, is shown in Figure 1.11. The key/value pairs generated by mappers have the following format, <reads' and references' indexes, *k*-mers of reads and references>. Reducers execute end-to-end alignments between reads and reference sequences sharing the same *k*-mers. Final results are converted into text files with the standard format as RMAP did, so CloudBurst can replace RMAP in other pipelines. CloudBurst's running time scales near linearly as the number of processors increases. In a configuration with 24-processor cores, CloudBurst achieved up to 30 times faster than RMAP executed on a single core given an identical set of alignments as input (39).

CloudAligner   Nguyen et al. (41) developed CloudAligner, a MapReduce-based application to handle long read mapping. It only uses map phase to locate the reads on references. The workflow is shown in Figure 1.12. CloudAligner accepts two types of input files, that is, read files and reference files. CloudAligner first splits read files into a bunch of chunks with limited size and distributes them to multiple mappers. Mappers then align the reads to the references. CloudAligner is able to deal with multiple mapping types, including mismatch mapping, bisulfite mapping, and pair-end mapping, and it is also compatible with various types of input, such as fastq and SAM. The experimental results demonstrated that CloudAligner can achieve significant improvement in terms of efficiency by using the MapReduce model without reduce phase.

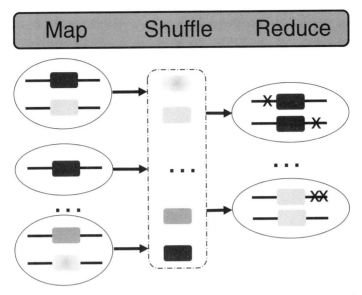

Figure 1.11 The overview of CloudBurst (39). Source: Schatz 2009 (39). Reproduced with permission of Oxford University Press.

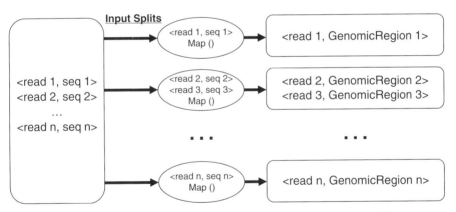

Figure 1.12 The workflow of CloudAligner (41). Source: Nguyen, http://bmcresnotes .biomedcentral.com/articles/10.1186/1756-0500-4-171. Used under CC BY 2.0 http:// creativecommons.org/licenses/by/2.0/.

1.4.3.4 SNP Detection SNP detection is used to scan for single-nucleotide poly-morphism (SNP), which is a DNA sequence variation occurring commonly within a population. A single nucleotide in the genome is defined as SNP when it differs between members of a biological species or paired chromosomes. These genetic variations are considered to be associated with many genetic diseases (56, 57). A central challenge of SNP detection based on NGS data is the sequence alignment that massive sequence reads must be compared to the references. Similar to the previous comparison problems, SNP detections on multiple sequence areas are independent of each other. In the following, an application, Crossbow, is used to show the cloud solution for SNP detection.

Crossbow   Langmead et al. (43) proposed a Hadoop-based tool, Crossbow, which uses alignment to detect SNP for the whole human genomes. Crossbow combines two methods together, that is, a short read aligner, Bowtie (58), and an SNP caller, SOAPsnp (59). Crossbow has three phases: Map phase, Sorting & merging phase, and Reduce phase. In Map phase, mappers align short reads to the reference sequences by using Bowtie to produce a stream of alignment pairs. There are two keys for each pair: primary key and secondary key. The primary key is the identifier of the chromosome, and the secondary key is the beginning location of the queried sequence on the chromosome. The value of each pair is the aligned sequence and quality scores. In Sorting & merging phase, Hadoop platform automatically groups these pairs according to the primary key and sorts them in a particular order based on the secondary key. In Reduce phase, reducers simply call SOAPsnp to perform SNP detection. As shown in the experiments, Crossbow consumed only 3 hours to finish the alignments and SNP detection for a Han Chinese male genome with 38-fold coverage by using a compute cluster with 320 cores.

1.4.4   Data as a Service (DaaS)

DaaS aims to deliver data from multiple sources in various formats as a service, which can be accessed through the Internet. Bioinformatic tools heavily depend on the data, especially those analyses about the genome, which are fundamentally crucial for downstream analyses. DaaS enables dynamic data access on demand, so users can remotely obtain the data via Internet as if they are using the data in a local file system. In addition, the sharing of data accelerates the progress of large projects for the real-time collaboration with other researchers. An example of DaaS is the AWS, which provides a centralized repository of public data, including archives of GenBank, Ensembl, 1000 Genomes, Model Organism Encyclopedia of DNA Elements, UniGene, Influenza Virus, and so on. In fact, AWS also contains many public data sets for other scientific fields, such as astronomy, chemistry, climate, and economics. All public data sets in AWS are delivered as services, and thus they can be seamlessly integrated into cloud-based applications (27).

1.5   CONCLUSIONS AND FUTURE DIRECTIONS

In this chapter, we first discussed some challenges proposed by NGS technologies in terms of storage, transportation, analysis, and economy. All these challenges are caused by the tsunami of genomic data generated by NGS platforms. It is becoming impossible for a single desktop to run sequential approaches to obtain results in an acceptable time. Cloud computing is a promising solution to rescue researchers for computationally intensive problems. Two categories of cloud programming models, MapReduce and task programming models, were elaborated to illustrate the parallel idea of current cloud platforms. A summary of cloud resources for NGS data analysis was listed in terms of four cloud services, that is, HaaS, PaaS, SaaS, and DaaS. Six cloud-based methods of SaaS for four NGS data analysis problems were revised, which were used as an inspiration for how to perform transformation of sequential programs onto cloud platforms. However, not all data-intensive algorithms can be easily transferred to cloud environments. Therefore, suitable parallel strategies, more powerful programming models, frameworks, and cloud platforms are urgently needed. There are three basic parallel tactics commonly used in developing cloud-based applications:

1. Straightforward Parallel. Many bioinformatics tasks, such as BLAST, are explicitly data independent. So they can run on cloud platforms with only minor modifications. The speedup of this parallel type is near linear with respect to the computation power of a computing cluster.
2. Iterative Parallel. Multiple MapReduce rounds or executions of a program are another solutions for some complicated jobs, like RSD. Additional operations about how to store, allocate, and collect intermediate temporary data may also be carefully considered. How to design an efficient method is a fundamental challenge, since temporary data could be too huge to be processed during the execution.
3. Data Transformation Parallel. For some sophisticated problems, the conversion of data and algorithm itself are needed to take the advantage of cloud computing. An example is the one in Reference 60. Although we did not introduce it in this chapter, it is still worthy to know that we can change the format of inputs wisely to facilitate the development of cloud-based solutions. The challenge lying in parallel tactic is how to design a suitable data formation without any or with few compromises of accuracy or efficiency of legacy programs.

Concerning bioinformatics, not limited in the analysis of NGS data, cloud computing is still at its early stage. More powerful and readily usable cloud platforms and programming models are aspired to be invented and invested to conquer complex scientific issues. Ultimately, all scientists will benefit by taking advantages of this new computation power in all areas.

REFERENCES

1. Sanger F, Nicklen S, Coulson AR. DNA sequencing with chain-terminating inhibitors. Proc Natl Acad Sci U S A 1977;74(12):5463–5467.
2. Quail MA, Smith M, Coupland P, Otto TD, Harris SR, Connor TR, Bertoni A, Swerdlow HP, Gu Y. A tale of three next generation sequencing platforms: comparison of ion Torrent, Pacific Biosciences and Illumina MiSeq sequencers. BMC Genomics 2012;13(1):341.
3. Shanker A. Genome research in the cloud. OMICS 2012;16(7-8):422–428.
4. Benson DA, Cavanaugh M, Clark K, Karsch-Mizrachi I, Lipman DJ, Ostell J, Sayers EW. GenBank. Nucleic Acids Res 2012;41(D1):D36–D42. DOI: 10.1093/nar/gks1195.

5. Brooksbank C, Camon E, Harris MA, Magrane M, Martin MJ, Mulder N, O'Donovan C, Parkinson H, Tuli MA, Apweiler R et al. The European Bioinformatics Institute's data resources. Nucleic Acids Res 2003;31(1):43–50.

6. Shumway M, Cochrane G, Sugawara H. Archiving next generation sequencing data. Nucleic Acids Res 2010;38 Suppl 1:870–871.

7. Stein LD et al. The case for cloud computing in genome informatics. Genome Biol 2010;11(5):207.

8. Moore GE et al. *Cramming More Components Onto Integrated Circuits*. New York; McGraw-Hill; 1965.

9. Walter C. Kryder's law. Sci Am 2005;293(2):32–33.

10. Reynolds C. As we may communicate. ACM SIGCHI Bull 1998;30(3):40–44.

11. Margulies M, Egholm M, Altman WE, Attiya S, Bader JS, Bemben LA, Berka J, Braverman MS, Chen Y-J, Chen Z et al. Genome sequencing in microfabricated high-density picolitre reactors. Nature 2005;437(7057):376–380.

12. Bennett S. Solexa Ltd. Pharmacogenomics 2004;5(4):433–438.

13. McKernan KJ, Peckham HE, Costa GL, McLaughlin SF, Fu Y, Tsung EF, Clouser CR, Duncan C, Ichikawa JK, Lee CC et al. Sequence and structural variation in a human genome uncovered by short-read, massively parallel ligation sequencing using two-base encoding. Genome Res 2009;19(9):1527–1541.

14. Internet Archive. Available at http://www.archive.org/. Accessed 2016 Mar 19.

15. Illumina. Available at http://www.illumina.com/. Accessed 2016 Mar 19.

16. Pacific Biosciences. Available at http://www.pacificbiosciences.com/. Accessed 2016 Mar 19.

17. Helicos Biosciences Corporation. Available at http://www.helicosbio.com/. Accessed 2016 Mar 19.

18. Ion Torrent. Available at http://www.iontorrent.com/. Accessed 2016 Mar 19.

19. The 1000 Genomes Project. Available at http://www.1000genomes.org/. Accessed 2016 Mar 19.

20. NCBI News. Available at http://www.ncbi.nlm.nih.gov/About/news/16feb2011. Accessed 2016 Mar 19.

21. Mediawiki. Available at http://sour-ceforge.net/apps/mediawiki/jnomics. Accessed 2016 Mar 19.

22. Zerbino DR, Birney E. Velvet: algorithms for de novo short read assembly using de Bruijn graphs. Genome Res 2008;18(5):821–829.

23. Buyya R, Yeo CS, Venugopal S, Broberg J, Brandic I. Cloud computing and emerging it platforms: vision, hype, and reality for delivering computing as the 5th utility. Future Gener Comput Syst 2009;25(6):599–616.

24. Vaquero LM, Rodero-Merino L, Caceres J, Lindner M. A break in the clouds: towards a cloud definition. ACM SIGCOMM Comput Commun Rev 2008;39(1):50–55.

25. O'Driscoll A, Daugelaite J, Sleator RD. 'Big data', Hadoop and cloud computing in genomics. J Biomed Inform 2013;46(5):774–781.

26. Amazon Elastic Compute Cloud (EC2). Available at https://aws.amazon.com/ec2/.

27. Fusaro VA, Patil P, Gafni E, Wall DP, Tonellato PJ. Biomedical cloud computing with Amazon Web Services. PLoS Comput Biol 2011;7(8):1002147.

28. Google App Engine. Available at http://appengine.google.com. Accessed 2016 Mar 19.

29. Windows Azure. Available at http://www.windowsazure.com/. Accessed 2016 Mar 19.

30. Jin C, Buyya R. MapReduce programming model for. NET-based cloud computing. In: Sips H, Epema D, Lin H-X, editors. *Euro-Par 2009 Parallel Processing*. Volume 5704 of Lecture Notes in Computer Science. Berlin Heidelberg: Springer-Veralg; 2009. p 417–428.

31. Guo X, Ding X, Meng Y, Pan Y. Cloud computing for de novo metagenomic sequence assembly. In: Cai Z, Eulenstein O, Janies D, Schwartz D, editors. *Bioinformatics Research and Applications*. Volume 7875 Lecture Notes in Computer Science. Berlin Heidelberg: Springer-Veralg; 2013. p 185–198.

32. Hadoop. Available at http://hadoop.apache.org. Accessed 2016 Mar 19.

33. Gunarathne T, Wu T-L, Choi JY, Bae S-H, Qiu J. Cloud computing paradigms for pleasingly parallel biomedical applications. Concurr Comput Pract Exp 2011;23(17): 2338–2354.

34. Redkar T, Guidici T, Meister T. *Windows Azure Platform*. Volume 1. Berkeley (CA): Apress; 2011.

35. Altschul SF, Gish W, Miller W, Myers EW, Lipman DJ. Basic local alignment search tool. J Mol Biol 1990;215(3):403–410.

36. Matsunaga A, Tsugawa M, Fortes J. CloudBlast: combining MapReduce and virtualization on distributed resources for bioinformatics applications. *eScience, 2008. eScience'08. IEEE 4th International Conference On*; 2008. p 222–229.

37. Schatz MC. *BlastReduce: High Performance Short Read Mapping with MapReduce*. University of Maryland; 2008. Available at http://cgis.cs.umd.edu/Grad/scholarlypapers/papers/MichaelSchatz.pdf. Accessed 2016 Mar 19.

38. Wall DP, Kudtarkar P, Fusaro VA, Pivovarov R, Patil P, Tonellato PJ. Cloud computing for comparative genomics. BMC Bioinformatics 2010;11(1):259.

39. Schatz MC. CloudBurst: highly sensitive read mapping with MapReduce. Bioinformatics 2009;25(11):1363–1369.

40. Li Y, Zhong S. SeqMapReduce: software and web service for accelerating sequence mapping. Critical Assessment of Massive Data Anaysis (CAMDA) 2009; 2009.

41. Nguyen T, Shi W, Ruden D. CloudAligner: a fast and full-featured MapReduce based tool for sequence mapping. BMC Res Notes 2011;4(1):171.

42. Pireddu L, Leo S, Zanetti G. Seal: a distributed short read mapping and duplicate removal tool. Bioinformatics 2011;27(15):2159–2160.

43. Langmead B, Schatz MC, Lin J, Pop M, Salzberg SL. Searching for SNPs with cloud computing. Genome Biol 2009;10(11):134.

44. Hung C-L, Lin Y-L, Hua G-J, Hu Y-C. CloudTSS: A TagSNP selection approach on cloud computing. In: Kim TH et al., editors. *Grid and Distributed Computing*. Volume 261. Berlin Heidelberg: Springer-Verlag; 2011. p 525–534.

45. Schatz M, Sommer D, Kelley D, Pop M. Contrail: assembly of large genomes using cloud computing. CSHL Biology of Genomes Conference; 2010.

46. Chang Y-J, Chen C-C, Chen C-L, Ho J-M. A de novo next generation genomic sequence assembler based on string graph and MapReduce cloud computing framework. BMC Genomics 2012;13 Suppl 7:28.

47. Langmead B, Hansen KD, Leek JT et al. Cloud-scale RNA-sequencing differential expression analysis with Myrna. Genome Biol 2010;11(8):83.

48. Jourdren L, Bernard M, Dillies M-A, Le Crom S. Eoulsan: a cloud computing-based framework facilitating high throughput sequencing analyses. Bioinformatics 2012;28(11):1542–1543.

49. Hong D, Rhie A, Park S-S, Lee J, Ju YS, Kim S, Yu S-B, Bleazard T, Park H-S, Rhee H et al. FX: an RNA-Seq analysis tool on the cloud. Bioinformatics 2012;28(5):721–723.

50. Niemenmaa M, Kallio A, Schumacher A, Klemelä P, Korpelainen E, Heljanko K. Hadoop-BAM: directly manipulating next generation sequencing data in the cloud. Bioinformatics 2012;28(6):876–877.

51. O'Connor BD, Merriman B, Nelson SF. SeqWare Query Engine: storing and searching sequence data in the cloud. BMC Bioinformatics 2010;11 Suppl 12:2.

52. McKenna A, Hanna M, Banks E, Sivachenko A, Cibulskis K, Kernytsky A, Garimella K, Altshuler D, Gabriel S, Daly M et al. The genome analysis toolkit: a MapReduce framework for analyzing next-generation DNA sequencing data. Genome Res 2010;20(9):1297–1303.

53. Lu W, Jackson J, Barga R. AzureBlast: a case study of developing science applications on the cloud. Proceedings of the 19th ACM International Symposium on High Performance Distributed Computing. ACM; 2010. p 413–420.

54. Darling A, Carey L, Feng W-C. The design, implementation, and evaluation of mpi-BLAST. Proceedings of ClusterWorld 2003; 2003.

55. Smith AD, Xuan Z, Zhang MQ. Using quality scores and longer reads improves accuracy of Solexa read mapping. BMC Bioinformatics 2008;9(1):128.

56. Patil N, Berno AJ, Hinds DA, Barrett WA, Doshi JM, Hacker CR, Kautzer CR, Lee DH, Marjoribanks C, McDonough DP et al. Blocks of limited haplotype diversity revealed by high-resolution scanning of human chromosome 21. Science 2001;294(5547):1719–1723.

57. Guo X, Meng Y, Yu N, Pan Y. Cloud computing for detecting high-order genome-wide epistatic interaction via dynamic clustering. BMC Bioinformatics 2014;15(1):102.

58. Langmead B, Trapnell C, Pop M, Salzberg SL et al. Ultrafast and memory-efficient alignment of short DNA sequences to the human genome. Genome Biol 2009;10(3):25.

59. Li R, Li Y, Fang X, Yang H, Wang J, Kristiansen K, Wang J. SNP detection for massively parallel whole-genome resequencing. Genome Res 2009;19(6):1124–1132.

60. Estrada T, Zhang B, Cicotti P, Armen R, Taufer M. A scalable and accurate method for classifying protein–ligand binding geometries using a MapReduce approach. Comput Biol Med 2012;42(7):758–771.

2

INTRODUCTION TO THE ANALYSIS OF ENVIRONMENTAL SEQUENCE INFORMATION USING METAPATHWAYS

NIELS W. HANSON

Graduate Program in Bioinformatics, University of British Columbia, Vancouver, Canada

KISHORI M. KONWAR

Department of Microbiology and Immunology, University of British Columbia, Vancouver, Canada

SHANG-JU WU

Department of Computer Science, University of British Columbia, Vancouver, Canada

STEVEN J. HALLAM

Graduate Program in Bioinformatics, Department of Microbiology and Immunology, University of British Columbia, Vancouver, Canada

2.1 INTRODUCTION & OVERVIEW

The rise of next-generation sequencing technologies has created a veritable tsunami of environmental sequence information sourced from a variety of natural and engineered ecosystems. While this tidal wave of information is changing our understanding of microbial community structure and function in the world around us, a

Computational Methods for Next Generation Sequencing Data Analysis, First Edition.
Edited by Ion I. Măndoiu and Alexander Zelikovsky.
© 2016 John Wiley & Sons, Inc. Published 2016 by John Wiley & Sons, Inc.
Companion website: www.wiley.com/go/Mandoiu/NextGenerationSequencing

general lack of versatile and scalable software tools to process, organize, and interact with environmental data sets limits knowledge creation and translation. In this chapter, we describe the capabilities of MetaPathways, a modular pipeline designed for metabolic pathway reconstruction and comparative analysis of environmental sequence information.

2.2 BACKGROUND

We live in a world dominated by microorganisms (1). Interconnected members of this invisible microbial majority form distributed metabolic networks linking genomic potential and phenotypic expression from single cells to earth systems (2, 3). Such networks are continuously changing through mutation, gene transfer, and habitat selection creating functionally redundant and resilient metabolic modules driving matter and energy conversion in natural and engineered ecosystems (4–8). Plurality sequencing, also known as environmental genomics, ecological genomics, or metagenomics, generates vast amounts of environmental sequence information with the potential to transform our worldview (4, 9–11) and enable technological and therapeutic innovations through perception and manipulation of these networks (12–18).

Although plurality and single-cell genomic sequencing technologies are rapidly expanding our capacity to generate environmental sequence information (19–22), there are a number of computational and analytical challenges that limit knowledge creation and translation (23–25). Collins and colleagues observe that similar problems are common to many areas of modern experimental research: "Data comes in all scales and shapes, covering large international experiments; cross-laboratory, single-laboratory, and individual observations; and potentially individuals lives. The discipline and scale of individual experiments and especially their data rates make the issue of tools a formidable problem" (26). In particular, computational and visual analytic bottlenecks present powerful limits to integration across different levels of biological information (DNA, RNA, protein, and metabolites) and scalable comparisons between replicate samples or communities. While good progress has been made in developing tools to inventory and to a lesser extent compare microbial community structure and function, user-friendly softwares that integrate novel statistical and distributed computing techniques and efficient data structures across multiple hierarchical levels are needed to fully realize the power and promise of environmental sequence information.

Several online services integrate environmental sequence analysis and information storage. The Metagenome Rapid Annotation using Subsystem Technology (MG-RAST http://metagenomics.nmpdr.org/) is a web-based portal using the SEED annotation framework of functional genes and subsystems that provides annotation and comparative analysis services (27). The Integrated Microbial Genomes and Metagenomes (IMG/M, http://img.jgi.doe.gov/m) (28, 29) and Community Cyberinfrastructure for Advanced Microbial Ecology Research and Analysis (CAMERA http://camera.calit2.net/) (30) archive data sets and provide annotation and comparative analysis services for selected sequencing projects. Other information resources such as the Genomes OnLine Database (GOLD) (31) and Marine Ecological Genomics (MEGX) (32) curate and compile information about microbial genome and metagenome projects worldwide.

In addition to online services and information resources, a number of specific software tools have been developed to support microbial community structure and function studies. For example, the Metagenome Analyzer (MEGAN) provides a stand-alone tool that parses and projects the Basic Local Sequence Alignment Tool (BLAST) searches onto various taxonomic or functional gene hierarchies (33–35). Other tools such as Qiime, MLTreeMap (http://mltreemap.org), and the Interactive Tree-of-Life (iTOL) focus on taxonomic identification through marker gene binning. Qiime uses phylogenetic distance of 16S rRNA genes combined with multivariate statistical methods to infer relationship among and between data sets (36). MLTreeMap uses selected single-copy clusters of orthologous groups (COGs) to discern the taxonomic composition of environmental data sets (37). iTOL enables the interactive display of phylogenetic relationships constructed from user-defined hierarchies (38). While these services and tools enable users to compare environmental sequence information based on gene abundance patterns, they also come with idiosyncratic bandwidth and formatting restrictions that can limit downstream analyses and data product generation.

In the following sections, we present MetaPathways (39, 40), a modular annotation and analysis pipeline, that enables construction of environmental pathway/genome databases (ePGDBs) from environmental sequence information using Pathway Tools (41, 42) and MetaCyc (43, 44). MetaPathways provides a pathway-centric view of microbial community metabolism that overcomes several data processing and integration bottlenecks. We begin with an overview of pipeline processing stages, introduce a refined task management system for high-performance computing on grids and clouds, describe a graphical user interface (GUI) and "Knowledge Engine" data structure enabling comparative community analysis on embarrassingly large data sets, and outline data product generation using selected R packages including VennDiagram (45), pvclust (46), and ggplot2 (http://ggplot2.org/) (47).

2.3   METAPATHWAYS PROCESSES

In this section, we review pipeline processes with special emphasis on design principles and algorithm selection. MetaPathways can be compartmentalized into five operational stages currently encompassing 16 processing steps whose order may vary incrementally between different pipeline versions (Figure 2.1). After initial quality control (QC) to remove poor quality sequences, MetaPathways implements open reading frame (ORF) prediction and conceptual translation. Translated ORFs are functionally annotated against protein sequence databases, using a seed-and-extend algorithm. Following functional annotation, transfer RNA (tRNA) and ribosomal RNA (rRNA) gene sequences are identified. The taxonomic affiliation of predicted ORFs is determined based on queries against reference protein databases. The lowest common ancestor (LCA) is placed onto the NCBI taxonomic hierarchy, and an automated tree building application is used to profile community structure based on

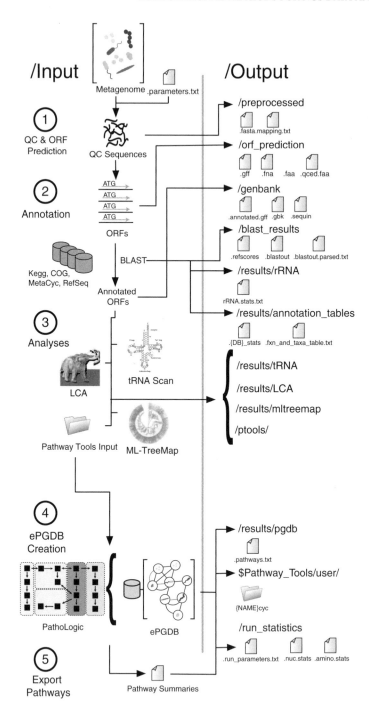

the recovery of selected phylogenetic and functional anchor genes. Annotated ORFs are refined and formatted for pathway prediction using the PathoLogic algorithm implemented in Pathway Tools to generate ePGDBs. Finally, gene, reaction, and pathway information are extracted for downstream comparison and data product generation.

2.3.1 Open Reading Frame (ORF) Prediction

The metabolic potential of an organism or community can be inferred from primary sequence information with the aid of computational methods that search for patterns or motifs representing ORFs based on identification of common genetic parts including promoters, translation initiation sites (TIS), start and stop codons or ribosomal binding sites (RBS) (Figure 2.2a) (48). A number of algorithms have been developed that accurately predict ORFs from assembled genomic sequences including GeneMark (49–51), Glimmer (52), and fgenesb (http://linux1.softberry.com/). Because these algorithms were developed to predict full-length genes in individual genomes, their performance declines on unassembled environmental data sets containing reads less than 1000 bp in length (average prokaryotic gene length) (53) (Figure 2.2b). Moreover, while algorithm accuracy can be increased through a training process that tunes parameters to common sequence patterns among closely related taxa, the genomic complexity of environmental sequence information presents an "anonymous sequence problem" (54). Given these constraints, short-read gene prediction algorithms need to sort incoming anonymous sequences into taxonomic bins or successively modify internal parameters for each sequence. Such algorithms also need to be robust to sequencing errors associated with short-read sequencing platforms (55).

A number of machine learning and statistical models have been developed to address the anonymous sequence problem (Figure 2.3). For example, MetaGene trains a heuristic scoring model based on log-odds frequencies of GC content, codon frequencies, ORF length, intra-start codon length, and orientation and distances between genes trained for Bacteria and Archaea, using the highest scoring model for each sequence (56). MetaGene Annotator (57), an updated version of MetaGene, adds a viral-trained model, as well as ribosomal binding site features to

---

Figure 2.1 Beginning with nucleotide sequences as input, the MetaPathways pipeline processes input sequences in five operational stages: (i) QC & ORF prediction where minor quality control and open reading frame (ORF) prediction are performed via Prodigal; (ii) predicted ORFs are annotated using a seed-and-extend algorithm against a collection of reference databases (e.g., KEGG, COG, MetaCyc, and RefSeq); (iii) secondary taxonomic analysis include MEGAN's lowest common ancestor (LCA) taxonomy, tRNA scan, and MLTreeMap; (iv) sequences and annotations are combined into a Pathway Tools compatible format and ePGDBs are constructed; and (v) genes, reactions, and pathways are extracted for downstream analysis (39). Source: Konwar, http://bmcbioinformatics.biomedcentral.com/articles/10.1186/1471-2105-14-202. Used under CC BY 2.0 http://creativecommons.org/licenses/by/2.0/.

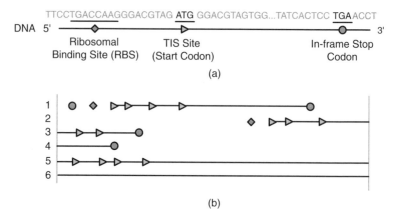

Figure 2.2 Short-read ORF prediction. (a) Three classic sequence features used to define an ORF are start codons or transcription initiation sites (TIS), for example, ATG, in-frame stop codons (e.g., TGA), and the presence of a 5′ upstream ribosomal binding site (RBS). (b) Current short-read sequencing length is smaller than most bacterial genes, thus many ORF sequences will truncate outside of the sequenced short read window; ORFs in short reads have a number of possible incomplete signals: (i) multiple TIS sites with upstream in-frame stop codon; (ii) in-frame stop codon absent; (iii) no valid TIS site present; (iv) no start but potential stop codon present; (v) start but no potential stop codon present; and (vi) no start, RBS, or potential stop present.

improve anonymous sequence identification. Glimmer-MG employs a sophisticated interpolated hidden Markov Model scheme using k-mer frequencies for model selection (58). Orphelia is a two-stage algorithm that first extracts relevant genetic features via linear discriminants and then combines these features in a neural network to create a gene prediction model (59). MetaGUN sorts environmental sequences into taxonomic bins based on k-mer frequencies prior to training a support vector machine (SVM) to identify binned ORFs (60). The Prokaryotic Dynamic Programming Gene-finding Algorithm (Prodigal) provides an efficient heuristic capable of identifying ORFs using alternative genetic codes from anonymous input sequences (61).

MetaProdigal (62), an optimized version of Prodigal developed for environmental sequence information, uses 50 training files derived from a pair-wise distance matrix composed of 1415 clustered NCBI RefSeq genomes. By calculating the GC content of each input sequence, MetaProdigal uses only those training profiles within a given GC range increasing computational efficiency. Moreover, a Bonferroni-like correction (63) process reduces false-positive identification on short reads by combining penalties for input sequence length, number of training files used, and length of the predicted ORF. When benchmarked against other ORF prediction methods tuned for anonymous and short read sequences, MetaProdigal performs exceptionally well with improved start site identification (60). Given these outcomes, open source code and widespread use at major sequencing centers Prodigal/MetaProdigal was integrated into MetaPathways although alternative prediction algorithms can readily be introduced based on user preference. MetaProdigal produces conceptually

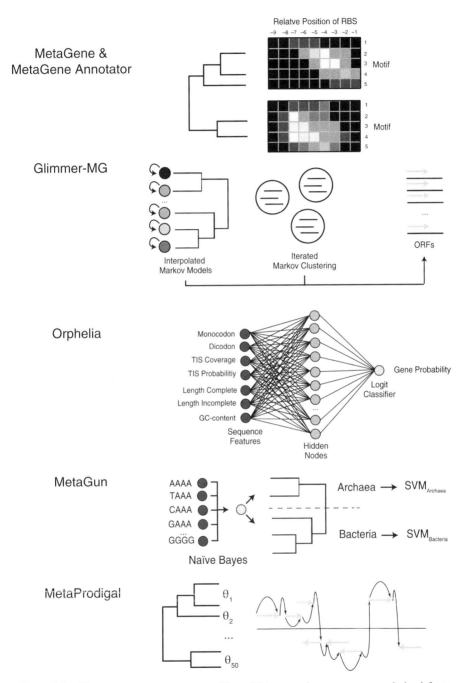

Figure 2.3 The anonymous sequence problem. Metagenomic sequences are derived from many different donor genotypes from a wide range of taxonomic groups creating a parameter training problem for many ORF prediction algorithms. Please see www.wiley.com/go/Mandoiu/NextGenerationSequencing for a color version of this figure.

translated protein sequences using NCBI Translation Table Code 11 (Standard Bacteria/Archaea) by default, but this parameter can be changed in a user-defined manner if alternative translation tables are needed. The pipeline collects MetaProdigal outputs and exports nucleotide and protein sequences as fasta files (.fna and .faa). ORFs below a default length of 180 nucleotides or 60 amino acids are removed and exported (.qc.fna and .qc.faa), and distribution summaries before and after quality control are generated (.nuc.stats and amino.stats).

2.3.2   Functional Annotation

Following ORF prediction, MetaPathways uses either BLAST or LAST seed-and-extend algorithms (33, 64) to query user defined protein sequence databases for functional annotations. One of the most comprehensive and rapidly expanding public databases for automated functional annotation is the National Center for Biotechnology Information (NCBI) nonredundant Reference Sequence (RefSeq) database. For example, the RefSeq Version 66 database released in July 2014 contained almost 60 million reference sequences with 20 million entries added in the last 12 months alone (Figure 2.4). Ironically, RefSeq expansion represents one

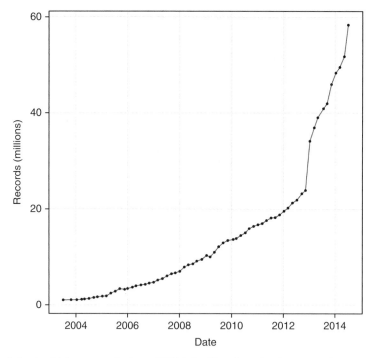

Figure 2.4 A tidal wave of data: NCBI RefSeq Protein Sequence Record growth (2004–2014). The steady increase in the number of protein reference sequences presents computational challenges to functional annotation of environmental sequence information using seed-and-extend algorithms such as BLAST or LAST.

of the biggest computational challenges when processing environmental sequence information due to run-time complexity of all-versus-all BLAST searches.

In addition to RefSeq, more specialized protein databases exist for metabolic reconstruction including KEGG and MetaCyc albeit with reduced taxonomic coverage. These databases have more stringent curation requirements based on controlled vocabularies and defined organizational schemas. For example, the Kyoto Encyclopedia of Genes and Genomes (KEGG) provides an integrated database of genes, enzymes, metabolites, and pathway maps based on enzyme commission (EC) numbers (65). The MetaCyc collection of genes, pathways, and metabolites contains more EC numbers than KEGG and encompasses an extensive collection of evolutionarily conserved pathways and variants specific to individual organisms or taxonomic groups represented in the BioCyc collection of PGDBs (44, 66–68). MetaPathways contains example versions of both databases that will automatically be formatted for BLAST/LAST queries. Users can add additional databases to simply search by copying an unformatted fasta file into the database directory. This file will automatically be formatted prior to the next seed-and-extend search.

Database queries can be run locally or externalized onto high-performance computing resources via an *ad hoc* distributed system (see Section 2.4). BLAST score ratios are calculated by dividing the query score against the self-BLAST or reference score (69). BLAST score ratios greater than 0.4 are considered reasonable for parsing functional annotations. KEGG and MetaCyc annotations are returned with their declination on their respective functional hierarchies with results exported to the results/annotation_tables/ folder along with RefSeq annotations. Functional annotations for each read are summarized in tabular format in the results/annotation_tables/ directory or viewed directly in the MetaPathways GUI. In subsequent processing steps, this information is combined with the annotated.gff file to generate input files for ePGDB construction and a standard .gbk file for GenBank submission.

2.3.3   Analysis Modules

The third stage of pipeline operation consists of modular analyses currently focused on transfer RNA (tRNA) and rRNA gene identification as well as taxonomic annotation and functional gene profiling.

2.3.3.1   tRNA Gene Identification   Transfer RNA genes comprise an extensive multicopy gene family tuned to codon usage and gene expression patterns in viral and cellular genomes. MetaPathways implements tRNA scan (version 1.4) using default parameters to identify tRNAs from QC nucleotide sequences (70). Resulting tRNA identifications are exported to the results/tRNA/ directory.

2.3.3.2   rRNA Gene Identification   All cellular organisms encode ribosomal RNA genes (Figure 2.5a and b). These ribonucleoprotein machines are integral to protein synthesis within the cell providing a translational mechanism for biological information flow between DNA, RNA, and protein. As universal marker genes, rRNA sequences can be used to construct phylogenetic trees unifying all three domains

(a)

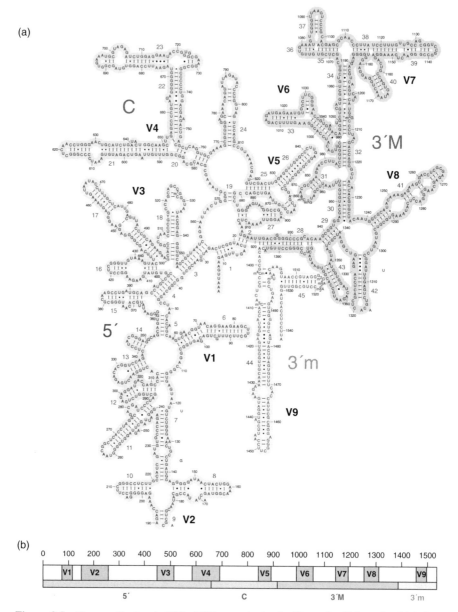

(b)

Figure 2.5 The small-subunit SSU rRNA gene is the "gold standard" for microbial diversity studies. (a) The *E. coli* SSU rRNA transcript and its folding structure contain four main domains (5′, C, 3′M, and 3′m) and nine hyper-variable regions (V1–V9) that correspond to areas with higher mutation rates Source: Dr Harry Noller, RNA Center, University of California, Santa Cruz, USA. Reproduced with permission of Dr Harry Noller. (b) A linearized diagram of the SSU rRNA showing the relative position of each domain and hyper-variable region. (c) By aligning rRNA sequences from multiple organisms, a three-domain phylogenetic tree can be generated revealing that the largest proportion of genetic diversity is represented by microorganisms. Please see www.wiley.com/go/Mandoiu/NextGenerationSequencing for a color version of this figure.

(c)

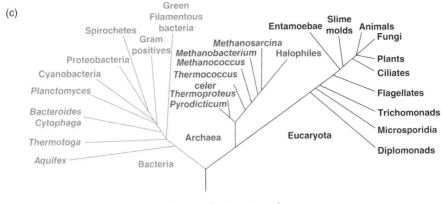

Figure 2.5 (*Continued*)

of life (Figure 2.5c) (71, 72). Indeed, prokaryotic small and large subunit ribosomal RNA (SSU/16S and LSU/23S rRNA) genes represent "gold standards" in the
study of microbial diversity (71, 72). Cultivation-independent rRNA gene surveys
over the past two decades have revolutionized our perception of the microbial world
and revealed the presence of numerous candidate divisions, the so-called microbial
dark matter, known solely on the basis of rRNA gene sequence information.

Today rRNA gene surveys represent a growth industry relevant to environmental
monitoring and human health and disease. High-throughput amplicon sequencing
projects based on polymerase chain reaction amplification with the so-called
"universal" primer sets (73) are expanding sequence databases used for taxonomic
profiling in natural and engineered ecosystems (GreenGenes, Silva, RDP2, etc.)
(74–76). While community composition profiling can provide quantitative insights
into microbial ecology, they are limited in their capacity to determine metabolic
pathways within uncultivated microbial communities. Although methods to extrap-
olate functional potential based on rRNA gene abundance in amplicon data sets
have been developed (77), identification of rRNA gene abundance directly from
metagenomic data sets provides a more accurate representation of donor genotypes
within a sample that can be directly related to metabolic pathway information based
on functional gene annotations (24).

MetaPathways uses BLAST to query assembled or unassembled nucleotide
sequences against reference nucleotide databases including SILVA (75) and Green-
Genes (74) to identify rRNA genes. Resulting rRNA identifications as summary
tables containing BLAST e-values, percent identities, bit scores, lengths, and
taxonomic identity are exported for each reference database (.rRNA.stats.txt),
and identified rRNA sequences can be exported as a feature in the MetaPathways
v2.0 GUI.

2.3.3.3 Clusters of Orthologous Groups The use of conserved single-copy functional genes provides a powerful adjunct or alternative to rRNA amplicon sequencing
when profiling microbial community structure. Indeed, COGs provide collections of

anchor genes useful in estimating genome completion or coverage and in the construction of phylogenetic trees (78, 79). Each COG contains orthologous proteins shared between at least three lineages. Because orthologous proteins tend to share equivalent functions, this information can be propagated throughout a cluster providing a rapid route for functional annotation of individual genomes or metagenomic data sets (Figure 2.6a). Interestingly, environmental sequencing projects have generated a large number of hypothetical ORFs that are conserved across taxonomic groups. While incorporating these novel sequences into COGs have traditionally relied on manual annotation, the sheer number of incoming sequences has driven the development of automated alternatives including the EggNOG database (80). Now in its second version, EggNOG uses reciprocal BLAST and the "triangle homology" method to build new clusters and append to existing ones (81). While the COG database is included in MetaPathways, the EggNOG database can be added as an unformatted fasta file to the database directory. This file will automatically be formatted prior to the next seed-and-extend search and results will be combined with other database outputs.

MetaPathways uses BLAST to query predicted protein sequences against reference databases including RefSeq and COG for taxonomic annotation. BLAST score ratios are calculated and reported as described in the functional annotations section. COG annotations are returned with their declination on the COG functional hierarchy with results exported to the results/annotation_tables/ folder. COG annotations for each read are summarized in a tabular format in the results/annotation_tables/ directory or viewed directly on screen in the MetaPathways v2.0 GUI.

MetaPathways also supports functional gene profiling using MLTreeMap (37). MLTreeMap automatically places marker genes onto a highly resolved reference phylogeny made from a multiple sequence alignment or "supermatrix" of 40 universal COGs (82). Taxonomic assignment proceeds in three steps. First marker genes are identified using nucleotide or protein sequences as input using BLAST and GeneWise (33, 83). Detected marker genes are added to curated reference alignments using hmmalign and Gblocks (84, 85). These aligned sequences are then placed into annotated reference phylogenies using RAxML (86, 87) (Figure 2.6b). In addition to universal COG reference phylogenies, user-defined trees can be appended to MLTreeMap supporting both phylogenetic and functional anchor screening.

2.3.3.4 Lowest Common Ancestor Taxonomic annotation of environmental sequence information is sensitive to read length and assembly. Although the presence of phylogenetic anchors such as rRNA genes on assembled contigs enables direct taxonomic assignment of linked ORFs, individual ORF assignments based on BLAST or LAST searches can be confounded by shared sequence homology between a number of different taxonomic groups. The LCA algorithm originally incorporated into MEGAN (34) uses BLAST scores to conservatively place sequences onto the NCBI taxonomic hierarchy. The LCA represents the lowest node in the NCBI taxonomy to which a sequence can be assigned based on the expected taxonomic range of its corresponding BLAST hits (Figure 2.6c). MEGAN LCA uses several user-defined criteria for determining this range, including a bit score cutoff, best hit cutoff, and minimum support percent. MetaPathways implements the LCA

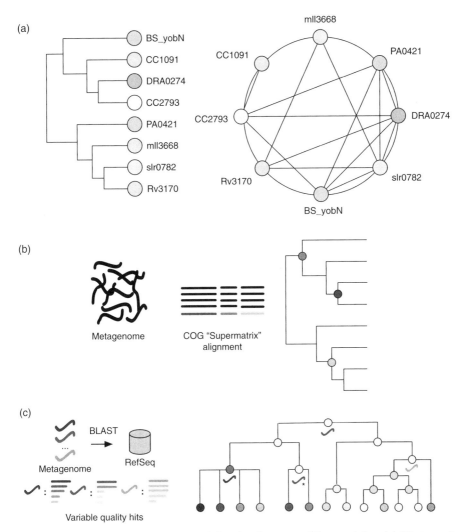

Figure 2.6 Taxonomic assignment and functional gene profiling modules: (a) Clusters of orthologous groups (COGs) are constructed using a "triangle homology" method. Similar to rRNA genes, many COGs can be used as phylogenetic anchors. (b) MLTreeMap leverages a subset of 40 universal COGs aligned and concatenated into a "supermatrix." Metagenomic reads can be added to this alignment and placed on the tree of life using a maximum likelihood method. (c) MEGAN parses BLAST outputs and projects this information onto the NCBI taxonomic hierarchy using the lowest common ancestor (LCA) ancestor algorithm. MEGAN also supports KEGG and SEED subsystems mapping. Please see www.wiley.com/go/Mandoiu/ NextGenerationSequencing for a color version of this figure.

algorithm for taxonomic annotation and exports the resulting taxonomic hierarchy information to results/annotation_tables/folder. Taxonomic annotations for each read are summarized in a tabular format in the results/annotation_tables/directory or viewed directly in the MetaPathways GUI.

2.3.4   ePGDB Construction

Pathway-centric approaches that predict metabolic networks using defined biochemical rules offer a more robust route for quantitative and predictive metabolic insights. Recently, pathway prediction and metabolic flux algorithms have been developed for microbial communities including the Human Microbiome Project Unified Metabolic Analysis Network (HUMAnN) and Predicted Relative Metabolic Turnover (PRMT), respectively. The HUMAnN combines gene prediction with MinPath (88), a parsimony approach for metabolic reconstruction based on KEGG pathway mapping (http://www.genome.jp/kegg/) (12, 65). PRMT predicts metabolic flux using KEGG pathways across multiple metagenomes (89). Neither HUMAnN nor PRMT support scalable data processing steps needed to annotate large environmental data sets, and neither offers a coherent data structure for exploring and interpreting predicted metabolic pathways. In addition to HUMAnN and PRMT, alternative algorithms for metabolic flux modeling within engineered microbial consortia are emerging although their application to environmental sequence information remains to be determined (90, 91).

MetaPathways uses Pathway Tools, a production-quality software environment developed at SRI International that supports Pathway/Genome database (PGDB) construction (41, 42). A PGDB contains information on the genes, reactions, and pathways that define metabolic and regulatory networks within individual organisms. The PathoLogic algorithm implemented in Pathway Tools allows users to construct new PGDBs from an annotated genome using MetaCyc (http://metacyc.org/) (43, 44), a highly curated, nonredundant, and experimentally validated database of metabolic pathways representing all domains of life. Unlike KEGG or SEED subsystems, MetaCyc emphasizes smaller, evolutionarily conserved or co-regulated units of metabolism and contains the largest collection (over 2000) of experimentally validated metabolic pathways (68). Navigable and extensively commented enzyme properties, pathway descriptions, and literature citations combined within a PGDB provide a coherent structure for exploring and interpreting metabolic networks. Although initially conceived for cellular organisms, MetaPathways extends the PGDB concept to environmental sequence information enabling ePGDB construction compatible with the editing, search, and navigation features of Pathway Tools (39).

MetaPathways concatenates predicted ORFs into a single "chromosomal" element as input to PathoLogic. Gene product descriptions are also harmonized across different protein databases using a scoring system based on keyword annotations and valid EC numbers to improve prediction performance. When constructing PGDBs for cellular organisms, PathoLogic uses a process called taxonomic pruning to constrain pathway predictions within a curated taxonomic range. Thus, inferring a pathway in a bacterium that is constrained to animals is contraindicated. Although this improves prediction accuracy for individual genomes of known origin (92), it can severely underestimate the metabolic potential of microbial communities composed of many interacting taxa (3). MetaPathways disables taxonomic pruning by default opening ePGDBs to the structural and functional diversity of microbial communities. A weighted taxonomic distance (WTD) is calculated for each pathway predicted from environmental sequence information providing a measure of taxonomic agreement between the realized and the expected taxonomic range associated with a given pathway (3). The WTD can be used to reduce false discovery or point to potentially unexpected or novel metabolic potential.

Once loaded within Pathway Tools, users can interact with and compare ePGDBs at the gene, reaction, pathway, or cellular levels (Figure 2.7). Because each pathway is typically composed of coding sequences from multiple genotypes, each pathway is a distributed function unless otherwise indicated by uniform taxonomic distribution or sequence coverage information. Direct comparisons between ePGDBs can be made on the cellular overview revealing similarities and differences in the abundance and taxonomic range of predicted metabolic pathways. In addition to interactive search and visualization features, Pathway Tools provides a metabolic encyclopedia, based on primary literature citations encompassing annotated MetaCyc pathways. MetaPathways can export this information in a tabular format where it can be viewed directly on screen or in the results/annotation_tables/directory.

2.4   BIG DATA PROCESSING

As indicated in the introductory paragraph, an expanding wave of environmental sequence information is inundating the user community with the potential to inform new scientific insights and biotechnological innovations. Processing and integrating this information is very much a Big Data problem that many researchers are not entirely prepared to solve. Typical environmental sequencing projects produce gigabases to terabases of sequence information with associated metadata that requires high-performance computing resources on grids or clouds to process and integrate. However, the use of HPC resources presents an additional layer of logistical and operational complexity. Most federated grid systems limit the number of jobs a user can submit into a batch-processing queue. Thus, job management can be improved with algorithms that split and load balance tasks across improved task environments. The most recent MetaPathways v2.0 release provides a number of improvements that directly address common data processing and integration challenges. From the standpoint of data processing, multiple grid management allows computationally intensive sections of the pipeline to be performed by multiple compute grids simultaneously in an ad hoc distributed system, accommodating dynamic availability, addition, and removal of compute clusters. From the standpoint of data integration, usability is improved through the development of a GUI for parameter setup, run monitoring, and result management. A custom Knowledge Engine enables real time and efficient interaction with summary tables, visualization modules, and data export features.

2.4.1   A Master–Worker Model for Grid Distribution

To simultaneously manage multiple dynamic grids with different processing capabilities, MetaPathways v2.0 implements a BlastBroker class that communicates and submits jobs to worker grids on the user's behalf. Specifically, the Broker monitors job submission and run status and manages worker throughput and job migration to improve processing resilience (Figure 2.8). The MetaPathways' Broker model

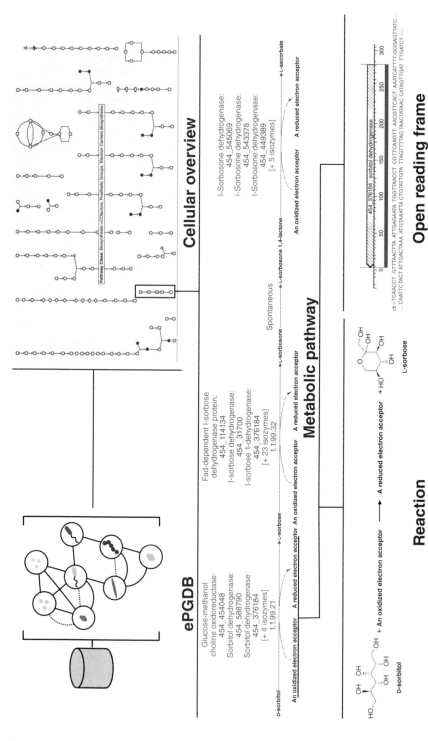

ePGDB

Glucose-methanol
choline oxidoreductase:
454_454048
Sorbitol dehydrogenase:
454_588790
Sorbitol dehydrogenase:
454_376184
[+ 4 isozymes]
1.1.99.21

Fad-dependent l-sorbose
dehydrogenase protein:
454_114134
l-sorbose dehydrogenase:
454_31700
l-sorbose 1-dehydrogenase:
454_376184
[+ 23 isozymes]
1.1.99.32

l-Sorbosone dehydrogenase:
454_545069
l-Sorbosone dehydrogenase:
454_543378
l-Sorbosone dehydrogenase:
454_449389
[+ 5 isozymes]

Cellular overview

D-sorbitol → L-sorbose → L-sorbosone → L-sorbosone 1,4-lactone → L-ascorbate

Spontaneous

An oxidized electron acceptor A reduced electron acceptor An oxidized electron acceptor A reduced electron acceptor An oxidized electron acceptor A reduced electron acceptor

Metabolic pathway

An oxidized electron acceptor ⟶ A reduced electron acceptor + HO

L-sorbose

OH + An oxidized electron acceptor ⟶ A reduced electron acceptor

D-sorbitol

Reaction

454_376184 : sorbitol dehydrogenase

0   50   100   150   200   250   300

ctt =TCAACCT GTTTAACTTA ATTGAGAATA TGGTTAACCT CGTTCAAGTT AACGTTCACT AAATCATTTT GGGAGTTATC...
         ... CAATTCTACT ATTGACTAAA ATGTAAATTA CTGTATTGTA TTAGTTTTAG TAACTATAAC CATAGTTGAT TTGATCT

Open reading frame

Figure 2.7 ePDGB navigation. An ePGDB can be interactively queried at multiple levels of biological organization from higher level cellular and pathway views down to individual coding sequences and enzymatic reactions (39). Source: Konwar, http://bmcbioinformatics.biomedcentral.com/
articles/10.1186/1471-2105-14-202. Used under CC BY 2.0 http://creativecommons.org/licenses/by/2.0/.

assumes an adversary that can sabotage the distributed setup in three ways: (i) a grid core can fail or become ineffective, causing a loss of jobs currently being computed on that core, (ii) an entire grid can fail or become ineffective, requiring all jobs to be migrated to other grids, or (iii) a sporadic or failed Internet connection increases the asynchronicity of the system, affecting the reliable submission and harvesting of results. The Broker handles each scenario in an intuitive way through a combination of job re-submission, job migration to other worker grids, and an escalating back-off in job submission and job harvesting from problematic grids.

2.4.2 GUI and Data Integration

MetaPathways v2.0 provides an interactive GUI improving parameter setup, run monitoring, and result management. Pipeline activities are configured via a configuration window that allows specification of input and output files, quality control parameters, target databases, executable stages, and grid computation. Grid systems and credentials are added and stored in an Available Grids window, allowing the user to add additional grids when credentials become available. Moreover, MetaPathways v2.0 implements a custom Knowledge Engine data structure and file indexing scheme that connects millions of data primitives, such as reads, ORFs, and metadata, and projects them onto a specified classification or hierarchy (e.g., KEGG, COG, Meta-Cyc, and the NCBI Taxonomy database) (Figure 2.9). The benefit of this structure is customizability and performance. Indeed, data primitives are extensible to the addition of new data types including transcriptome and proteome sequence information. From the standpoint of performance, tables of millions of rows of annotations can now be quickly queried (sorted, searched, overviewed, and exported) much faster than typical database schemas including MySQL or Oracle. Additionally, the use of the Knowledge Engine data structure enables the development of high-performance visualization modules for users to interactively query their data including tree-based taxonomy and contig viewing (Figure 2.10). This structure also supports versatile export features including subsets of sequences or annotations in a variety of formats compatible with downstream analysis in different software environments.

2.5 DOWNSTREAM ANALYSES

Comparative community analysis integrating pathway and environmental parameter information is a primary objective of most environmental sequencing projects. MetaPathways facilitates comparative community analysis through a uniform directory structure enabling programmatic integration of output files. The "Large Table" data feature available in MetaPathways 2.0 facilitates browsing through millions of functional and taxonomic annotations from one or more samples in real time and then to export this information in a user-defined manner. At the pathway level, Pathway Tools supports ePGDB comparisons via the cellular overview feature. Moreover, ORF abundance and taxonomic assignment information can be incorporated into the cellular overview as metadata providing a more nuanced perspective of microbial

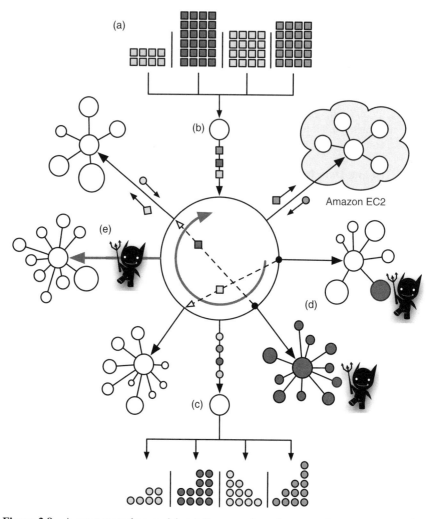

Figure 2.8 A master–worker model. (a) Sequence homology searches present an embarrassingly parallel problem. (b) The MetaPathways Broker distributes a BLAST job into equal-sized subtasks. The broker establishes blast services (with all the required executables and databases) on worker grids available to accept incoming tasks. Jobs (squares) are submitted in a round-robin manner to each worker grid. (c) The broker intermittently harvests results (circles) from each grid as they become available, demultiplexing if there are multiple samples being run. (d) An adversary can cause nodes or entire grids to fail at random. The broker provides fault tolerance by migrating lost jobs (dashed lines) to alternative grids. (e) An adversary can also cause intermittent or failed Internet connections (the line with the cartoon of a demon). The broker uses exponential back-off to determine job migration to other girds if latency becomes excessive. Source: Hanson et al. 2014 (40). Reproduced with permission of IEEE.

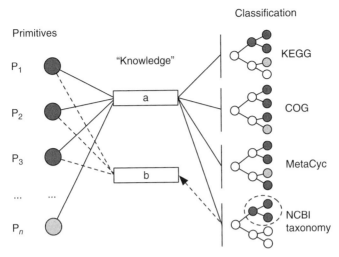

Figure 2.9   The "Knowledge Engine" data structure. This data structure considers sequence reads or predicted ORFs as data primitives that can be summarized by projection onto a series of classification schemes using pointer following. Pointer following is a computationally efficient operation so the identification and enumeration of data primitives are robust to large environmental data sets. The connection between primitives and classification schemes are called Knowledge objects. Through the exploration of data projected on classification schemes, new Knowledge objects can be created (dashed lines). Once created, Knowledge objects can be projected onto tables or other visualization modes that can in turn be used to create new Knowledge objects, enabling iterative and interactive data exploration. Source: Hanson et al. 2014 (40). Reproduced with permission of IEEE.

community metabolism. Finally, pathways and ORFs can be exported from MetaPathways and Pathway Tools to flat files that can be formatted for downstream statistical analyses in the R statistical programming environment.

2.5.1   Large Table Comparisons

MetaPathways supports display and interactive comparison of functional and taxonomic annotations from multiple samples via the Large Table feature. Tabulated annotations can be compared in a supermatrix composed of product descriptions, taxonomic assignments, and predicted pathways, each declined onto the KEGG, COG, SEED, NCBI Taxonomy, and MetaCyc hierarchies, using the Knowledge Engine data structure. In addition to supporting standard search, sort, and export functions, the Knowledge Engine data structure allows users to dynamically compare and export subsets of genes or pathways across hierarchical levels even when samples each contain millions of annotations.

2.5.2   Pathway Tools Cellular Overview

The features of Pathway Tools Cellular Overview provide an holistic representation of metabolism for single cells and microbial communities (93). The feature displays

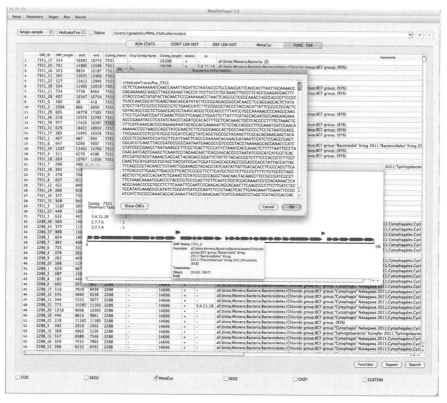

Figure 2.10 Knowledge objects projections. A Large Table view enables efficient query, look-up, and sub-setting of reads, ORFs, translated ORFs into amino acid sequences, statistics, hierarchical annotations, while a Contig View enables genome context navigation based on ORF positions with functional and taxonomic annotations for the ORFs, on both strands, appearing as tool-tips pop-ups. Source: Hanson et al. 2014 (40). Reproduced with permission of IEEE.

pathways predicted by the PathoLogic algorithm as glyphs, highlighting annotations mapped to reactions in each pathway. While this is a useful view for identifying pathways of interest in individual ePGDBs, it can also support comparisons between multiple ePGDBs using Pathway Tools highlighting features. For example, pathways from any two ePGDBs can be highlighted in the Cellular Overview, a list of Gene or Pathway Names, or via count data mapped to the MetaCyc schema via the Omics Data feature (Figure 2.11). The Omics Data feature has the additional benefit of being integrated with tool-tip visualizations on both the Cellar Overview and the individual pathway pages. Tool-tips can be used to highlight ORF abundance within predicted pathways or the taxonomic distribution of reactions within a pathway using a workaround that represents names as numbers. Future versions of Pathway Tools should enable qualitative variables to be displayed on the Cellular Overview or individual pathway pages more easily.

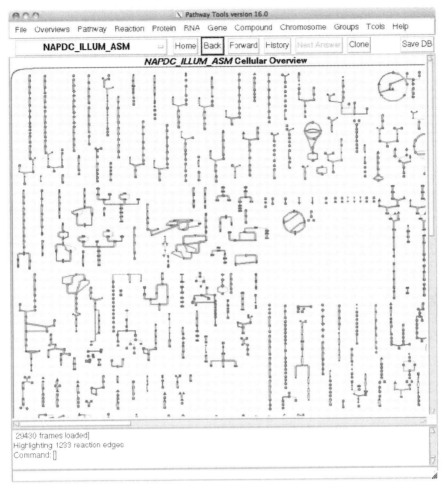

Figure 2.11 Comparative pathway analysis. Two or more ePGDBs can be compared using the Cellular Overview and Omics Data features (glyphs represent pathways predicted for one sample using three different sequencing methods) (39). Source: Konwar, http://bmcbioinformatics.biomedcentral.com/articles/10.1186/1471-2105-14-202. Used under CC BY 2.0 http://creativecommons.org/licenses/by/2.0/.

2.5.3 Statistical Analysis with R

The R statistical environment provides a versatile toolkit for conducting comparative analyses and generating useful data products. R is well adopted by scientists and statisticians alike, having many useful packages for data analysis and publication quality visualization. However, there is also a significant activation barrier to becoming R proficient. Given that MetaPathways supports export file formats compatible with downstream analyses using R, it is useful to examine specific use cases that lower this activation barrier.

(a)   Wide

Sample	Bacteria	Archaea	Eukaryotes	Salinity	Depth	DOC
1	30	2	3	35.08	10	78
2	25	5	5	35.21	70	79
3	2	20	4	35.04	130	69
4	3	22	6	34.07	200	63

melt(wide, id.vars=c("Sample", "Depth"))   dcast(long, Sample + Depth ~ variable)

(b)   Long

Sample	Depth	Variable	Value
1	10	Bacteria	30
2	70	Bacteria	25
3	130	Bacteria	2
4	200	Bacteria	3
1	10	Archaea	2
2	70	Archaea	5
3	130	Archaea	20
4	200	Archaea	22
1	10	Eukaryotes	3
2	70	Eukaryotes	5
3	130	Eukaryotes	4
4	200	Eukaryotes	6
1	10	Salinity	35.08
2	70	Salinity	35.21
3	130	Salinity	35.04
4	200	Salinity	34.07
1	10	DOC	78
2	70	DOC	79
3	130	DOC	69
4	200	DOC	63

One of the primary barriers for biologists using statistical code is data formatting. Input file formats must be delimited correctly and free of noncompliant characters or fields. This often requires a nontrivial and sometimes byzantine task of (re)processing output data files known as data cleaning or wrangling. Fortunately, Pathway Tools allows for ePGDBs to be programmatically accessed for downstream analysis via its API mode through the perlCyc and javaCyc packages (http://arabidopsis.org/biocyc/perlcyc/index.jsp) for the Perl and Java programming languages, respectively. For example, the script extract_pathway_table_from_pgdb.pl included with MetaPathways is able to extract pathways from ePGDBs in Pathway Tools in an R-parsable, tab-delimited format.

Two common ways data tables are formatted for statistical software are the so-called wide and long table formats; a part of a collection of data organization concepts that Hadley Wickham describes as "tidy data" (http://vita.had.co.nz/papers/tidy-data.pdf). The wide table format is intuitive to most users (it is sometimes referred to as a "master table"). In wide format, each variable has its own column. While compact and appropriate for some analytical tasks (principal component analysis (PCA), Clustering, etc.), for other packages such as ggplot2 a long table format is adopted where each row represents one individual observation on a particular variable (Figure 2.12). The long table format is useful for sophisticated plotting packages such as lattice and ggplot2. Additionally, the melt() and dcast() functions in the reshape2 package (94) serve as a way to swap between the two formats; melt() converts wide tables to long format; and dcast() converts long tables to wide format, using the R formula operator ($\sim$).

2.5.4   Venn Diagrams

When analyzing sets of observed objects such as functional annotations, taxonomic assignments, or pathway predictions, one might ask what is common or unique between sets. Once a user has loaded a list of pathways into R, it becomes possible to isolate the list of pathways found in each sample and compare differences setdiff(a,b) and similarities intersect(a,b) between sets. The lists generated by these set operations will be useful for selecting and setting up plots and identifying pathways of interest for more in-depth exploration. A common way to display set differences is to use a Venn Diagram (Figure 2.13a). The VennDiagram package provides an easy to use solution (45). However, this script assumes that you have figured out all possible combinations of sets, which can quickly get complicated

---

Figure 2.12 Tables of long and wide format. A key "tidy data" concept is the format of long and wide tables. Algorithms implementing hierarchical clustering and cdimensionality-reduction techniques such as principal component analysis (PCA) and non-metric multidimensional scaling (NMDS) often require the wide format (a), while many other plotting packages such as ggplot2 and lattice require the long format (b). It is important to understand how to use the melt() and dcast() functions in the reshape2 R package to convert between the two formats.

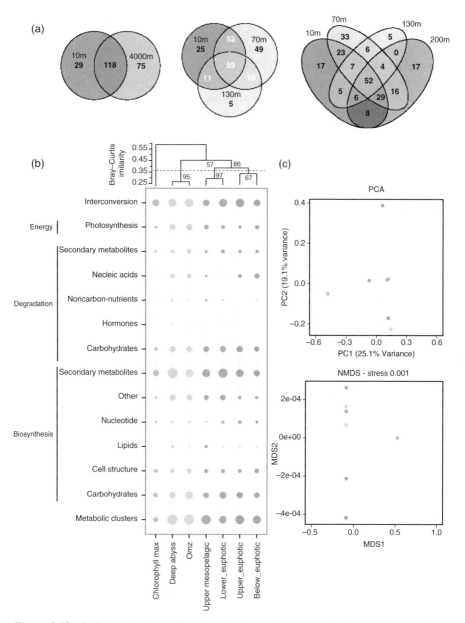

Figure 2.13 R data products. (a) Shared and unique pathways can be identified using a Venn diagram. (b) Hierarchical clustering or (c) dimensionality reduction methods such as PCA or NMDS can be used to determine the extent to which pathway profiles are shared between environmental samples. The grid visualizations framework ggplot2 can be used to create a wide variety of plots based on the Grammar of Graphics framework. In (b), pathways have been declined to the second level of the MetaCyc hierarchy. The areas of each circle indicate pathway abundance. Please see www.wiley.com/go/Mandoiu/NextGenerationSequencing for a color version of this figure.

when comparing multiple data sets as the number of nonempty sets is $2^n - 1$, where n is the total number of sets to compare.

2.5.5   Clustering and Relating Samples by Pathways

Comparing ORF counts across predicted pathway profiles for multiple samples provides is another common analysis mode amenable to hierarchical clustering. A distance or similarity measure is used to iteratively group samples together generating hierarchical groups in the process (Figure 2.13b). A point to consider is the distance measure and data transform being used, environmental data sets often have many zero entries (the so-called double-zero problem) and a wide range of observed count data. To deal with these two problems, different similarity measures such as Manhattan distance or Bray–Curtis similarity have been adopted to focus on differences of observed nonzero data, and the log or square-root transforms are used to pull-in data with a large dynamic range. Clustering with different transforms and distance measures can be conducted using hclust and pvclust packages in R, with bootstrapped subsamples providing a measure of statistical confidence in resulting cluster groups (46, 95).

Another way to highlight similarity or differences between environmental data sets involves dimensionality reduction techniques such as PCA or nonmetric multidimensional scaling (NMDS) (Figure 2.13c). Intuitively one can think of PCA as being motivated from the fact that all multidimensional data sets can be viewed as a high-dimensional ellipse. PCA rotates and shifts this ellipse such that the longest axis, the first principal component, aligns with the direction of highest variance in the data, the next components following orthogonally based on this first axis. NMDS takes objects in a high-dimensional space, for example, pathway profiles, and places them in a lower dimensional space while maintaining their relative distances as best as possible. A number of operational settings need to be considered when generating PCA and NMDS plots, the details of which are beyond the scope of this chapter. Readers are encouraged to familiarize themselves with ordination and dimensionality reduction techniques described in numerical ecology texts (96–98).

2.5.6   Faceting Variables with ggplot2

Finally, another approach to analyzing environmental sequence information is to isolate interesting subsets of pathways in relation to specific physical, chemical, or process variables, a process known as faceting. Two frameworks for doing this in the R statistical environment are ggplot2 and lattice (47, 99), allowing subsets of data to be displayed in small multiple plots based on user-defined variables. ggplot2 in particular is a full-featured and flexible framework based on the principles outlined in the Grammar of Graphics (100), which is able to automatically construct a variety of data plots based on drawing primitives, including bar plots, heatmaps, bubble plots, and so on. Readers are encouraged to familiarize themselves with this framework in more detail given its ability to produce publication quality figures.

2.6 CONCLUSIONS

In this chapter, we presented conceptual background and contemporary methods for analysis of environmental sequence information using MetaPathways. Beginning with methods for ORF prediction and annotation using an ever-growing ecosystem of algorithms and reference databases, we then described taxonomic and functional gene profiling using the LCA algorithm and MLTreeMap and the construction and exploration of ePGDBs using Pathway Tools before presenting useful comparative analysis methods in the R statistical environment. The design and iterative development of MetaPathways focuses a spotlight on Big Data challenges associated with environmental sequence analysis. These include computationally intensive pipeline steps related to scalable seed-and-extend searches, data integration across multiple information levels, and interactive visualization of embarrassingly large data sets. We highlighted recent pipeline innovations related to several of these challenges including a grid and cloud distribution system, GUI, and Knowledge Engine data structure that integrates millions of functional and taxonomic annotations across multiple samples. Future versions of MetaPathways will build on distributed computing and interactive visualization themes and append modules for single-cell genomic and population genome assembly and binning. At the same time, parallel improvements to import/export features of Pathway Tools promise to enhance Cellular Overview and Omics Data features enabling more effective metabolic pathway reconstruction and comparative community analysis.

REFERENCES

1. Whitman WB, Coleman DC, Wiebe WJ. Prokaryotes: the unseen majority. Proc Natl Acad Sci U S A 1998;95(12):6578–6583.
2. Falkowski PG, Fenchel T, Delong EF. The microbial engines that drive Earth's biogeo-chemical cycles. Science 2008;320(5879):1034–1039.
3. Hanson NW, Konwar KM, Hawley AK, Altman T, Karp PD, Hallam SJ. Metabolic pathways for the whole community. BMC Genomics 2014;15:619.
4. Delong EF, Preston CM, Mincer T, Rich V, Hallam SJ, Frigaard N-U, Martinez A, Sullivan MB, Edwards R, Brito BR, Chisholm SW, Karl DM. Community genomics among stratified microbial assemblages in the ocean's interior. Science 2006;311(5760): 496–503.
5. Walsh DA, Zaikova E, Howes CG, Song YC, Wright JJ, Tringe SG, Tortell PD, Hallam SJ. Metagenome of a versatile chemolithoautotroph from expanding oceanic dead zones. Science 2009;326(5952):578–582.
6. Shi Y, Tyson GW, Eppley JM, Delong EF. Integrated metatranscriptomic and metagenomic analyses of stratified microbial assemblages in the open ocean. ISME J 2011;5(6):999–1013.
7. Fierer N, Jackson RB. The diversity and biogeography of soil bacterial communities. Proc Natl Acad Sci U S A 2006;103(3):626–631.
8. Warnecke F, Luginbühl P, Ivanova N, Ghassemian M, Richardson TH, Stege JT, Cayouette M, McHardy AC, Djordjevic G, Aboushadi N, Sorek R, Tringe SG, Podar M, Martín HG, Kunin V, Dalevi D, Madejska J, Kirton E, Platt D, Szeto E, Salamov A, Barry K,

Mikhailova N, Kyrpides NC, Matson EG, Ottesen EA, Zhang X, Hernández M, Murillo C, Acosta LG, Rigoutsos I, Tamayo G, Green BD, Chang C, Rubin EM, Mathur EJ, Robertson DE, Hugenholtz P, Leadbetter JR. Metagenomic and functional analysis of hindgut microbiota of a wood-feeding higher termite. Nature 2007;450(7169):560–565.

9. Yooseph S, Sutton G, Rusch DB, Halpern AL, Williamson SJ, Remington K, Eisen JA, Heidelberg KB, Manning G, Li W. The Sorcerer II Global Ocean Sampling expedition: expanding the universe of protein families. PLoS Biol 2007;5(3):e16.

10. Tyson GW, Chapman J, Hugenholtz P, Allen EE, Ram RJ, Richardson PM, Solovyev VV, Rubin EM, Rokhsar DS, Banfield JF. Community structure and metabolism through reconstruction of microbial genomes from the environment. Nature 2004;428(6978):37–43.

11. Turnbaugh PJ, Ley RE, Hamady M, Fraser-Liggett CM, Knight R, Gordon JI. The human microbiome project. Nature 2007;449(7164):804–810.

12. Abubucker S, Segata N, Goll J, Schubert AM, Izard J, Cantarel BL, Rodriguez-Mueller B, Zucker J, Thiagarajan M, Henrissat B, White O, Kelley ST, Methe B, Schloss PD, Gevers D, Mitreva M, Huttenhower C. Metabolic reconstruction for metagenomic data and its application to the human microbiome. PLoS Comput Biol 2012;8(6):e1002358.

13. Alcalde M, Ferrer M, Plou FJ, Ballesteros A. Environmental biocatalysis: from remediation with enzymes to novel green processes. Trends Biotechnol 2006;24(6):281–287.

14. Human Microbiome Project Consortium. Structure, function and diversity of the healthy human microbiome. Nature 2012;486(7402):207–214.

15. Cho I, Blaser MJ. The human microbiome: at the interface of health and disease. Nat Rev Genet 2012;13(4):260–270.

16. Taupp M, Mewis K, Hallam SJ. The art and design of functional metagenomic screens. Curr Opin Biotechnol 2011;22(3):465–472.

17. Mewis K, Armstrong Z, Song YC, Baldwin SA, Withers SG, Hallam SJ. Biomining active cellulases from a mining bioremediation system. J Biotechnol 2013;167(4):462–471.

18. Strachan CR, Singh R, VanInsberghe D, Ievdokymenko K, Budwill K, Mohn WW, Eltis LD, Hallam SJ. Metagenomic scaffolds enable combinatorial lignin transformation. Proc Natl Acad Sci U S A 2014;111(28):10 143–10 148.

19. Margulies M, Egholm M, Altman WE, Attiya S, Bader JS, Bemben LA, Berka J, Braverman MS, Chen Y-J, Chen Z, Dewell SB, Du L, Fierro JM, Gomes XV, Godwin BC, He W, Helgesen S, Ho CH, Irzyk GP, Jando SC, Alenquer MLI, Jarvie TP, Jirage KB, Kim J-B, Knight JR, Lanza JR, Leamon JH, Lefkowitz SM, Lei M, Li J, Lohman KL, Lu H, Makhijani VB, McDade KE, McKenna MP, Myers EW, Nickerson E, Nobile JR, Plant R, Puc BP, Ronan MT, Roth GT, Sarkis GJ, Simons JF, Simpson JW, Srinivasan M, Tartaro KR, Tomasz A, Vogt KA, Volkmer GA, Wang SH, Wang Y, Weiner MP, Yu P, Begley RF, Rothberg JM. Genome sequencing in microfabricated high-density picolitre reactors. Nature 2005;437(7057):376–380.

20. Quail MA, Kozarewa I, Smith F, Scally A, Stephens PJ, Durbin R, Swerdlow H, Turner DJ. A large genome center's improvements to the Illumina sequencing system. Nat Methods 2008;5(12):1005–1010.

21. Leung K, Zahn H, Leaver T, Konwar KM, Hanson NW, Pagé AP, Lo C-C, Chain PS, Hallam SJ, Hansen CL. A programmable droplet-based microfluidic device applied to multiparameter analysis of single microbes and microbial communities. Proc Natl Acad Sci U S A 2012;109(20):7665–7670.

22. Bayley H. Nanotechnology: holes with an edge. Nature 2010;467(7312):164–165.

23. Wooley JC, Ye Y. Metagenomics: facts and artifacts, and computational challenges. J Comput Sci Technol 2009;25(1):71–81.

24. Kunin V, Copeland A, Lapidus A, Mavromatis K, Hugenholtz P. A bioinformatician's guide to metagenomics. Microbiol Mol Biol Rev 2008;72(4):557–578.

25. O'Driscoll A, Daugelaite J, Sleator RD. 'Bigdata', Hadoop and cloud computing in genomics. J Biomed Inform 2013;46(5):774–781.

26. Collins JP, Gates B, Francis R, Kell D, Bell G. *The Fourth Paradigm: Data-Intensive Scientific Discovery*. Redmond, VA: Microsoft Research; 2009.

27. Meyer F, Paarmann D, D'Souza M, Olson R, Glass EM, Kubal M, Paczian T, Rodriguez A, Stevens R, Wilke A, Wilkening J, Edwards RA. The metagenomics RAST server - a public resource for the automatic phylogenetic and functional analysis of metagenomes. BMC Bioinformatics 2008;9:386.

28. Markowitz VM, Ivanova NN, Szeto E, Palaniappan K, Chu K, Dalevi D, Chen I-MA, Grechkin Y, Dubchak I, Anderson I, Lykidis A, Mavromatis K, Hugenholtz P, Kyrpides NC. IMG/M: a data management and analysis system for metagenomes. Nucleic Acids Res 2008;36(Database issue):D534–D538.

29. Markowitz VM, Chen I-MA, Chu K, Szeto E, Palaniappan K, Grechkin Y, Ratner A, Jacob B, Pati A, Huntemann M, Liolios K, Pagani I, Anderson I, Mavromatis K, Ivanova NN, Kyrpides NC. IMG/M: the integrated metagenome data management and comparative analysis system. Nucleic Acids Res 2012;40(Database issue):D123–D129.

30. Seshadri R, Kravitz SA, Smarr L, Gilna P, Frazier M. CAMERA: a community resource for metagenomics. PLoS Biol 2007;5(3):e75.

31. Bernal A, Ear U, Kyrpides N. Genomes OnLine Database (GOLD): a monitor of genome projects world-wide. Nucleic Acids Res 2001;29(1):126–127.

32. Kottmann R, Kostadinov I, Duhaime MB, Buttigieg PL, Yilmaz P, Hankeln W, Waldmann J, Glockner FO. Megx.net: integrated database resource for marine ecological genomics. Nucleic Acids Res 2009;38(Database):D391–D395.

33. Altschul SF, Gish W, Miller W, Myers EW, Lipman DJ. Basic local alignment search tool. J Mol Biol 1990;215(3):403–410.

34. Huson DH, Auch AF, Qi J, Schuster SC. MEGAN analysis of metagenomic data. Genome Res 2007;17(3):377–386.

35. Huson DH, Mitra S, Ruscheweyh H-J, Weber N, Schuster SC. Integrative analysis of environmental sequences using MEGAN4. Genome Res 2011;21(9):1552–1560.

36. Caporaso JG, Kuczynski J, Stombaugh J, Bittinger K, Bushman FD, Costello EK, Fierer N, Pe na AG, Goodrich JK, Gordon JI, Huttley GA, Kelley ST, Knights D, Koenig JE, Ley RE, Lozupone CA, McDonald D, Muegge BD, Pirrung M, Reeder J, Sevinsky JR, Turnbaugh PJ, Walters WA, Widmann J, Yatsunenko T, Zaneveld J, Knight R. QIIME allows analysis of high-throughput community sequencing data. Nat Methods 2010;7(5):335–336.

37. Stark M, Berger SA, Stamatakis A, von Mering C. MLTreeMap–accurate Maximum Likelihood placement of environmental DNA sequences into taxonomic and functional reference phylogenies. BMC Genomics 2010;11:461.

38. Letunic I, Bork P. Interactive Tree Of Life v2: online annotation and display of phylogenetic trees made easy. Nucleic Acids Res 2011;39(Web Server issue):W475–W478.

39. Konwar KM, Hanson NW, Pagé AP, Hallam SJ. MetaPathways: a modular pipeline for constructing pathway/genome databases from environmental sequence information. BMC Bioinformatics 2013;14(1):202.

40. Hanson NW, Konwar KM, Wu S-J, Hallam SJ. MetaPathways v2.0: a master-worker model for environmental Pathway/Genome Database construction on grids and clouds. Computational Intelligence in Bioinformatics and Computational Biology, 2014 IEEE Conference on; 2014. p 1–7.

41. Karp PD, Paley S, Romero P. The pathway tools software. Bioinformatics 2002;18 Suppl 1:S225–S232.

42. Karp PD, Paley SM, Krummenacker M, Latendresse M, Dale JM, Lee TJ, Kaipa P, Gilham F, Spaulding A, Popescu L, Altman T, Paulsen I, Keseler IM, Caspi R. Pathway tools version 13.0: integrated software for pathway/genome informatics and systems biology. Brief Bioinform 2010;11(1):40–79.

43. Karp PD, Riley M, Saier M, Paulsen IT, Paley SM, Pellegrini-Toole A. The EcoCyc and MetaCyc databases. Nucleic Acids Res 2000;28(1):56–59.

44. Caspi R, Altman T, Dreher K, Fulcher CA, Subhraveti P, Keseler IM, Kothari A, Krummenacker M, Latendresse M, Mueller LA, Ong Q, Paley S, Pujar A, Shearer AG, Travers M, Weerasinghe D, Zhang P, Karp PD. The MetaCyc database of metabolic pathways and enzymes and the BioCyc collection of pathway/genome databases. Nucleic Acids Res 2012;40(Database issue):D742–D753.

45. Chen H, Boutros PC. VennDiagram: a package for the generation of highly-customizable Venn and Euler diagrams in R. BMC Bioinformatics 2011;12:35.

46. Suzuki R, Shimodaira H. Pvclust: an R package for assessing the uncertainty in hierarchical clustering. Bioinformatics 2006;22(12):1540–1542.

47. Hadley W. ggplot2: Elegant graphics for data analysis; 2009.

48. Mathé C, Sagot M-F, Schiex T, Rouzé P. Current methods of gene prediction, their strengths and weaknesses. Nucleic Acids Res 2002;30(19):4103–4117.

49. Borodovsky M, McIninch JD. GeneMark: parallel gene recognition for both DNA strands. Comput Chem 1993;17(2):123–133.

50. Lukashin AV, Borodovsky M. GeneMark.hmm: new solutions for gene finding. Nucleic Acids Res 1998;26(4):1107–1115.

51. Besemer J, Borodovsky M. GeneMark: web software for gene finding in prokaryotes, eukaryotes and viruses. Nucleic Acids Res 2005;33(Web Server):W451–W454.

52. Delcher AL, Harmon D, Kasif S, White O, Salzberg SL. Improved microbial gene identification with GLIMMER. Nucleic Acids Res 1999;27(23):4636–4641.

53. Hoff KJ, Tech M, Lingner T, Daniel R, Morgenstern B, Meinicke P. Gene prediction in metagenomic fragments: a large scale machine learning approach. BMC Bioinformatics 2008;9:217.

54. Zhu W, Lomsadze A, Borodovsky M. Ab initio gene identification in metagenomic sequences. Nucleic Acids Res 2010;38(12):e132.

55. Hoff KJ. The effect of sequencing errors on metagenomic gene prediction. BMC Genomics 2009;10:520.

56. Noguchi H, Park J, Takagi T. MetaGene: prokaryotic gene finding from environmental genome shotgun sequences. Nucleic Acids Res 2006;34(19):5623–5630.

57. Noguchi H, Taniguchi T, Itoh T. MetaGeneAnnotator: detecting species-specific patterns of ribosomal binding site for precise gene prediction in anonymous prokaryotic and phage genomes. DNA Res 2008;15(6):387–396.

58. Kelley DR, Liu B, Delcher AL, Pop M, Salzberg SL. Gene prediction with Glimmer for metagenomic sequences augmented by classification and clustering. Nucleic Acids Res 2011;40(1):e9.

59. Hoff KJ, Lingner T, Meinicke P, Tech M. Orphelia: predicting genes in metagenomic sequencing reads. Nucleic Acids Res 2009;37(Web Server):W101–W105.

60. Liu Y, Guo J, Hu G, Zhu H. Gene prediction in metagenomic fragments based on the SVM algorithm. BMC Bioinformatics 2013;14 Suppl 5:S12.

61. Hyatt D, Chen G-L, LoCascio PF, Land ML, Larimer FW, Hauser LJ. Prodigal: prokaryotic gene recognition and translation initiation site identification. BMC Bioinformatics 2010;11(1):119.

62. Hyatt D, LoCascio PF, Hauser LJ, Uberbacher EC. Gene and translation initiation site prediction in metagenomic sequences. Bioinformatics 2012;28(17):2223–2230.

63. Bonferroni CE. In: Carboni SO, editor. *Il calcolo delle assicurazioni su gruppi di teste.* Rome: Tipografia del Senato; 1935. p 13–60.

64. Kiełbasa SM, Wan R, Sato K, Horton P, Frith MC. Adaptive seeds tame genomic sequence comparison. Genome Res 2011;21(3):487–493.

65. Kanehisa M, Goto S. KEGG: kyoto encyclopedia of genes and genomes. Nucleic Acids Res 2000;28(1):27–30.

66. Karp PD, Ouzounis CA, Moore-Kochlacs C, Goldovsky L, Kaipa P, Ahrén D, Tsoka S, Darzentas N, Kunin V, Lopez-Bigas N. Expansion of the BioCyc collection of pathway/genome databases to 160 genomes. Nucleic Acids Res 2005;33(19):6083–6089.

67. Caspi R, Foerster H, Fulcher CA, Hopkinson R, Ingraham J, Kaipa P, Krummenacker M, Paley S, Pick J, Rhee SY, Tissier C, Zhang P, Karp PD. MetaCyc: a multiorganism database of metabolic pathways and enzymes. Nucleic Acids Res 2006;34(Database issue):D511–D516.

68. Altman T, Travers M, Kothari A, Caspi R, Karp PD. A systematic comparison of the MetaCyc and KEGG pathway databases. BMC Bioinformatics 2013;14:112.

69. Rost B. Twilight zone of protein sequence alignments. Protein Eng 1999;12(2):85–94.

70. Lowe TM, Eddy SR. tRNAscan-SE: a program for improved detection of transfer RNA genes in genomic sequence. Nucleic Acids Res 1997;25(5):0955–0964.

71. Fox GE, Pechman KR, Woese C. Comparative cataloging of 16S ribosomal ribonucleic acid: molecular approach to procaryotic systematics. Int J Syst Bacteriol 1977;27(1):44–57.

72. Woese CR. Bacterial evolution. Microbiol Rev 1987;51(2):221.

73. Klindworth A, Pruesse E, Schweer T, Peplies J, Quast C, Horn M, Glöckner FO. Evaluation of general 16S ribosomal RNA gene PCR primers for classical and next-generation sequencing-based diversity studies. Nucleic Acids Res 2013;41(1):e1.

74. DeSantis TZ, Hugenholtz P, Larsen N, Rojas M, Brodie EL, Keller K, Huber T, Dalevi D, Hu P, Andersen GL. Greengenes, a chimera-checked 16S rRNA gene database and workbench compatible with ARB. Appl Environ Microbiol 2006;72(7):5069–5072.

75. Pruesse E, Quast C, Knittel K, Fuchs BM, Ludwig W, Peplies J, Glöckner FO. SILVA: a comprehensive online resource for quality checked and aligned ribosomal RNA sequence data compatible with ARB. Nucleic Acids Res 2007;35(21):7188–7196.

76. Cole JR. The Ribosomal Database Project (RDP-II): previewing a new autoaligner that allows regular updates and the new prokaryotic taxonomy. Nucleic Acids Res 2003;31(1):442–443.

77. Langille MGI, Zaneveld J, Caporaso JG, McDonald D, Knights D, Reyes JA, Clemente JC, Burkepile DE, Vega Thurber RL, Knight R, Beiko RG, Huttenhower C. Predictive functional profiling of microbial communities using 16S rRNA marker gene sequences. Nat Biotechnol 2013;31(9):814–821.

78. Tatusov RL, Koonin EV, Lipman DJ. A genomic perspective on protein families. Science 1997;278(5338):631–637.

79. Tatusov RL, Fedorova ND, Jackson JD, Jacobs AR, Kiryutin B, Koonin EV, Krylov DM, Mazumder R, Mekhedov SL, Nikolskaya AN, Rao BS, Smirnov S, Sverdlov AV, Vasudevan S, Wolf YI, Yin JJ, Natale DA. The COG database: an updated version includes eukaryotes. BMC Bioinformatics 2003;4:41.

80. Jensen LJ, Julien P, Kuhn M, von Mering C, Muller J, Doerks T, Bork P. eggNOG: automated construction and annotation of orthologous groups of genes. Nucleic Acids Res 2007;36(Database):D250–D254.

81. Muller J, Szklarczyk D, Julien P, Letunic I, Roth A, Kuhn M, Powell S, von Mering C, Doerks T, Jensen LJ, Bork P. eggNOG v2.0: extending the evolutionary genealogy of genes with enhanced non-supervised orthologous groups, species and functional annotations. Nucleic Acids Res 2010;38(Database issue):D190–D195.

82. Ciccarelli FD. Toward automatic reconstruction of a highly resolved tree of life. Science 2006;311(5765):1283–1287.

83. Birney E. GeneWise and genomewise. Genome Res 2004;14(5):988–995.

84. Durbin R. *Biological Sequence Analysis: Probabilistic Models of Proteins and Nucleic Acids*. Cambridge: University Press; 1998.

85. Castresana J. Selection of conserved blocks from multiple alignments for their use in phylogenetic analysis. Mol Biol Evol 2000;17(4):540–552.

86. Stamatakis A. RAxML-VI-HPC: maximum likelihood-based phylogenetic analyses with thousands of taxa and mixed models. Bioinformatics 2006;22(21):2688–2690.

87. Stamatakis A, Hoover P, Rougemont J. A rapid bootstrap algorithm for the RAxML Web servers. Syst Biol 2008;57(5):758–771.

88. Ye Y, Doak TG. A parsimony approach to biological pathway reconstruction/inference for genomes and metagenomes. PLoS Comput Biol 2009;5(8):e1000465.

89. Larsen PE, Collart FR, Field D, Meyer F, Keegan KP, Henry CS, McGrath J, Quinn J, Gilbert JA. Predicted Relative Metabolomic Turnover (PRMT): determining metabolic turnover from a coastal marine metagenomic dataset. Microb Inf Exp 2011;1(1):4.

90. Chiu H-C, Levy R, Borenstein E. Emergent biosynthetic capacity in simple microbial communities. PLoS Comput Biol 2014;10(7):e1003695.

91. Harcombe WR, Riehl WJ, Dukovski I, Granger BR, Betts A, Lang AH, Bonilla G, Kar A, Leiby N, Mehta P, Marx CJ, Segrè D. Metabolic resource allocation in individual microbes determines ecosystem interactions and spatial dynamics. Cell Rep 2014;7(4):1104–1115.

92. Karp PD, Latendresse M, Caspi R. The pathway tools pathway prediction algorithm. Stand Genomic Sci 2011;5(3):424–429.

93. Paley SM, Karp PD. The Pathway Tools cellular overview diagram and Omics Viewer. Nucleic Acids Res 2006;34(13):3771–3778.

94. Wickham H. Reshaping data with the reshape package. J Stat Softw 2007;21(12):1–20.

95. Efron B, Halloran E, Holmes S. Bootstrap confidence levels for phylogenetic trees. Proc Natl Acad Sci U S A 1996;93(23):13 429–13 434.

96. Ramette A. Multivariate analyses in microbial ecology. FEMS Microbiol Ecol 2007;62(2):142–160.

97. Legendre P, Legendre LF. *Numerical Ecology*. Volume 20. Amsterdam, The Netherlands: Elsevier; 2012.

98. Borcard D, Gillet F, Legendre P. *Numerical Ecology with R*. New York: Springer-Verlag; 2011.

99. Sarkar D. *Lattice: Multivariate Data Visualization with R*. New York: Springer-Verlag; 2008.

100. Wilkinson L, Wills D, Rope D, Norton A, Dubbs R. *The Grammar of Graphics*. New York: Springer-Verlag; 2006.

3

POOLING STRATEGY FOR MASSIVE VIRAL SEQUENCING

PAVEL SKUMS

Division of Viral Hepatitis, Centers of Disease Control and Prevention, Atlanta, GA, USA

ALEXANDER ARTYOMENKO AND OLGA GLEBOVA

Department of Computer Science, Georgia State University, Atlanta, GA, USA

SUMATHI RAMACHANDRAN, DAVID S. CAMPO, AND ZOYA DIMITROVA

Division of Viral Hepatitis, Centers of Disease Control and Prevention, Atlanta, GA, USA

ION I. MĂNDOIU

Department of Computer Science and Engineering, University of Connecticut, Storrs, CT, USA

ALEXANDER ZELIKOVSKY

Department of Computer Science, Georgia State University, Atlanta, GA, USA

YURY KHUDYAKOV

Division of Viral Hepatitis, Centers of Disease Control and Prevention, Atlanta, GA, USA

3.1 INTRODUCTION

Next-generation sequencing (NGS) allows generation of very large sets of viral sequences carried in samples of infected individuals. It offers previously unapproachable opportunities for studying microbial populations and understanding pathogen evolution and epidemiology. In particular, NGS provides means to implement a global molecular surveillance of infectious diseases for detailed description of disease dynamics in host population, evaluation of structures of transmission networks, monitoring of epidemic progress, and providing of informed guidance for planning public health interventions.

Although NGS offers a significant increase in throughput, sequencing of a large number of viral samples still is prohibitively expensive and extremely time consuming. Therefore, massive molecular surveillance requires development of a strategy for simple, rapid, and cost-effective sequencing of microbial populations from a large number of specimens.

Highly mutable viruses, such as hepatitis C virus (HCV) and human immunodeficiency virus (HIV), exist in infected hosts as large populations of genetically related variants. Such intrahost variants are commonly referred to as *quasispecies* in virological literature. The most preferable way of assessment of intrahost viral populations in each sample is the analysis of whole-genome sequences. However, NGS usually generates short reads, which should be assembled into whole-genome sequences. Assembly of viral quasispecies and estimation of their frequencies is extremely complex task, and currently even most advanced computational tools for whole-genome quasispecies reconstruction often only allows inference of most prevalent intrahost variants, with minority variants being frequently undetectable (1–5). Alternatively, genetic viral variants can be detected using highly variable subgenomic regions that can be easily amplified and sequenced. Although genetic information presented in such regions does not allow for identification of all viral variants, it is usually sufficient for inferring transmission networks (6–8), detecting drug-resistant variants, predicting therapy outcome (9–11), and studying intrahost viral evolution (12–14).

Cost of sequencing of multiple viral samples can be reduced using multiplexing through barcoding. Although this is probably the simplest approach to a simultaneous sequencing of a large number of specimen, it requires individual handling of each sample starting from nucleic acid extraction to polymerase chain reaction (PCR) and library preparation, which increases the sequencing costs (15, 16). Additionally, bias in amplification of different viral variants using PCR primers with different barcodes may affect the distribution of reads (16, 17). Moreover, maintaining a large library of barcodes is daunting (15, 16).

Combinatorial pooling provides an alternative approach to sequencing costs reduction. Applications of pooling to diagnostic testing goes back to the 1940s (18). Commonly, it is used for tests producing binary results; for example, positive or negative, as in group testing (19–22). Recently, several pooling strategies were proposed for more complex assays based on DNA sequencing, SNP calling, and a rare alleles detection (23–28). In particular, recent application of combinatorial pooling protocol to selective genome sequencing using NGS (15) should be mentioned.

In this chapter, we describe a framework for a cost-effective NGS of heterogeneous viral populations, which combines barcoding and pooling recently proposed by Skums et al. (29). This framework includes the following steps (Figure 3.1):

(i)   mixing samples in a specially designed set of pools in such a way that the identity of each sample is encoded in the composition of pools;

(ii) sequencing pools using barcoding;

(iii) deconvolution of samples, that is, assignment of viral variants from the pools to individual samples.

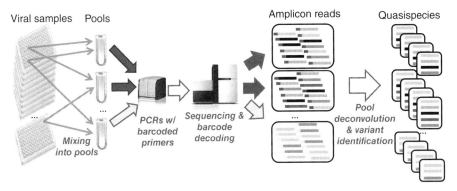

Viral samples Pools — Amplicon reads Quasispecies

PCRs w/ barcoded primers
Sequencing & barcode decoding
Mixing into pools
Pool deconvolution & variant identification

Figure 3.1 Combinatorial pooling strategy for viral samples sequencing. Source: Skums 2015 (29). Reproduced with permission of Oxford University Press. Please see www.wiley.com/go/Mandoiu/NextGenerationSequencing for a color version of this figure.

This approach significantly decreases the number of PCR and NGS runs, thus reducing the cost of testing and hands-on time. As an additional benefit, pooling provides opportunity for PCR amplification of viral variants from each sample in different mixtures of samples generated in each pool, thus introducing variation in amplification biases and contributing to sequencing of a more representative set of viral variants from each sample. In difference to most pooling methods and algorithms for human samples, which aim at SNP calling (i.e., the identification of positions in the sequenced region that differs from the reference), this approach allows for finding the whole *viral quasispecies spectra*, that is, viral sequences and their frequencies. However, application of the approach requires a careful designing of pools and significantly increases complexity of deconvolution of pools into individual samples, with the last task being especially demanding when applied to highly heterogeneous viral populations.

Sequence analysis of highly mutable RNA viruses is particularly difficult because of the complexity of their intrahost populations, the assessment of which can be distorted by PCR or sampling biases, presenting additional challenges for application of the pool-based sequencing to these viruses. The complex nature of viral samples imposes restrictions on the pool design and deconvolution. It is essential to detect not only major but also minor viral intrahost variants from pools, since minor variants may have important clinical implications and in many cases may define outcomes of therapeutic treatment (9, 30, 31). Mixing of a large number of specimens or specimens with significant differences in viral titers may contribute to under-representation of viral variants from some patients in pools, suggesting that size and composition of pools should be carefully designed.

Stochastic sampling from genetically diverse intrahost viral populations usually produces variability in compositions of sets of variants in different pools obtained from a single patient. Additionally, mixing specimens may differentially bias PCR amplification, contributing to mismatching between viral variants sampled from the same host in two pools with different specimen compositions. Thus, straightforward set-theoretical intersections among pools cannot be used for samples deconvolution, indicating that a more complex approach based on clustering techniques is needed. To increase the effectiveness of cluster-based deconvolution and minimize possible clustering errors, it is important to minimize mixing of genetically close samples as can be expected in epidemiologically related samples and samples collected from a small geographic region.

The rest of the chapter is organized as follows: in the next section, we consider the problem of pooling design for massive viral data. We formulate the pool design problem as an optimization problem on graphs, study the computational complexity of this and the associated problems, and describe heuristic algorithms to solve them. Then we review the method for deconvolution of individual samples from sequenced pools based on the maximum likelihood clustering algorithm for heterogeneous viral samples. And finally, we report the results of the framework evaluation using simulated data and experimentally obtained sequences of the HCV hypervariable region 1 (HVR1).

3.2   DESIGN OF POOLS FOR BIG VIRAL DATA

The basic idea of the overlapping pools strategy for sequencing n samples is to generate m pools (i.e., mixtures of samples) with $m \ll n$ in such a way that every sample is uniquely identified by the pools to which it belongs (23). Then, after sequencing of pools, the obtained amplicon reads can be assigned to samples by the sequence of set-theoretical intersections and differences of pools. Then the two examples show that a small number of pools can be used to uniquely identify a larger number of samples.

Example 1. Consider three samples S_1, S_2, S_3 and two pools $P_1 = S_1 \cup S_2$, $P_2 = S_2 \cup S_3$ (see Figure 3.2). These pools satisfy the separation requirement, and, therefore, each sample can be recovered, for example, $S_2 = P_1 \cap P_2$, $S_1 = P_1 \backslash P_2$ and $S_3 = P_2 \backslash P_1$. Thus, pooling sequencing of all three samples requires two sequencing runs.

Example 2. As a more complex example, consider eight samples $S_1, \ldots, S_8$ and four pools $P_1, \ldots, P_4$ defined as follows: $P_1 = S_1 \cup S_2 \cup S_3 \cup S_4$, $P_2 = S_5 \cup S_6 \cup S_7 \cup S_8$, $P_3 = S_1 \cup S_2 \cup S_5 \cup S_6$, $P_4 = S_1 \cup S_3 \cup S_5 \cup S_7$. These pools satisfy the separation requirement, and therefore each sample could be recovered by the sequence of intersections and differences of pools. For instance, $S_1 = P_1 \cap P_3 \cap P_4$, $S_2 = (P_1 \cap P_3) \backslash P_4, \ldots, S_8 = (P_2 \backslash P_3) \backslash P_4$. Therefore, sequencing of all eight samples may require four sequencing runs instead of eight.

The unique identification is possible if and only if for any two samples there is a pool *separating* them, that is, containing exactly one of the samples. Indeed, if any

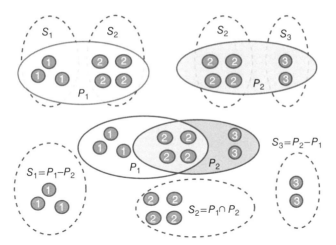

Figure 3.2 Two pools for three samples: S_1 has three, S_2 has four, and S_3 has two variants. All three samples can be reconstructed from these two pools by pool intersection and subtraction. Source: Skums 2015 (29). Reproduced with permission of Oxford University Press.

two samples are separated by a pool, then the intersection of all pools containing sample S minus the union of all pools not containing S coincides with S. On the other hand, if two samples S_1 and S_2 are not separated by any pool, then it is impossible to distinguish them from each other by set-theoretical operations. This fact leads to the following efficient pool design method described.

Theorem [23, 32]. If any subset of samples can form a single pool, then n samples can be reconstructed using $m = \lceil \log(n) \rceil + 1$ pools.

Proof. Assume for simplicity that n is a power of 2, that is, $n = 2^k$ (the proof is analogous for any n). Then apply induction by k. If $k = 1$, then $P = \{\{S_1\}, \{S_2\}\}$ clearly is a valid pool design with $m = 2$. Suppose that $P\prime = \{P'_1, \dots, P'_{m\prime}\}$ is a valid pool design with $S\prime = \{S'_1, \dots, S'_{m\prime}\}$, $n' = n/2 = 2^{k-1}$, $m\prime = \log(n') + 1 = k$. Construct a family $P = \{P_1, \dots, P_{k+1}\}$ as follows:

$$P_i = \{S_{2i-1}, S_{2i} : S'_i \in P'_i\}, \quad i = 1, \dots, k; \tag{3.1}$$

$$P_{k+1} = \{S_1, S_3, \dots, S_{n-1}\}. \tag{3.2}$$

The family P is a valid pool design. Indeed, it is clear that $\bigcup_{i=1}^{k+1} P_i = S$. Since P' is a feasible pool design for $n' = n/2$, for every $i, j \in \{1, n/2\}$, $i \neq j$, there exists $l \in \{1, \dots, k\}$ such that P'_l separates S'_i and S'_j. Thus, by the definition of the family P, the set P_l separates the sets $\{S_{2i-1}, S_{2i}\}$ and $\{S_{2j-1}, S_{2j}\}$. Finally, the set P_{k+1} separates the samples S_{2i-1}, S_{2i} for every $i = 1, n/2$. ∎

3.2.1 Pool Design Optimization Formulation

However, sequencing of heterogeneous RNA viral samples imposes the following additional restrictions on the pool composition: (i) the maximal number of samples that can be pooled without losing detection of many minority viral variants; and (ii) undesirability of mixing samples with drastically different viral titers or samples, which may be epidemiologically related. These restrictions make the pool design problem computationally much harder. Here, we formalize these restrictions and formulate the optimal pool design problem as an optimization problem on graphs.

Let $S = \{S_1, \ldots, S_n\}$ be a set of samples and $X \subseteq S$. The set X *separates* samples S_i and S_j, if X contains exactly one of the samples, that is, $|X \cap \{S_i, S_j\}| = 1$.

Restrictions on the pool composition can be represented by a *sample compatibility* graph $G = G(S)$ with $V(G) = S$ and $S_i S_j \in E(G)$ if and only if the samples S_i and S_j could be mixed in the same pool. So, every feasible pool is a clique of the graph G. Let T be an upper bound for the pool size. The problem of the optimal pool design for sequencing of viral samples can be formulated as follows:

Viral Sample Pool Design (VSPD) Problem. Given a sample compatibility graph $G = (V, E)$ and a number $T > 0$, find the set of cliques $\mathcal{P} = \{P_1, \ldots, P_m\}$ of G such that m is minimized and

(1) $\bigcup_{i=1}^{m} P_i = V$;
(2) for every $u, v \in V(G)$ there is a clique $P_i \in \mathcal{P}$ separating u and v;
(3) $|P_i| \leq T$ for every $i = 1, \ldots, m$.

Unlike the case when any subset of samples can be a pool, the general VSPD problem is more challenging.

Theorem [32]. Viral Sample Pool Design (VSPD) Problem is NP-hard, even for $T = 3$.

Proof. We will reduce to VSPD with $T = 3$, the following special case of the yes/no three-dimensional matching problem.

Problem A. *Given non-intersecting sets X, Y, Z, such that $|X| = |Y| = |Z| = q$; $M \subseteq X \times Y \times Z$, such that the following condition holds:*

(*) if $(a, b, w), (a, x, c), (y, b, c) \in M$, then $(a, b, c) \in M$.

Does M contain a subset $M' \subseteq M$ such that $|M'| = q$ and every two elements of M' do not have common coordinates?

The subset M' is called *three-dimensional matching*. It is known that the problem A is NP-complete (33). Let X, Y, Z, M, $|X| = |Y| = |Z| = q$, be the input of the problem A. Construct a graph G as follows:

$$V(G) = X \cup Y \cup Z \cup A, \tag{3.3}$$

where $A = \{a_v : v \in X \cup Y \cup Z\}$;

$$E(G) = \bigcup_{(a,b,c) \in M} \{ab, bc, ac\} \cup \{va_v : v \in X \cup Y \cup Z\}. \tag{3.4}$$

We will show that the set M contains three-dimensional matching if and only if the graph G contains a clique test collection $\mathcal{P} = \{P_1, \ldots, P_m\}$ of size $m = 4q$.

Let $\mathcal{P} = \{P_1, \ldots, P_m\}$ be a clique test collection of G, $m = 4q$. Let $R = X \cup Y \cup Z$. Let $\mathcal{P}' \subseteq \mathcal{P}$ be a set of cliques covering the vertices from the set A. For every $v \in R$ set, $\mathcal{P}'$ contains either clique $\{a_v\}$ or clique $\{v, a_v\}$ or both of them. Let $R = R_1 \cup R_2$, where $R_1 = \{v \in R : \{a_v\}, \{v, a_v\} \in \mathcal{P}\}$, $R_2 = R \backslash R_1$.

Consider an arbitrary vertex $v \in R_2$. Set $\mathcal{P}'$ contains either clique $\{a_v\}$ or clique $\{v, a_v\}$. If $\{a_v\} \in \mathcal{P}'$, then set $\mathcal{P}'' = \mathcal{P} \backslash \mathcal{P}'$ contains at least one clique covering the vertex v. If $\{v, a_v\} \in \mathcal{P}'$, then $\mathcal{P}''$ contains at least one clique, which separates v and a_v. Thus, every $v \in W_2$ is covered by a clique from the set $\mathcal{P}''$.

Let $r_1 = |R_1|$. We have $|R_2| = 3q - r_1$, $|\mathcal{P}'| = 3q + r_1$, $|\mathcal{P}''| = 4q - |\mathcal{P}'| = q - r_1$. So, $3q - r_1$ vertices from the set R_2 are covered by $q - r_1$ cliques from set $\mathcal{P}''$. Since sizes of cliques from $\mathcal{P}''$ are at most 3 (by construction of the graph G), it is possible only if $r_1 = 0$, all cliques from $\mathcal{P}''$ contain exactly three vertices and do not pairwise intersect. The condition (*) guarantees that every triangle of the graph G belongs to set M, and so $\mathcal{P}''$ is three-dimensional matching.

Conversely, if $M' \subseteq M$ is a three-dimensional matching, then $\mathcal{P} = M' \cup \{\{v, a_v\} : v \in R\}$ is a clique test collection of graph G. Indeed, $\mathcal{P}$ covers all vertices of G; for every $v \in R$, a clique $\{v, a_v\}$ separates sets $\{v, a_v\}$ and $V(G) \backslash \{v, a_v\}$ and vertices v and a_v are separated by the clique from M' which contains v. ∎

In practice, the condition (1) is not essential. Indeed, since every pair of vertices should be separated by some clique, at most one vertex $v \in V(G)$ is not covered by a clique from the set $\mathcal{P}$. Thus, any family of cliques satisfying (2) and (3) can be transformed into a family satisfying (1) by adding just one additional clique $\{v\}$. Therefore, we will consider the problem without the condition (1).

3.2.2 Greedy Heuristic for VSPD Problem

We propose a heuristic algorithm for the VSPD problem. For the algorithmic purposes, in addition to the graph G, consider the graph H with $V(H) = V(G) = V$ and $ij \in E(H)$ if and only if the pair of vertices (i, j) is not separated yet. Initially, H is a complete graph.

Let $A \subseteq V$ be a set of vertices. A *cut* in the graph H is the pair $(A, V \backslash A)$, the *size of the cut* $c(A, V \backslash A)$ is the number of edges with one end in A and the other end in $V \backslash A$.

The basic scheme of the heuristics is described in Algorithm GPDA. At each iteration, Algorithm GPDA finds and adds to the solution a locally optimal pool, that is, the pool that consists of compatible vertices and separates the maximal number of nonseparated samples.

Algorithm GPDA: Greedy Pool Design Algorithm

1:   $H \leftarrow$ complete graph on n vertices
2:   $\mathcal{P} \leftarrow \emptyset$
3:   **while** $E(H) \neq \emptyset$ **do**
4:      find a subset $A \subseteq V$ such that the followingconditions hold:
        (C1) A is a clique of the graph G;
        (C2) $|A| \leq T$;
        (C3) the size of the cut $(A, V \setminus A)$ in the graph H ismaximal.
5:      $\mathcal{P} \leftarrow \mathcal{P} \cup \{A\}$
6:      remove from H all edges uv with $u \in A$ and$v \in V \setminus A$.
7:   **end while**
8:   **return** $\mathcal{P}$

The crucial step of Algorithm GPDA finds locally optimal pool (step 4). It solves the following:

Optimal Clique Cut Bi-Graph (OCBG) Problem. Given a graph $H = (V, E)$ and a constant T, find a clique in G with the set of vertices A such that $|A| \leq T$ and the size of the cut $(A, V\setminus A)$ is maximized.

The OCBG problem is a previously unstudied discrete optimization problem. It is easy to see that this problem itself is NP-hard, and it is hard to approximate within a linear factor (29).

Theorem OCBG Problem is not approximable within $O(n^{1-\epsilon})$ for any $\epsilon > 0$, unless P=NP.

Proof. Let an n-vertex graph G' be the input of the maximum clique problem and $G = G' \cup O_n$. Without loss of generality, we assume that G is connected. Consider the instance of LOP problem with G and $H = K_{2n}$ as an input. Then for the value f_{opt} of the optimal solution of LOP, we have

$$f_{opt} = \max\{f(\omega) = \omega(2n - \omega) : \omega = |A|, a \text{ is a clique of } G\}.$$

Let us first show that the maximum clique size of G' is ω_{opt} if and only if $f_{opt} = \omega_{opt}(2n - \omega_{opt})$. Indeed, by construction $\omega = |A| \leq n$ for every clique A of G. The function $\omega(2n - \omega)$ increases monotonically on the segment $[1, n]$, and therefore f reaches its maximum on $\omega_{opt} = |A_{opt}|$, where A_{opt} is the maximal clique of graph G (and therefore of G').

Let $(A, V(G)\setminus A)$ be a solution of LOP, where A is a clique, $\omega = |A|$, and $f = \omega(2n - w)$. Suppose that

$$\frac{f_{opt}}{f} \leq \frac{1}{4}|V(G)|^{1-\epsilon} = \frac{1}{4}(2n)^{1-\epsilon}$$

for some $\epsilon > 0$. Then

$$\frac{1}{2}\frac{\omega_{opt}}{\omega} \le \frac{\omega_{opt}(2n - \omega_{opt}))}{\omega(2n - \omega)} = \frac{f_{opt}}{f} \le \frac{1}{4}(2n)^{1-\epsilon},$$

and therefore

$$\frac{\omega_{opt}}{\omega} \le n^{1-\epsilon}.$$

So, if LOP is approximable within $\frac{1}{4}|V(G)|^{1-\epsilon}$ for some $\epsilon > 0$, then Clique is approximable within $|V(G')|^{1-\epsilon}$. The latter is impossible, unless P=NP (34). ∎

In Section 3.2.3, we will describe an efficient heuristic to solve the OCBG problem.

3.2.3 The Tabu Search Heuristic for the OCBG Problem

In this section, we propose tabu search heuristic to solve the OCBG problem.

Let $M = |E(H)| + 1$. OCBGP can be formulated as the following quadratic programming problem:

$$\text{maximize } f(x) = \frac{1}{2}\sum_{ij\in E(H)}(1 - x_ix_j)-\frac{1}{8}M\sum_{ij\notin E(G)}(x_i + x_j)(x_1 + x_j + 2) \quad (3.5)$$

subject to

$$\frac{1}{2}\sum_{i\in V}(x_i + 1) \le T; \quad (3.6)$$

$$x_i \in \{-1, 1\}, i \in V. \quad (3.7)$$

There is a one-to-one correspondence between solutions x of the problems (3.5)–(3.7) and the pairs of sets (A_x, B_x), $A_x \cup B_x = V$, where $A_x = \{i : x_i = 1\}$ and $B_x = \{i : x_i = -1\}$. Next, we will indicate the solutions of (3.5)–(3.7) either by x or by (A_x, B_x).

The term $\frac{1}{2}\sum_{ij\in E(H)}(1 - x_ix_j)$ is equal to the size of the cut (A_x, B_x) in H. The term $\frac{1}{8}\sum_{ij\in E(G)}(x_i + x_j)(x_1 + x_j + 2)$ is equal to the number of nonadjacent pairs of vertices in the induced subgraph $G[A_x]$; in particular, it is equal to 0 if A_x is a clique. So, $f(x) \ge 0$ if and only if A_x is a clique. Therefore, for any optimal solution of the problems (3.5)–(3.7), the set A_x is a clique. The constraint (3.6) ensures that $|A_x| \le T$.

Initially, we relax the constraint (3.6). Suppose that (A_x, B_x) is a feasible solution of (3.5, 3.7). For a vertex $v \in A_x$, consider the solution $(A_{x'}, B_{x'})$, where $A_{x'} = A_x\backslash\{v\}, B_{x'} = B_x \cup \{v\}$. Then for $\Delta_1 = f(A_{x'}, B_{x'}) - f(A_x, B_x)$, we have

$$\Delta_1 = deg^H_{A_x}(v) - deg^H_{B_x}(v) + Mdeg^{\overline{G}}_{A_x}, \quad (3.8)$$

where $deg_U^H(v)$ denotes the number of vertices from the set $U \subseteq V$ adjacent to a vertex $v \in V$ in a graph H and $\overline{G}$ is a complement of a graph G. In particular, if v is nonadjacent to some vertex $u \in A_x$, then $\Delta_1 > 0$. Analogously, for $v \in B_x$, the solution $(A_{x'}, B_{x'})$ with $A_{x'} = A_x \cup \{v\}$, $B_{x'} = B_x \setminus \{v\}$ and $\Delta_2 = f(A_{x'}, B_{x'}) - f(A_x, B_x)$, we have

$$\Delta_2 = deg_{B_x}^H(v) - deg_{A_x}^H(v) - M deg_{A_x}^{\overline{G}}. \tag{3.9}$$

Thus, according to the relations (3.8) and (3.9), any initial solution (A, B) can be iteratively improved by moving vertices from one part of the partition to the other until a local optimum is reached, and the obtained solution cannot be further improved. According to (3.8), a local optimum A is a clique.

The major well-known general drawback of such local search strategies is that the value of the objective function in a local optimum may be far from the value of the globally optimal solution. Another problem, which is specific to our case, is that it is possible that the size of the locally optimal cut in H is 0. In this case, the solution found at the stage (4) of Algorithm GOPDA will not decrease the set $E(H)$, and, therefore, Algorithm GOPDA will go into an infinite loop. To overcome these problems, we use the variation of the tabu search strategy (35). The basic idea is that if after the moving of a vertex v the algorithm arrives at a local optimum, then the following actions are taken: the value of the local optimum is compared to the current best solution, v is placed back, and the moving of v is prohibited for the next k_t iterations of the algorithm. This idea is implemented in the following Algorithm OCBGP, which is described in more detail.

Let $\text{tabu}^i = \{\text{tabu}_1^i, \ldots, \text{tabu}_n^i\}$ be the tabu state at the iteration i, that is, a sequence of integers, where tabu_j^i is a current number of iterations during which it is not allowed to move a vertex j. Let optStates^i be the set of algorithm states, that is, the set of pairs $((A, B), t_{A,B})$, where (A, B) is a local optimum found by the algorithm at some iteration $j < i$ and $t_{A,B}$ is the tabu state at that iteration, that is, $t_{A,B} = \text{tabu}^j$. Let (A^*, B^*) be the current record cut, that is, the cut of the biggest size $c(A^*, B^*)$ found by the algorithm before the ith iteration. Let also moveList be the sequence of vertices moved by the algorithm from one part of the cut to another in the order of movement. This sequence is easier to implement as a stack, and it allows the algorithm to return to the previous solutions when the neighborhoods of solutions are completely explored. Let k_t denote the initial number of steps for which a move of a vertex is prohibited.

The steps of the algorithm are detailed in the following. Let $\text{tabu}^i = \{\text{tabu}_1^i, \ldots, \text{tabu}_n^i\}$ be the tabu state at the iteration i, that is, a sequence of integers, where tabu_j^i is a current number of iterations during which it is not allowed to move a vertex j. Let optStates^i be a set of algorithm states, that is, a set of pairs $((A, B), t_{A,B})$, where (A, B) is a local optimum found by the algorithm at certain iteration $j < i$ and $t_{A,B}$ is the tabu state at that iteration, that is, $t_{A,B} = \text{tabu}^j$. Let (A^*, B^*) be the current record cut, that is, the cut of the biggest size $c(A^*, B^*)$ found by the algorithm before the ith iteration. Let also moveList be the sequence of vertices moved by the algorithm from one part of the cut to another in the order of movement. This sequence is easier to implement as a stack, and it allows the algorithm to return to the previous solutions when neighborhoods of solutions are completely explored. Let k_t denote the initial number of steps for which a move of a vertex is prohibited.

Algorithm OCBGRP: Optimal Clique Cut Bi-Graph Problem Algorithm

1: Find solution (X, Y) of 0.5-approximation algorithm for Maximum Cut problem [36] applied to the graph H.

2: Apply 2-29 to initial solutions $(A^0, B^0) = (X, Y)$ and $(A^0, B^0) = (Y, X)$

3: $i \leftarrow 0$; $\text{tabu}_i \leftarrow (0, \dots, 0)$; $\text{optStates}_i \leftarrow \emptyset$; $\text{moveList} \leftarrow \emptyset$

4: **while true do**

5:     **for** $u \in V$ **do** Calculate a value δ_u using relations (3.8),(3.9):

$$\delta_u \leftarrow \begin{cases} \Delta_1, & \text{if } u \in A^i \text{ and } \text{tabu}_u^i = 0; \\ \Delta_2, & \text{if } u \in B^i \text{ and } \text{tabu}_u^i = 0; \\ 0, & \text{if } \text{tabu}_u^i > 0. \end{cases}$$

6:     **end for**

7:     $\delta^* \leftarrow \max_{i=1,\dots,n} \{\delta_i\}; u^* \leftarrow \arg \max_{i=1,\dots,n} \{\delta_i\}$

8:     Update the tabu state, that is,

$$\text{tabu}_j^{i+1} \leftarrow \begin{cases} \text{tabu}_j^i - 1, & \text{if } \text{tabu}_j^i > 0; \\ 0, & \text{otherwise .} \end{cases}$$

9:     **if** $\delta^* > 0$ **then** Move the vertex u^* to another part of the cut (A^i, B^i)

10:         Push u^* into the stack moveList

11:     **end if**

12:     **if** $\delta^* = 0$ **then**

13:         **if** $c(A^i, B^i) > c(A^*, B^*)$ **then** $(A^*, B^*) \leftarrow (A^i, B^i)$

14:         **end if**

15:         **if** $\text{moveList} \neq \emptyset$ **then** Pop v from moveList and move v back

16:             Forbid to move the vertex v for the next k_t iterations: $\text{tabu}_v^{i+1} \leftarrow k_t$.

17:         **end if**

18:         $s \leftarrow ((A^{i+1}, B^{i+1}), \text{tabu}^{i+1})$

19:         **if** $s \notin \text{optStates}$ **then** $\text{optStates} \leftarrow \text{optStates} \cup \{s\}; i \leftarrow i + 1$ **continue**

20:         **else**

21:             **while** $|A^*| > T$ **do** $a \leftarrow argmin\{deg_{B^*}^H(a) - deg_{A^*}^H | a \in A^*\}$

22:             $(A^*, B^*) \leftarrow (A^* \setminus \{a\}, B^* \cup \{a\})$

23:         **end while**

24:         **return** (A^*, B^*)

25:     **end if**

26:     **end if**

27: **end while**

At each iteration, Algorithm OCBGP tries to improve the current solution by moving one vertex from one part of the current cut to another part (steps 4–7). After the calculations for the cut improvement, the current tabu state is updated (step 8) and, if the current solution can be improved, the algorithm does it and proceeds to the next iteration (steps 9–12). If the current solution (A^i, B^i) cannot be improved, then it is a local optimum (stages 13–32). Then according to (3.8), A^i is a clique. In that case, the algorithm compares the obtained locally optimal solution with the record and updates it if necessary (steps 14–16). Then the algorithm returns to the previous solution (step 18) and forbids for the next k_t steps of moving the vertex, which leads to the previous local optimum (step 19). If the current algorithm state has not occurred previously, then the algorithm adds it to the set optStates and proceeds to the next iteration (steps 21–24). Otherwise, the algorithm reduces the current record (A^*, B^*) to the solution where $|A^*| \leq T$ and stops (steps 25-30).

The default value of k_t is 1. It is still possible that for the solution (A^*, B^*) found by Algorithm 2, we have $c(A^*, B^*) = 0$. If it happens, we increase k_t by one and repeat Algorithm 2.

3.3   DECONVOLUTION OF VIRAL SAMPLES FROM POOLS

According to Examples 1 and 2, deconvolution requires computing of intersections and differences of pools. In Section 3.3.1, we formally define generalized intersections and differences of pools and show how to use them for pool deconvolution. The challenges of implementation of generalized intersections and differences are addressed in Section 3.3.2

3.3.1   Deconvolution Using Generalized Intersections and Differences of Pools

Let $\mathcal{P}$ be the set of pools designed using a solution of the VSPD problem found by Algorithm GOPDA and sequenced using NGS. As discussed earlier, the obtained reads theoretically can be assigned to samples by the sequence of set-theoretical intersections and differences of pools (see Examples 1 and 2). However, owing to the high heterogeneity of viral populations and sampling bias, individual viral variants and even subpopulations of viral variants sequenced from a certain sample mixed with different pools may be different in each pool(see an example on Figure 3.3). It

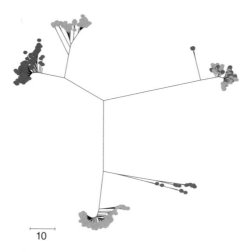

10

Figure 3.3   Phylogenetic tree representing a union of two pools: P_1 consisting of samples S_1, S_2, S_3 (shown in light grey) and P_3 consisting of samples S_1, S_4, S_5 (shown in dark grey). The intersection of two pools consists of the sample S_1 (upper right cluster in the tree); however, sequences sampled from S_1 in pools P_1 and P_2 are different. Source: Skums 2015 (29). Reproduced with permission of Oxford University Press.

hampers the usage of straightforward set-theoretical intersections and differences, and, therefore, "generalized" intersections and differences should be used instead.

For a pool P_i, let $S(P_i)$ be a set of samples mixed in it. In particular, for simplicity of notations, we can assume that each individual sample R_j is a special type of pool with $|S(R_j)| = 1$.

We define the *generalized intersection* of pools P_1 and P_2 as the pool $P_1 \overline{\cap} P_2$ with $S(P_1 \overline{\cap} P_2) = S(P_1) \cap S(P_2)$, consisting of sequences from $P_1 \cup P_2$ that belong to the samples from $S(P_1) \cap S(P_2)$. The *generalized difference* $P_1 \backslash P_2$ then can be defined as follows: $P_1 \overline{\backslash} P_2$ is the pool with $S(P_1 \overline{\backslash} P_2) = S(P_1) \backslash S(P_2)$ that contains sequences of the set $P_1 \backslash (P_1 \overline{\cap} P_2)$.

Individual samples can be inferred from pools by a sequence of generalized intersections and differences using Algorithm IS. By definition, generalized differences may be reduced to generalized intersections. For calculations of generalized intersections, we propose the scheme described in Algorithm GI, which is based on clustering techniques.

Algorithm PD: Pools Deconvolutionlgorithm

Require: The set of pools $\mathcal{P} = \{P_1, \dots, P_m\}$
Ensure: The set of individual samples $\mathcal{R} = \{R_1, \dots, R_n\}$

1: **procedure** IS($(\mathcal{P})$)
2:      define two queues Q and $\mathcal{R}'$ as follows: Q contains all pools and $\mathcal{R}' = \emptyset$.
3:      **while** $Q \neq \emptyset$ **do**
4:          $P' \leftarrow$ the first element of Q
5:          **if** $|S(P')| = 1$ **then**
6:             $\mathcal{R}' \leftarrow \mathcal{R}' \cup \{P'\}$
7:          **end if**
8:          $Q \leftarrow Q \backslash P'$
9:          **for** $P'' \in Q$ **do**
10:             **if** $S(P') \cap S(P'') \neq \emptyset$ **then**
11:                 $Q \leftarrow Q \cup \{P' \overline{\cap} P''\}$
12:                 **if** $S(P') \backslash S(P'') \neq \emptyset$ **then**
13:                     $Q \leftarrow Q \cup \{P' \overline{\backslash} P''\}$
14:                 **end if**
15:                 **if** $S(P'') \backslash S(P') \neq \emptyset$ **then**
16:                     $Q \leftarrow Q \cup \{P'' \overline{\backslash} P'\}$
17:                 **end if**
18:                 **break**
19:             **end if**
20:          **end for**
21:      **end while**
22:      **return** last n elements of $\mathcal{R}'$
23: **end procedure**

Algorithm GI: Generalized Intersection Algorithm

Require: Pools P_1, P_2, parameter $W \geq 1$
Ensure: The generalized intersection $P_1 \overline{\cap} P_2$
 1: **procedure** GI(P_1, P_2)
 2:     $K \leftarrow |S(P_1) \cup S(P_2)|$
 3:     Partition the union $P_1 \cup P_2$ into WK clusters using Maximum Likelihood
        k-Clustering of viral data described in the next subsection (Algorithm kGEM).
 4:     **return** the union of clusters, which contain reads from both pools P_1 and P_2.
 5: **end procedure**

Theoretically, Algorithm GI may be used with the parameter $W = 1$. However, viral populations of highly mutable viruses, such as HCV and HIV, may differ greatly in heterogeneity. In extreme cases, intrahost and interhost heterogeneity of certain samples may be comparable. If such samples belong to the same pool, it can lead to the effect when with $W = 1$, highly heterogeneous samples may be partitioned into multiple clusters while samples with lower heterogeneity will be joined into one cluster. Such clustering will lead to the incorrect detection of generalized intersections and consecutive loss of samples, which were not separated from other samples. To avoid this effect, higher values of W should be used. In our experiments, we used the default value $W = 2$. If certain samples are not found by Algorithm IS (i.e., the corresponding data sets are empty), we increase the value of W by one and repeat Algorithm IS.

3.3.2 Maximum Likelihood k-Clustering

In this section, we formulate the viral sample clustering problem and describe our solution, which is based on the probabilistic k-means approach (see Reference 37).

Sample Clustering Problem. Given a set R of NGS reads drawn from a mix of k' RNA viral samples, partition R into $k = Wk'$ subsets consisting of reads from a single sample.

The presence of numerous sequence variants in each viral sample, extreme heterogeneity of viral populations, and a very large number of reads make sample clustering problem challenging. Although a commonly used clustering objective is to minimize intracluster distances or distance to cluster centers (e.g., the k-means algorithm), we propose to use a statistically sound objective of maximizing likelihood. Our likelihood model estimates the probability of a certain read being emitted by a cluster consensus (or *centroid*).

Our algorithm receives a multiple sequence alignment of a given set of reads R as an input. We represent R as a matrix with columns corresponding to the consensus positions and rows corresponding to aligned reads. Our model assumes that each read in a cluster is emitted by a particular genotype (centroid). The proposed clustering (i) finds k genotypes $g_1, \ldots, g_k$, which most likely emit the observed set of reads, (ii) estimates probability $p_{i,r}$ that read r is emitted by a genotype g_i, and (iii) assigns a read r to a cluster that genotype most likely emits r.

Formally, given a set of reads C, a *genotype* $g(C)$ of C is a matrix with each column corresponding to a consensus position and five rows each corresponding to one of the alleles $\{a, c, t, g, d\}$. Each entry $f_m(e), e \in \{a, c, t, g, d\}$ is the frequency of allele e in mth position among all variants in C, $\sum_{e \in \{a,c,t,g,d\}} f_m(e) = 1$. In particular, every read can be considered as a genotype with a single 1 and 4 zeroes in each column. Given a set of reads R, a k-genotype is a set $G^* = \{g_1, \ldots, g_k\}$ of k distinct genotypes that most likely emitted R:

$$G^* = \arg\max_{|G|=k} \Pr(R|G),$$

where $\Pr(R|G)$ is the probability to observe R given a set of genotypes G, which is calculated as a product of probabilities to observe each read from R. The probability to observe read r equals to $\Pr(r) = \sum_{i=1}^{k} f_i \Pr(r|g = g_i)$, where

$$\Pr(r|g = g_i) = \prod_{m=1}^{L} f_{i,m}(r_m), \tag{3.10}$$

and $f_{i,m}(r_m)$ is the frequency of r_m, the mth character of read r, in the mth position of genotype g_i. Then the log-likelihood of the set of allele frequencies $\mathcal{F} = \{f_{i,m}(e) | i = 1, \ldots, k; m = 1, \ldots, L; e \in \{a, c, t, g, d\}\}$ equals to

$$\ell(\mathcal{F}) = \sum_{r \in R} o_r \log \Pr(r),$$

where o_r is the observed read frequency.

We iteratively estimate the missing data $p_{i,r}$, that is, the number of times the read r originated from the genotype g_i, and solve the easier optimization problem of maximizing the log-likelihood of the hidden model

$$\ell_{\text{hid}}(\mathcal{F}) = \sum_{r \in R} \sum_{i=1}^{k} p_{i,r} \log \left(f_i \Pr(r|g = g_i) \right).$$

Our clustering method is described in Algorithm kGEM. The initial set of genotypes $G^{(0)}$ is selected as follows: starting from the most frequent read, we iteratively select the read maximizing the minimum Hamming distance to the previously selected reads and add to $G^{(0)}$ the corresponding genotype.

3.4 PERFORMANCE OF POOLING METHODS ON SIMULATED DATA

3.4.1 Performance of the Viral Sample Pool Design Algorithm

The pool generation algorithm was evaluated using three sets of simulated data.

1. Complete graphs.
   Pools were generated for complete graphs with $n = 4, \ldots, 1024$ vertices without the threshold of pools' size. For every test instance, exactly $m = \lceil \log(n) \rceil + 1$ pools were constructed, coinciding with the theoretically justified estimation (23, 32). Hence, the VSPD algorithm produces optimal solutions for complete graphs.

Algorithm kGeEM: Maximum Likelihood k-Clustering Algorithm

1: For each genotype $g_i \in G^{(0)}, i = 1, \dots, k$, initialize allele frequencies in the mth position ($m = 1, \dots, L$) as follows:

$$f_{i,m}(e) = \begin{cases} 1 - 4\varepsilon & \text{if } r_{i,m} = e \\ \varepsilon & \text{otherwise,} \end{cases}$$

2: Initialize $h_{i,r}$ by the probability that genotype g_i emitted read r using formula (3.10): $h_{i,r} \leftarrow \Pr(r|g = g_i)$

3: **repeat**

4:      $f_i^{(0)} \leftarrow \frac{1}{k}$ for all $i = 1, \dots k$

5:      **repeat**

6:          Compute the expected number of reads $e_{i,r}$ emitted by the ith genotype g_i that match read r as follows:

$$e_{i,r} \leftarrow o_r \cdot p_{i,r}$$

$$p_{i,r} \leftarrow \frac{f_i^{(\tau)} \cdot h_{i,r}}{\sum_{i'=1}^{k} f_{i'}^{(\tau)} \cdot h_{i',r}},$$

where f_i is the frequency of g_i and o_r is the observed frequency of r.

7:          Estimate the frequency $f_i^{(\tau+1)}$ of each $G_i^{(t)}$ as the portion of all reads emitted by $g_i^{(t)}$ as follows:

$$f_i^{(\tau+1)} \leftarrow \frac{\sum_{r \in R} e_{i,r}}{\sum_{i'=1}^{k} \sum_{r \in R} e_{i',r}}$$

8:      **until** $\sum_{i=1}^{k} (f_i^{(\tau)} - f_i^{(\tau+1)})^2 \geq \delta$

9:      Update allele frequencies $f_{i,m}(e)$ of each allele $e \in \{a, c, t, g, d\}$ in the mth position of $g_i^{(t+1)}$ as follows:

$$f_{i,m}(e) \leftarrow \begin{cases} 1 - 4\varepsilon & \text{if } e = \underset{e' \in \{a,c,t,g,d\}}{\arg\max} \sum_{r \in R : r_m = e'} p_{i,r} \\ \varepsilon & \text{otherwise} \end{cases}$$

10:     Update $h_{i,r}$ using (3.10): $h_{i,r} \leftarrow \Pr(r|g = g_i)$

11: **until** there exists i such that $g_i^{(t)} \neq g_i^{(t+1)}$

12: **for** each read $r \in R$ **do**

13:     $i' \leftarrow \arg\max_i p_{i,r}$.

14:     Assign read r to i'th cluster

15: **end for**

---

2. Random graphs, where each vertex v receives a random titer $w_v \in \{1, L\}$, and two vertices u and v are adjacent if and only if $|w_u - w_v| \leq R$. This family of test instances represents *titer compatibility model*, that is, it simulates the case in which two samples could be mixed into one pool only if their viral titers are not sufficiently different.

Twenty-five thousand test instances were generated with $n = 10, \ldots, 1000$, with parameters $L = 20$, $R = 4$, and thresholds of pools' sizes $T = n$ (i.e., without the threshold), $T = 55$, $T = 35$, and $T = 25$. For each n, the mean size of the set of pools constructed by the VSPD algorithm and the mean sequencing reduction coefficient (i.e., the number of pools divided by the number of samples) were calculated. The results are shown in Figure 3.4a.

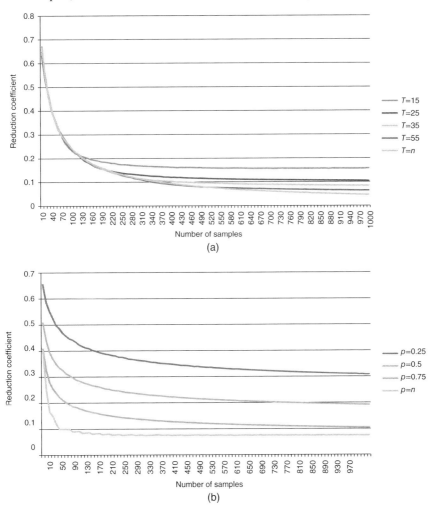

Figure 3.4 Sequencing reduction coefficient for the pools generated by the VSPD algorithm for (a) random titer compatibility model graphs and (b) random graphs. Source: Skums 2015 (29). Reproduced with permission of Oxford University Press.

For $n = 1000$ sets of pools generated by the VSPD algorithm, more than 21-fold reduction in the number of sequencing runs is achieved for $T = n$, 15-fold reduction for $T = 55$, 11-fold reduction for $T = 35$, 9-fold reduction for $T = 25$, and 6-fold reduction for $T = 15$. The reduction coefficient in all these cases is a decreasing function of n, which suggests a higher reduction for the larger n.

3. Random graphs, where each edge is chosen with probability $p = 0.25, 0.5, 0.75$, and 1 and sizes of pools are bounded by $T = 35$.

Twenty thousand test instances with $n = 10, \ldots, 1000$ were generated and processed by the VSPD algorithm. As above-mentioned, for each n, the mean sequencing reduction coefficient was calculated. The results are shown in Figure 3.4b. In this case, as well as in the previous one, pooling provides a great reduction in the number of sequencing runs, although it is generally lower than for the test instances (2) (from more than 13-fold reduction for $p = 1$ (complete graph) to more than 3-fold reduction for $p = 0.25$). The reduction coefficient is also a decreasing function of n.

3.4.2 Performance of the Pool Deconvolution Algorithm

Four hundred and fifty test instances with $n = 10, \ldots, 150$ samples and with thresholds of pools' sizes $T = 15, 25,$ and 35 were generated. Simulated pools were constructed using 155 HCV HVR1 samples previously sequenced in Molecular Epidemiology and Bioinformatics Laboratory, Division of Viral Hepatitis, Centers for Disease Control and Prevention using 454 GS Junior System (454 Life Sciences, Branford, CT) (9, 38–40). Reads from each sample were cleaned from sequencing errors using NGS error correction algorithms KEC and ET (41). Test instances were generated as follows:

1. n samples were chosen randomly.
2. A random sample's compatibility graph on n vertices was generated based on the titer compatibility model and pools were designed using the VSPD algorithm (Algorithm GOPDA).
3. Pools were created by taking D randomly selected reads from the samples composing each pool (in order to simulate a sampling bias). The number of reads per pool D was set as $D = 10,000$, which approximately corresponds to the sequencing settings, under which the data used for simulation were obtained (454 Junior System with 8–10 MIDs per sequencing run).

For all test instances, all samples were inferred, that is, all n data sets produced by Algorithm IS were nonempty. It is possible that some reads are not classified into samples and therefore are lost by the algorithm. However, the number of such reads was extremely low (Figure 3.5a): in average 99.996% of reads for $T = 15$, 99.993% for $T = 25$, and 99.984% for $T = 35$ were classified into samples.

An overwhelming majority of reads were classified correctly (Figure 3.5b): in average, 99.998% of reads for $T = 15$, 99.982% for $T = 25$, and 99.959% for $T = 35$ were assigned to the right samples. It should be noted that the percentages of classified and correctly classified reads, in general, do not depend on the number of samples if this number is large enough.

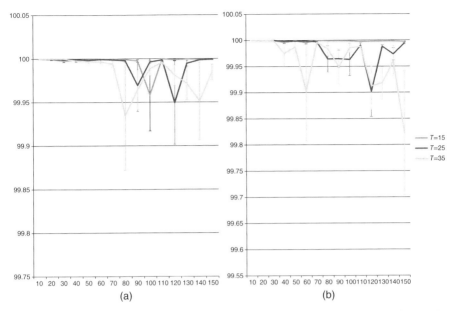

Figure 3.5 (a) Percentage of classified reads and (b) percentage of correctly classified reads. Bars represent a standard error. Source: Skums 2015 (29). Reproduced with permission of Oxford University Press.

We call an incorrect assignment of reads to the samples *in silico contamination*. The average percentage of samples without in silico contamination ranges from 100% to 98.13% for $T = 15$, from 100% to 96.13% for $T = 25$, and from 100% to 93.8% for $T = 35$ (Figure 3.6a); the percentage of in silico contaminated samples increases with the total number of samples. In silico contaminants constitute a small minority within contaminated samples: in average 0.163% of all reads for $T = 15$, 0.545% for $T = 25$, and 0.892% for $T = 35$ (Figure 3.6b).

Root mean square error of the frequency estimations of deconvoluted haplotypes is in average 0.031–0.107% for $T = 15$, 0.025–0.139% for $T = 25$, and 0.028–0.174% for $T = 35$; it is an increasing function of the number of samples (Figure 3.7).

According to all measures considered earlier, the accuracy of the samples deconvolution is affected by the number of allowed samples per pool. The algorithm is more accurate for smaller pools although the accuracy remains high even for larger pools.

3.5 EXPERIMENTAL VALIDATION OF POOLING STRATEGY

3.5.1 Experimental Pools and Sequencing

Serum specimens collected from HCV-positive cases (6) were used to sequence HCV HVR1. Seven serum samples $S_1, \dots, S_7$ were mixed to form four pools $P_1, \dots, P_4$ using the VSPD algorithm with the parameter $T = 7$ as follows: P_1 was created by mixing samples S_1, S_2, S_3, P_2 – samples S_4, S_5, S_6, S_7, P_3 – samples S_1, S_4, S_5 and P_4 - samples S_2, S_4, S_6. Then the seven specimens and four pools were sequenced

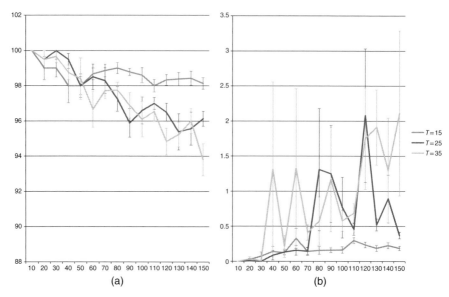

(a)   (b)

Figure 3.6   (a) Percentage of samples without in silico contamination. (b) Total frequency of in silico contaminants within contaminated samples. Bars represent a standard error. Source: Skums 2015 (29). Reproduced with permission of Oxford University Press.

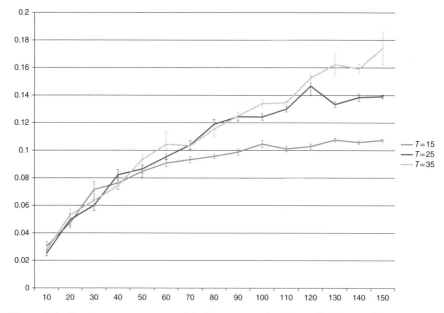

Figure 3.7   Root mean square error of the frequency estimations of haplotypes. Bars represent a standard error. Source: Skums 2015 (29). Reproduced with permission of Oxford University Press.

using 454 GS Junior System (454 Life Sciences, Branford, CT). Extraction of the total nucleic acids was performed using MagNA Pure LC Total Nucleic Acid Isolation Kit (Roche Diagnostics, Mannheim, Germany) and reverse-transcribed using the SuperScript Vilo cDNA synthesis kit (Invitrogen, Carlsbad, CA).

The HVR1 amplification was accomplished using two rounds of PCR. For the first round of amplification, regular region-specific primers were used. Forward and reverse tag sequences consisting of primer adaptors and multiple identifiers (MID – 454A and 454B) were added to the HVR1-specific nested primers. For the high-throughput purpose, pools were processed as a single specimen, tagged with a single MID for deep sequencing. PCR products were pooled and amplified by emulsion PCR using the GS FLX Titanium Series Amplicon kit and were bidirectionally sequenced. The sequenced reads were identified and separated using sample-specific MID tag identifiers. Low-quality reads were removed using the GS Run Processor v2.3 (Roche, 2010).

3.5.2 Results

The algorithmic approach described in Section 3.3 was used to reconstruct seven samples from experimental data described in Section 3.5.1. Before applying algorithms for samples recovery, the data were preprocessed in order to get rid of sequencing errors and PCR chimeras.

The reads from each pool were separated into clusters using Algorithm kGEM, each cluster was processed using NGS error correction algorithms KEC and ET (41) and the corrected reads were merged back. Then the samples were deconvoluted using Algorithm IS. The obtained samples will be further referred as *pooling samples*

For the verification of pooled samples, we compared them with the individually sequenced samples. The sequences were compared using pairwise alignment; insertions and deletions were ignored (since indels are rare in HVR1 and, therefore, are rather sequencing artifacts; moreover, some indels in alignment of sequences from individually sequenced and pooling samples may be introduced due to the inaccurate correction of homopolymer errors for the samples). For each sample, 10 reference sequences were taken from the set of individually sequenced variants, and the correctness of samples reconstruction was assessed using alignment of sequences in the pooled samples with these references. For alignment, Muscle (42) was used.

In average, 259 unique haplotypes per sample from a pool were obtained (from 23 haplotypes in Sample S_2 to 548 haplotypes in Sample S_4), which exceeds the number of HCV haplotypes obtained in other studies (43–45) after the standard individual sequencing using 454 Junior System and subsequent error correction. A total of 99.9634% (5463 of 5465) of all analyzed sequence reads were correctly classified into samples. Two reads assigned to sample S_7 showed a higher similarity with the reference sequence from sample S_6. However, the subsequent analysis showed that these reads are highly different from each other and from other sequences that belong to samples S_6 and S_7.. Therefore, these to reads are likely to be sequencing artifacts, which were not removed by the error correction algorithm.

In general, the percentage of haplotypes from individually sequenced samples found in pooled samples was not high (Figure 3.8a), with an average of 14.66%.

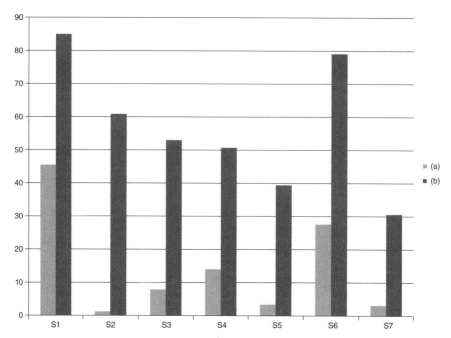

Figure 3.8 (a) Percentage of haplotypes from individually sequenced samples found in pooling experiment. (b) Total frequency of haplotypes from individually sequenced samples found in pooling experiment. Source: Skums 2015 (29). Reproduced with permission of Oxford University Press.

However, when the frequencies of these haplotypes were considered, the level of agreement between samples was much higher, with an average total haplotype frequency of 56.94% (Figure 3.8b). In particular, all individually sequenced haplotypes with frequencies greater than 10% and 72.73% of haplotypes with frequencies greater than 5% were found in pooled samples.

The differences between haplotype frequency distributions for individually sequenced and pooled samples were measured using Jensen–Shannon Divergence (JSD) (46) and correlation coefficient (Table 3.1). JSD varies from 0.15% for the sample S_1 to 0.65% for the sample S_7. There is a statistically significant positive correlation between frequency distributions for samples S_1–S_6. The only exception is the sample S_7, in which a large cluster of viral variants was not detected in the individually sequenced specimen but was found in the pooling experiment (see Figure 3.9).

Phylogenetic trees of viral populations from samples S_1–S_7 obtained by individual and pool sequencings are shown in Figure 3.9. Although haplotypes obtained from two different sequencing experiments are not completely matching, they cover the same areas of the sequence space. Some tree branches are formed by variants sequenced in one experiment but not the other. For instance, sequencing of individual samples S_1 and S_2 produced sequences forming branches that cannot be found when sequences from pooling experiments were considered. The opposite was observed for samples S_6 and S_7.

TABLE 3.1 Comparison of Frequency Distributions for Individually Sequenced and Pooled Samples

	JSD	Correlation (p-value)
S_1	0.15	0.95 (1.7×10^{-77})
S_2	0.57	0.30 (0.0023)
S_3	0.32	0.89 (2.7×10^{-173})
S_4	0.37	0.66 (4.14×10^{-99})
S_5	0.50	0.25 (8.6×10^{-7})
S_6	0.17	0.99 (0)
S_7	0.65	-0.07 (0.16)

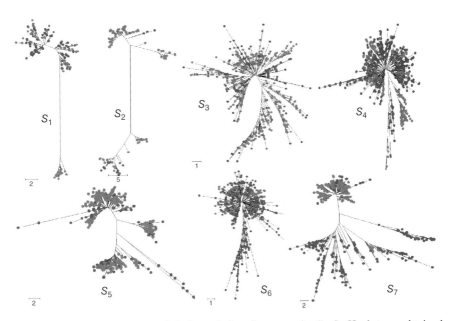

Figure 3.9   Phylogenetic trees of viral populations from samples S_1–S_7. Haplotypes obtained by individual sequencing of samples are shown in red, and haplotypes obtained from sequencing of pools are shown in blue. Source: Skums 2015 (29). Reproduced with permission of Oxford University Press. Please see www.wiley.com/go/Mandoiu/NextGenerationSequencing for a color version of this figure.

3.6   CONCLUSION

In this study, we present a novel framework for massive NGS of highly mutable RNA viruses, such as HCV and HIV. To the best of our knowledge, this is the first application of the pooling strategy to highly heterogeneous viral samples. The developed framework takes into account specific aspects of viral sequencing, such as the extensive heterogeneity of viral samples, the large number of distinct viral variants sequenced from each sample, and the effects of PCR and sampling biases. The proposed strategy is highly effective in reducing the number of sequencing runs, while still providing sufficient amount of information in support of molecular surveillance and numerous other applications of viral sequences in clinical and epidemiological settings. The novel clustering algorithm developed here significantly facilitates assignment of intrahost viral variants from massive sequence data sets obtained by pooling specimens to individual patients. The strategy of overlapping pools drastically reduces the cost of sequencing per specimen, especially when a large number of specimens require to be tested. This computational framework is applicable to viral agents infecting humans and animals, and, with further development of the experimental protocols, it should serve as a cost-effective foundation for accurate molecular surveillance of infectious diseases.

Ultradeep sequencing of viral samples produces a wide range of intrahost viral variants and allows for detecting minority variants, some of which have been shown to have important clinical implications such as drug resistance (9, 30, 31). Pooling of numerous specimens reduces the depth of sequencing for each specimen. However, this reduction is not as detrimental for identifying minor viral variants since each specimen is usually used in more than one pool in the strategy developed here. As specimen is tested more than once, the number of sequenced variants is increased, so representative sampling of viral subpopulations infecting each patient can be improved. The experiments conducted here showed that comparable number of haplotypes were recovered from individual specimens and from pools (Figure 3.9), at least at the pooling scale used in this study. Both individual sequencing and pooling produce sequences covering approximately the same areas of the sequence space, thus providing a consistent structure of a viral population.

Repeat sampling from the same complex viral population results frequently in poorly matched sets of viral sequences, thus presenting a significant challenge to the assignment of all sequences obtained by pool sequencing to each patient. Such stochastic sampling has a potential to diminish the effectiveness of pool sequencing and usefulness of the obtained sequences by impeding the correct allocation of sequences to samples, leaving some samples without sequences assigned or allocating only a fraction of the obtained sequences to samples. The clustering-based approach to finding generalized intersections of pools developed in this study significantly improves identification of sequences that belong to a patient and, thus, not only substantially overcomes the aforementioned potential pitfalls, but converts stochastic sampling into an advantage.

The cost of sequencing and accuracy of pool deconvolution are two major measures of quality of our computational framework. However, these two measures are in conflict with each other. While increase in pool size improves cost-effectiveness of sequencing by reducing the number of sequencing runs, it reduces accuracy of deconvolution. Considering that accuracy of deconvolution significantly depends on the genetic complexity of intrahost viral populations, an optimal pool size should be carefully selected for each virus and each genomic region.

In conclusion, success of the pool-based mass sequencing of viral populations depends to a significant degree on the efficacy of sequence assignments and on the risk of underrepresentation of viral variants from some patients, owing to PCR and sample biases. The pool design and clustering algorithms presented here substantially minimize the detrimental effect of these biases on quality of the mass sequencing. However, the further reduction of the biases using generalizations of error-correcting codes and optimization of experimental conditions should further improve the strategy, facilitating its application to molecular surveillance and study of infectious diseases.

REFERENCES

1. Zagordi O, Bhattacharya A, Eriksson N, Beerenwinkel N. ShoRAH: estimating the genetic diversity of a mixed sample from next-generation sequencing data. BMC Bioinformatics 2011;12(1):119.

2. Astrovskaya I, Tork B, Mangul S, Westbrooks K, Mandoiu II, Balfe P, Zelikovsky A. Inferring viral quasispecies spectra from 454 pyrosequencing reads. BMC Bioinformatics 2011;12 Suppl 6:S1.

3. Prosperi MC, Salemi M. QuRe: software for viral quasispecies reconstruction from next-generation sequencing data. Bioinformatics 2012;28(1):132–133.

4. Mancuso N, Tork B, Skums P, Ganova-Raeva L, Mandoiu II, Zelikovsky A. Reconstructing viral quasispecies from NGS amplicon reads. In Silico Biol 2012;11(5):237–249.

5. Skums P, Mancuso N, Artyomenko A, Tork B, Mandoiu I, Khudyakov Y, Zelikovsky A. Reconstruction of viral population structure from next-generation sequencing data using multicommodity flows. BMC Bioinformatics 2013;14 Suppl 9:S2.

6. Holodniy M, Oda G, Schirmer PL, Lucero CA, Khudyakov YE, Xia G, Lin Y, Valdiserri R, Duncan WE, Davey VJ, Cross GM. Results from a large-scale epidemiologic look - back investigation of improperly reprocessed endoscopy equipment. Infect Control Hosp Epidemiol 2012;33(7):649–656.

7. Vaughan G, Xia G, Forbi JC, Purdy MA, Rossi LM, Spradling PR, Khudyakov YE. Genetic relatedness among hepatitis A virus strains associated with food-borne outbreaks. PLoS ONE 2013;8(11):e74546.

8. Wertheim J, Leigh Brown A, Hepler N, Mehta S, Richman D et al. The global transmission network of HIV-1. J Infect Dis 2014;209(2):304–313.

9. Campo DS, Skums P, Dimitrova Z, Vaughan G, Forbi J, Teo C-G, Khudyakov Y, Lau DT-Y. Drug-resistance of a viral population and its individual intra-host variants during the first 48 hours of therapy. Clin Pharmacol Ther 2014;95(6):627–635.

10. Wang W, Zhang X, Xu Y, Weinstock GM, Di Bisceglie AM et al. High- resolution quantification of hepatitis C virus genome-wide mutation load and its correlation with the outcome of peginterferon-alpha2a and ribavirin combination therapy. PLoS ONE 2014;9(6):e100131.

11. Dierynck I, Thy K, Ghys A, Sullivan JC, Kieffer TL, Aerssens J, Picchio G, De Meyer S. Deep sequencing analysis of the HCV ns3-4a region confirms low prevalence of telaprevir-resistant variants at baseline and end of the REALIZE study. J Infect Dis 2014;210(12):1871–1880.

12. Ramachandran S, Campo DS, Dimitrova Z, Xia GL, Purdy MA, Khudyakov YE. Temporal variations in the hepatitis C virus intrahost population during chronic infection. J Virol 2011;85(13):6369–6380.

13. Palmer BA, Moreau I, Levis J, Harty C, Crosbie O, Kenny-Walsh E, Fanning LJ. Insertion and recombination events at hypervariable region 1 over 9.6 years of hepatitis C virus chronic infection. J Gen Virol 2012;93:2614–2624.

14. Culasso ACA, Bare P, Aloisi N, Monzani MC, Corti M, Campos RH. Intra- host evolution of multiple genotypes of hepatitis C virus in a chronically infected patient with HIV along a 13-year follow-up period. Virology 2014;449:317–327.

15. Lonardi S, Duma D, Alpert M, Cordero F, Beccuti M et al. Combinatorial pooling enables selective sequencing of the barley gene space. PLoS Comput Biol 2013;9(4):e1003010.

16. Duma D, Wootters M, Gilbert AC, Ngo HQ, Rudra A, Alpert M, Close TJ, Ciardo G, Lonardi Stefano. Accurate decoding of pooled sequenced data using compressed sensing. Lect Notes Comput Sci 2013;8126:70–84.

17. Alon S, Vigneault F, Eminaga S et al. Barcoding bias in high-throughput multiplex sequencing of miRNA. Genome Res 2011;21(9):1506–1511.

18. Dorfman R. The detection of defective members of large population. Ann Math Stat 1943;14(4):436–440.

19. Du D-Z, Hwang FK. *Pooling Design and Nonadaptive Group Testing: Important Tools for DNA Sequencing*. Volume 18, Series on Applied Mathematics. Singapore: World Scientific Publishing Company; 2006.

20. Weili WU, Huang Y, Huang X, Li Y. On error-tolerant DNA screening. Discrete Appl Math 2006;154:1753–1758.

21. Wu W, Li Y, Huang C-H, Du D-Z. Molecular biology and pooling design. In: *Data Mining in Biomedicine*. Volume 7, Springer Optimization and Its Applications. New York: Springer Science and Business Media LLC; 2007. p 133–139.

22. Berman P, DasGupta B, Kao M-Y. Tight approximability results for test set problems in bioinformatics. In: *Algorithm Theory - SWAT 2004*. Volume 3111, of Lecture Notes in Computer Science. Heidelberg; Springer Berlin; 2004. p 39–50.

23. Prabhu S, Pe'er I. Overlapping pools for high-throughput targeted resequencing. Genome Res 2009;19(7):1254–1261.

24. He D, Zaitlen N, Pasaniuc B, Eskin E, Halperin E. Genotyping common and rare variation using overlapping pool sequencing. Proceedings of the 1st Annual RECOMB Satellite Workshop on Massively Parallel Sequencing (RECOMB-seq), Volume 12. BMC Bioinformatics; 2011.

25. Erlich Y, Chang K, Gordon A, Ronen R, Navon O, Rooks M, Hannon GJ. DNA Sudoku - harnessing high-throughput sequencing for multiplexing specimen analysis. Genome Res 2009;19:1243–1253.

26. Shental N, Amir A, Zuk O. Identification of rare alleles and their carriers using compressed se(que)nsing. Nucleic Acids Res 2010;38:1–22.

27. Golan D, Erlich Y, Rosset S. Weighted pooling-practical and cost-effective techniques for pooled high-throughput sequencing. Bioinformatics 2012;28(12):i197–i206.

28. Bansal V. A statistical method for the detection of variants from next-generation resequencing of DNA pools. Bioinformatics 2010;26(12):i318–i324.

29. Skums P, Artyomenko A, Glebova O, Ramachandran S, Mandoiu I, Campo DS, Dimitrova Z, Zelikovsky A, Khudyakov Y. Computational framework for next-generation sequencing of heterogeneous viral populations using combinatorial pooling. Bioinformatics 2015;31(5):682–690.

30. Skums P, Campo DS, Dimitrova Z, Vaughan G, Lau DT, Khudyakov Y. Numerical detection, measuring and analysis of differential interferon resistance for individual HCV intra-host variants and its influence on the therapy response. In Silico Biol 2012;11(5):263–269.

31. Metzner KJ et al. Minority quasispecies of drug-resistant HIV-1 that lead to early therapy failure in treatment-naive and -adherent patients. Clin Infect Dis 2009;48(2):239–247.

32. Skums P, Glebova O, Zelikovsky A, Mandoiu I, Khudyakov Y. Optimizing pooling strategies for the massive next-generation sequencing of viral samples. 3rd Workshop on Computational Advances for Next Generation Sequencing (CANGS); 2013.

33. Garey MR, Johnson DS. *Computers and Intractability. A Guide to the Theory of NP-Completeness.* San Francisco (CA); W. H. Freeman; 1979.

34. Zuckerman D. Linear degree extractors and the inapproximability of max clique and chromatic number. Proceedings of the 38th ACM Symposium Theory of Computing; 2006. p 681–690.

35. Glover F, Kochenberger G. *Handbook of Metaheuristics.* Dordrecht: Kluwer Academic Publishers; 2003.

36. Khuller S, Raghavachari B, Young NE. Greedy methods. In: *Handbook of Approximation Algorithms and Metaheuristics.* Boca Raton (FL): Chapman and Hall/CRC; 2007.

37. Artyomenko A, Mancuso N, Skums P, Mandoiu I, Zelikovsky A. kGEM: an em-based algorithm for local reconstruction of viral quasispecies. 2013 IEEE 3rd International Conference on Computational Advances in Bio and Medical Sciences (ICCABS); 2013.

38. Dimitrova Z, Campo DS, Ramachandran S, Vaughan G, Ganova-Raeva L, Lin Y, Forbi JC, Xia G, Skums P, Honisch C, Pearlman B, Khudyakov Y. Assessments of intra- and inter-host diversity of hepatitis C virus using next generation sequencing and mass spectrometry. In Silico Biol 2012;11(5):183–192.

39. Lara J, Tavis JE, Donlin MJ, Yuan HJ, Lee WM, Pearlman BL, Vaughan G, Forbi JC, Xia GL, Khudyakov YE. Coordinated evolution among hepatitis C virus genomic sites is coupled to host factors and resistance to interferon. In Silico Biol 2011-2012;11(5-6):213–224.

40. Campo DS, Dimitrova Z, Yamasaki L, Skums P, Lau D, Vaughan G, Forbi JC, Teo C-G, Khudyakov Y. Next-generation sequencing reveals large connected networks of intra-host HCV variants. BMC Genomics 2014;15(5):1.

41. Skums P, Dimitrova Z, Campo DS, Vaughan G, Rossi L, Forbi JC, Yokosawa J, Zelikovsky A, Khudyakov Y. Efficient error correction for next-generation sequencing of viral amplicons. BMC Bioinformatics 2012;13 Suppl 10:S6.

42. Edgar RC. MUSCLE: multiple sequence alignment with high accuracy and high throughput. Nucleic Acids Res 2004;32(5):1792–1797.

43. Bull RA, Luciani F, McElroy K, Gaudieri S, Pham ST, Chopra A, Cameron B, Maher L, Dore GJ, White PA, Lloyd AR. Sequential bottlenecks drive viral evolution in early acute hepatitis C virus infection. PLoS Pathogens 2011;7(9):e1002243.

44. Cortes KC, Zagordi O, Laskus T, Ploski R, Bukowska-Osko I, Pawelczyk A, Berak H, Radkowski M. Ultradeep pyrosequencing of hepatitis C virus hypervariable region 1 in quasispecies analysis. Biomed Res Int 2013;2013:626083.

45. Gregori J, Esteban JI, Cubero M, Garcia-Cehic D, Perales C, Casillas R et al. Ultra-deep pyrosequencing (UDPS) data treatment to study amplicon HCV minor variants. PLoS ONE 2013;8(12):e83361.

46. Lin J. Divergence measures based on the shannon entropy. IEEE Trans Inform Theory 1991;37:145–151.

4

APPLICATIONS OF HIGH-FIDELITY SEQUENCING PROTOCOL TO RNA VIRUSES

SERGHEI MANGUL

Department of Computer Science, University of California, Los Angeles, CA, USA

NICHOLAS C. WU

Department of Integrative Structural and Computational Biology, The Scripps Research Institute, La Jolla, CA, USA

EKATERINA NENASTYEVA, NICHOLAS MANCUSO AND ALEXANDER ZELIKOVSKY

Department of Computer Science, Georgia State University, Atlanta, GA, USA

REN SUN AND ELEAZAR ESKIN

Department of Molecular and Medical Pharmacology, University of California, Los Angeles, CA, USA

4.1 INTRODUCTION

RNA viruses have high genomic diversity within an infected host. It effects many clinically important phenotypic traits such as escape from vaccine-induced immunity, virulence, and response to antiviral therapies (1). Sequencing technologies must be sensitive enough in order to accurately characterize an intrahost RNA virus population (2, 3). Next-generation sequencing (NGS) technologies offer deep coverage of genomic data in the form of millions of sequencing reads allowing to capture rare variants (4). But the full picture of viral diversity in a population remains undiscovered due to errors produced by sequencing platforms. The presence of sequencing errors

Computational Methods for Next Generation Sequencing Data Analysis, First Edition.
Edited by Ion I. Măndoiu and Alexander Zelikovsky.
© 2016 John Wiley & Sons, Inc. Published 2016 by John Wiley & Sons, Inc.
Companion website: www.wiley.com/go/Mandoiu/NextGenerationSequencing

makes it difficult to distinguish between variants and sequencing errors. Additionally, low viral population variability (i.e., pairs of individual viral genomes that have small genetic distance) and the presence of individual variants having low abundance complicate accessing viral diversity and assembling full-length viral variants (5).

Current sequencing technologies use different underlying chemistry and offer trade-offs among throughput, read length, and cost (4). Although the sequencing platforms can potentially detect point mutations, error rates may result in false-positive single-nucleotide variant (SNV) calls or wrong genome variant sequences. Current assembly methods (6–10) are not able to differentiate true biological mutations from sequencing artifacts, thus significantly limiting the possibility of a method to assemble the underlying viral population. Computational error correction approaches are able to partially correct the sequencing error, but they are not well suited for mixed viral samples and may lead to filtering out true biological mutations. As a result, the low abundant variants remain undiscovered. Additional difficulty is the genomic architecture of viruses. Long conserved regions shared across viral population introduce ambiguity in the assembly process because they may have multiple crossovers. In contrast to repeats in genome assembly, conserved regions may be phased based on relative abundances of viral variants.

In Reference 11, a high-fidelity sequencing protocol was proposed. It overcomes above limitations and eliminates sequencing errors. We propose to couple this protocol with an accurate method, referred to as Viral Genome Assembler (VGA) (5). It allows to assemble a heterogeneous viral population. The advantage of the method is that it does not need a reference genome. This feature makes our method applicable to newly emerged viruses with unknown genome sequences (5). The ability to discover rare viral variants makes our tool applicable for monitoring and quantifying an RNA virus population structure in order to dissect its evolutionary landscape and study genomic interaction. In particular, our approach allows to discover rare mutations and variants for HIV that are of particular interest because of their potential influence on drug resistance and treatment failure (12–14). Another application of the protocol can be an evaluation of error correction methods for NGS reads. The result of the protocol is erroneous reads that can be considered as a ground truth for any error correction method.

The rest of the chapter is organized as follows. In Section 4.2, the used high-fidelity sequencing protocol is described. Section 4.3 introduces the approach for viral genome assembly VGA based on high-fidelity sequencing data. Sections 4.4 and 4.5 present the results of performance of VGA and some other viral assemblers on simulated data, while in Section 4.6, we describe the performance of VGA on real HIV data. In Section 4.7, we compare different aligners to investigate the effect of their alignment on mapping statistics. Finally, in Section 4.8, we discuss the application of the high-fidelity protocol for the evaluation of error correction methods.

4.2 HIGH-FIDELITY SEQUENCING PROTOCOL

To eliminate errors from sequencing data, we apply a high-fidelity sequencing protocol (Figure 4.1a–d). The protocol, known as the Safe-Sequencing System

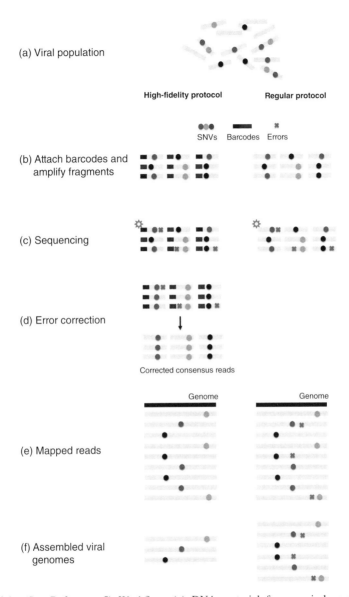

Figure 4.1 (See Reference 5) Workflow. (a) DNA material from a viral population is cleaved into sequence fragments using any suitable restriction enzyme. (b) Individual barcode sequences are attached to the fragments. Each tagged fragment is amplified by the polymerase chain reaction (PCR). (c) Amplified fragments are then sequenced. (d) Reads are grouped according to the fragment of origin based on their individual barcode sequence. An error-correction protocol is applied for every read group, correcting the sequencing errors inside the group and producing corrected consensus reads. (e) Error-corrected reads are mapped to the population consensus. (f) SNVs are detected and assembled into individual viral genomes. The ordinary protocol lacks steps (b) and (d). Source: Mangul 2014 (5). Reproduced with permission of Oxford University Press.

("safe-SeqS"), has been proposed in Reference 11 and applied to detect rare somatic mutations, but its application on detecting rare viral mutations has been neglected. The protocol involves two steps. The first is the assignment of a unique barcode to each DNA template molecule to be analyzed. The second is the amplification of each uniquely tagged template so that many copies of a molecule with the preexisted mutation are generated. Polymerase chain reaction (PCR) fragments with the same barcode are considered mutant ("supermutants") only if $\geq 90\%$ of them contain the identical mutation. Errors occurring during the amplification steps or sequencing errors in base calling should not give rise to supermutants.

In our work (5) similar to Safe-SeqS, we apply a special library preparation technique that eliminates sequencing errors during the de-multiplexing step. The proposed approach attaches individual barcode sequences for every fragment and then amplifies each tagged fragment. After the fragments are sequenced, reads belonging to the original fragment are clustered based on the barcode. It follows that every sequenced position of the fragment would have multiple independent evidence, suitably promoting highly accurate consensus reads. Given that many reads are required to sequence each fragment, we are trading off an increase in sequence coverage for a reduction in error rate. By applying an error-correction procedure of the protocol, we are able to address both sequencing and PCR errors, which leads to high assembly accuracy. Deep coverage and highly accurate data allow for accurate estimation of the underlying diversity of a viral population. Importantly, in order to detect ultra-rare variants, the low per-base sequencing cost of the Illumina platform makes it realistic to greatly increase coverage.

4.3 ASSEMBLY OF HIGH-FIDELITY SEQUENCING DATA

4.3.1 Consensus Construction

We apply the *de novo* consensus reconstruction tool, Vicuna (15) prior to assembly to produce a linear consensus directly from the sequence data. As far as our sequencing method does not contain any particularly low coverage region, we can reconstruct population consensus for viral sample. Vicuna produces multiple contigs rather than a complete consensus, so we use BLAST to merge contigs. Specifically, we require 50*nt* overlap to merge any pair of contigs. This approach offers more flexibility for samples that do not have "close" reference sequences available. Traditional assembly methods (16–18) aim to reconstruct a linear consensus sequence and are not well suited for assembling a large number of highly similar but distinct viral genomes. We instead take our ideas from haplotype assembly methods (19, 20), which aim to reconstruct two closely related haplotypes. However, these methods are not applicable for assembly of a large (*a priori* unknown) number of individual genomes. Many existing viral assemblers estimate local population diversity and are not well suited for assembling full-length quasispecies variants spanning the entire viral genome. Available genome-wide assemblers able to reconstruct full-length quasispecies variants are originally designed for low throughput and are impractical for high-throughput technologies containing millions of sequencing reads. In the next step, the population consensus is used as a reference genome to map reads. Building the reference genome from actual sequencing data rather than using an annotated genome provides us with an accurate and unique mapping.

4.3.2   Reads Mapping

As with many viral population analyses, the first step of assembly is to map the reads (Figure 4.1e). We map reads onto the *de novo* consensus using InDelFixer (21) with default parameters. False read alignments were filtered out using fragment length distribution inferred from the mapping data. Assuming that the fragment length follows a normal distribution (22), we only keep reads with fragment length within three standard deviations from the mean. In total, 1.2% of reads have been filtered out versus expected 0.3% according to the three-sigma rule.

4.3.3   Viral Genome Assembler (VGA)

In Reference 5, we introduce a viral population assembly method (Figure 4.2) working on highly accurate sequencing data able to detect rare variants and tolerate conserved regions shared across the population. The method includes a postassembly procedure to detect and resolve ambiguity raised from long conserved regions using expression profiles. After a consensus has been reconstructed directly from the sequence data (Figure 4.2a), our method detects SNVs from the aligned sequencing reads. Read overlapping is used to link individual SNVs and distinguish between genome variants in the population. The viral population is condensed in a conflict graph built from aligned sequencing data (Figure 4.2b). Two reads are originated from different viral genomes if they share different SNVs in the overlapping region. Viral variants are identified from the graph as independent sets of nonconflicting reads. Noncontinuous coverage of rare viral variants may limit assembly capacities, indicating that increase in coverage is required to increase the assembly accuracy. Frequencies of identified variants are then estimated using an expectation–maximization algorithm. Compared with existing approaches, we are able to detect rare population variants while achieving high assembly accuracy.

   The viral population assembly starts with determining pairs of mapped reads conflicting with each other in the overlapping region. The combination of deep coverage with high accuracy provides an unprecedented opportunity for estimating genomic diversity in a viral population. Following (23), we construct the conflict graph $G = (V, E)$ with vertices corresponding paired-end reads, that is, $V = R$, and edges connecting conflicting pairs of paired-end reads.

   Obviously, any true viral genome corresponds to a maximal independent set in the conflict graph (i.e., a maximal set of pairwise nonadjacent vertices), although, not every maximal independent set necessarily corresponds to a true viral genome. In Reference 5, we adopt a parsimonious approach requiring to cover the conflict graph with the minimum number of maximal independent sets. This problem is equivalent to MIN-GRAPH-COLORING, which is NP-hard. There exists many heuristics for solving this problem (see, e.g., References 24, 25) based on greedy selection of a maximal independent set. Unfortunately, our attempts to build even a single viral genome failed since it is difficult to arrange paired-end reads into a connected single

(a)

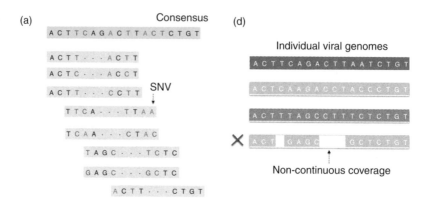

(d)

(b)   Conflict graph

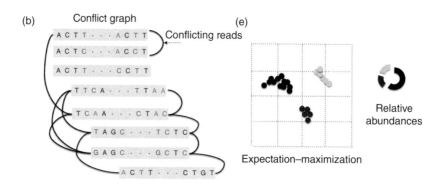

(e)

(c)   Graph coloring

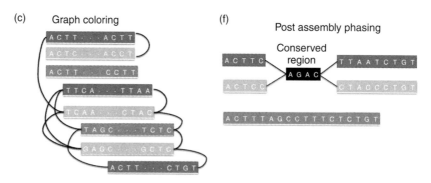

(f)

Colors = independent sets of non-conflicting reads

path. Indeed, a greedy algorithm runs out of any possible extension after just a few steps while concatenating paired-end reads from left to right.

Instead, in Reference 5, we apply an alternative "top-down" approach of recursive graph partitioning along the maximum cut (Max Cut), which has been previously successfully applied for human haplotyping (26). Given a graph $G = (V, E)$, the Max-Cut problem asks for partitioning of the vertices into two components $V = V_1 \cup V_2$ maximizing the total number of edges, which have one endpoint in V_1 and the other in V_2. The Max-Cut problem though NP-hard is well approximated by a simple 0.5 approximation algorithm, which randomly assigns vertex to one of the two components (27). Our Max-Cut heuristic starts with alternatively assigning left-to-right sorted mapped reads to two components and then repeatedly moves one vertex at a time from one component to another, improving the solution at each step, until no more improvements of this type can be made.

Our coloring heuristic recursively partitions the conflict graph until each component becomes independent (Figure 4.2c). If reads of a given color completely cover the consensus genome, then the resulted sequence is accepted as the next viral genome. Otherwise, if assembled genome contains gaps, we add nonconflicting reads from other color classes in left-to-right order in attempt to fill the gaps (Figure 4.2d). If all SNV positions are covered, then a newly reconstructed viral genome is added to the set $\mathcal{VG}$. Finally, the genomes whose gaps cannot be filled with the above procedure are dropped.

4.3.4 Viral Population Quantification

In the final step of the workflow (Figure 4.2e and f), an expectation–maximization algorithm is used to infer the relative abundances of assembled viral quasispecies

---

Figure 4.2 Overview of VGA (see Reference 5). (a) The algorithm takes as input paired-end reads that have been mapped to the population consensus. (b) The first step in the assembly is to determine pairs of conflicting reads that share different SNVs in the overlapping region. Pairs of conflicting reads are connected in the "conflict graph." Each read has a node in the graph, and an edge is placed between each pair of conflicting reads. (c) The graph is colored into a minimal set of colors to distinguish between genome variants in the population. Colors of the graph correspond to independent sets of nonconflicting reads that are assembled into genome variants. In this example, the conflict graph can be minimally colored with four colors (red, green, violet, and turquoise), each representing individual viral genomes. (d) Reads of the same color are then assembled into individual viral genomes. Only fully covered viral genomes are reported. (e) Reads are assigned to assembled viral genomes. Read may be shared across two or more viral genomes. VGA infers relative abundances of viral genomes using the expectation–maximization algorithm. (f) Long conserved regions are detected and phased based on expression profiles. In this example, red and green viral genomes share a long conserved region(colored in black). There is no direct evidence how the viral subgenomes across the conserved region should be connected. In this example, four possible phasings are valid. VGA uses the expression information of every subgenome to resolve ambiguous phasing. Source: Mangul 2014 (5). Reproduced with permission of Oxford University Press.

Algorithm VGA Assembly Algorithm (5)

Input : Set of reads R aligned to the consensus genome
Build conflict graph G = (V, E) from set R
Recursively color G into color classes C using Max-Cut
Initialize the set of complete viral genomes $\mathcal{VG} \leftarrow \emptyset$
for each color class $c_i \in C$ **do**
    Compute maximal independent set in $G = (V, E)$ containing c_i
    Assemble reads in c_i into viral genome g_i
    if g_i covers all positions in the consensus genome **then**
        $\mathcal{VG} \leftarrow \mathcal{VG} \cup \{g_i\}$
    end if
end for
Output: Set of complete viral genomes $\mathcal{VG}$

---

similar to what is described in Reference 28. In Reference 5, we extend the previous EM and likelihood formulation to incorporate a prior probability for the viral population and compute the *maximum a posteriori* estimate rather than the MLE. Let H be a random variable over the set of viral variant genomes $H = \mathcal{VG}$, and let R be a random variable over the set of reads $\mathcal{R}$. Let $p[H] \sim \mathrm{Dir}(\alpha_1, \ldots, \alpha_{|H|})$ be the prior probability of observing a given set of variants and denote $p_h = \Pr[H = h]$ to be the probability of observing a particular variant h. The probability of observing read $r \in \mathcal{R}$ is given by marginalizing over all variants

$$\Pr[R = r] = \sum_{h \in H} \Pr[R = r | H = h] \cdot p_h,$$

where

$$\Pr[R = r | H = h] = \begin{cases} 1/K_h & \text{if } r \text{ is consistent with } h \\ 0 & \text{otherwise} \end{cases}$$

and K_h is the number of reads consistent with h. We can now define the log-posterior as

$$\log \Pr[H|R] = \sum_{r \in R} n_r \cdot \log \Pr[R = r] + \sum_{h \in H} (\alpha_h - 1) \log p_h - C_R,$$

where C_R is a constant and n_r is the number of reads r. As this function is nonconvex and difficult to globally optimize for, we solve the easier problem of maximizing its lower bound,

$$\sum_{r \in R} \sum_{h \in H} n_{rh} \cdot \log (\Pr[R = r | H = h] \cdot p_h) + \sum_{h \in H} (\alpha_h - 1) \log p_h,$$

where n_{rh} is the expected number of reads r generated by variant h. The EM algorithm computes this by

$$n_{rh} = n_r \cdot \frac{p_h \cdot \Pr[R = r | H = h]}{\Pr[R = r]},$$

and subsequently maximizes the log-posterior with the MAP estimate given by

$$\hat{p}_h = \left(\alpha_h - 1 + \sum_{r \in R} n_{rh} \right) \Big/ \left(|\mathcal{R}| + \sum_{h \in \mathcal{H}} \alpha_h - |\mathcal{H}| \right).$$

We see that by setting a uniform prior, that is, $\alpha_h = 1$ for all $h \in \mathcal{H}$, we obtain the original MLE formulation.

4.4 PERFORMANCE OF VGA ON SIMULATED DATA

Since the ground truth is unknown for sequenced viral populations, in Reference 5 we used simulations as a standardized way to assess the performance of viral assembly tools. The proposed high-fidelity protocol allows to correct sequencing errors, thus giving access to highly accurate sequencing data. Post-sequencing error correction techniques are available for reads obtained by regular protocol offering the possibility to partially correct sequencing errors trading off for real biological mutations. Grinder (29) is used to generate reads from both the high fidelity and regular sequencing protocol. Reads are generated from both real and synthetic viral variants with different sequencing parameters and viral expression profiles. Grinder is a state-of-the-art sequencing read simulator able to produce shotgun sequencing data from a viral population with different expression profiles. We mapped the simulated paired-end reads onto the consensus using Mosaik (30). The consensus was constructed using Vicuna (15), a *de novo* assembly tool able to produce a linear consensus from deep paired-end sequencing data (see Section 4.3.1 for details).

In Reference 5, we use sensitivity and positive predictive value (PPV) to evaluate the quality of viral genomes assembled by VGA. We consider fully assembled viral genome without errors. Sensitivity is defined as the portion of assembled quasispecies that match true quasispecies, that is, $Sensitivity = TP/(TP + FN)$. PPV is defined as the portion of true sequences among assembled sequences, that is, $PPV = TP/(TP + FP)$. Additionally, we evaluate the ability of our method to estimate population size (i.e., number of viral genomes in the population). Accuracy of population size prediction is defined as a ratio between estimated and true population sizes. Finally, we use Jensen–Shannon divergence (JSD) to measure the accuracy of frequency estimation. Given two probability distributions, JSD measures the "distance" between them or, in other words, the quality of approximation of one probability distribution by the other distribution. It is defined as the Kullback–Leibler divergence from distributions P and Q to their mixture. Formally, the JSD between true distribution P and approximation distribution Q is given by the formula

$$\text{JSD}(P\|Q) = \frac{1}{2}D_{KL}(P\|M) + \frac{1}{2}D_{KL}(Q\|M)$$

where Kullback–Leibler divergence D_{KL} is

$$D_{KL}(P\|Q) = \sum_{i=1}^{n} P(i) \log \frac{P(i)}{Q(i)}$$

and $M = \frac{1}{2}(P + Q)$. The motivation for using JSD is a consequence of KL divergence being undefined when assembly methods fail to reconstruct some variant i, hence forcing $Q(i)$ to be 0. JSD averts this by measuring the distance to the mixture, which contains all true and called variants (TP and FP).

Our first simulated study compares the assembly accuracy across different virus species. In Reference 5, we focus on the effect of read length and throughput on assembly quality for different types of viruses. Paired-end reads of various length corresponding to high fidelity and regular sequencing protocols are simulated from HIV and HCV populations assuming uniform and power-law distributions. A power-law distribution (i.e., frequency of an individual viral genome is a power of the previous one) corresponds to a population with several dominant variants and many rare variants. The uniform distribution has equal frequencies for all viral genomes. HCV population is presented by 1739-bp long fragment from the E1E2 region of 44 real HCV sequences. HIV population consists of a mixture of 10 real intrahost viral variants from 1.3 kb-long HIV-1 region, which included pol protease and part of the pol reverse transcriptase (31).

The genomic architecture across virus species was investigated and its influence on assembly accuracy was studied (5). HCV virus exhibits more complex genomic architecture with lower population diversity and longer conserved regions (Figure 4.3) than HIV. Conserved regions were present in both viruses although only HCV contains conserved regions longer than 450 bp. Conserved regions longer than the average fragment length (450 bp) may introduce ambiguity in the assembly process due to a lack of direct evidence of subgenomes phasing across the conserved region. We performed simulated sequencing experiment where the average fragment amplification rate is 5, resulting in a five-time decrease in throughput due to the consensus error correction performed by the high-fidelity sequencing protocol. Also the simulation experiments were adjusted to simulate a nonuniform amplification rate. Nonuniform amplification rate results in discarding fragments with insufficient amplification rate (less than 3). From real studies, it is known that around 10% of fragments are amplified less the three times. Sequencing errors produced by the regular protocol limited the ability of VGA to accurately assemble a viral population. All assembled variants contained a large number of mismatches, in addition VGA significantly overestimated population size.

As expected, short read lengths dramatically inhibit reconstruction, which is evidenced by VGA failing to produce any full-length genomes when given 2×36 bp reads (Figure 4.6). Since common regions for distinct HCV viral genomes are significantly longer than for HIV, it is not surprising that performance of VGA is worse on HCV data – for 3M 2 × 150 bp reads simulated from 44 1739 bp-long viral genomes, sensitivity is 50%, and PPV is 80%. Results on HCV data confirm that the lower mutation rate and the presence of conserved regions have a negative impact on the ability to accurately reconstruct individual viral genomes. Surprisingly, increasing the read length for HIV from 100 to 150 bp yields no benefits for reconstruction accuracy, suggesting that 100 bp read length is enough to distinguish between HIV viral variants with high mutation rate. Although further experiments are needed to determine optimum read length, our simulations suggest that 2 × 100 bp is recommended

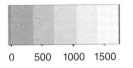

0   500   1000   1500

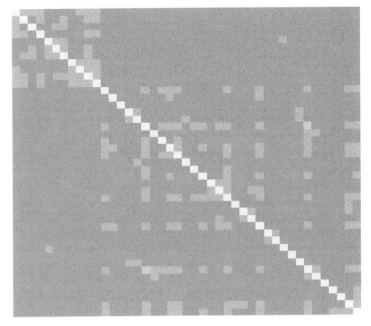

Figure 4.3   Genomic architecture of 44 real HCV viral genomes from 1739-bp long fragment of E1E2 region (see Reference 5). Length of longest common region shared between any two viral genomes is represented by color. Source: Mangul 2014 (5). Reproduced with permission of Oxford University Press.

for small HIV viral populations and 2×150 bp is recommended for medium HCV population with complex genomic architecture.

In Reference 5, we separately analyzed the ability of our method to estimate the viral population size (i.e., number of genomic variants present in the population). Noncontinuous coverage limits the ability of the method to assemble full-length viral variants. To evaluate the accuracy of population size estimation, we compared the true population size known from simulated data with estimated results. Continuous coverage of each individual viral genome present in the sample has a strong impact on the quality of population assembly. The probability of noncontinuous coverage increases dramatically for viral genomes with low abundance. Thus, the presence of coverage gaps for rare variants introduce additional challenges in the assembly process, making rare genomes unreachable by assembly tools. The number of problematic genomes can be reduced by increasing sequencing depth; however, it does not guarantee complete elimination. While complete assembly of all such genomes is unrealistic, it is still possible to estimate the number of viral genomes present in the sample (population size). The number of independent sets reported by VGA provides us with

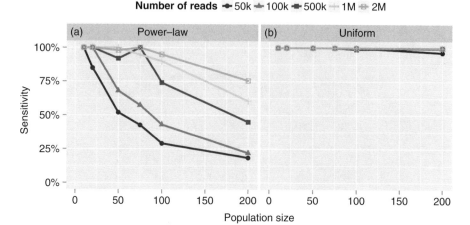

Figure 4.4 Accuracy of population size prediction (see Reference 5). Up to 200 viral genomes were generated from the Gag/Pol 3.4 kb HIV region. The population diversity is 5–10%. Variant abundances follow power-law (a) and uniform (b) distributions. Highly accurate 100×2 bp paired-end reads were simulated from HIV population. Source: Mangul 2014 (5). Reproduced with permission of Oxford University Press.

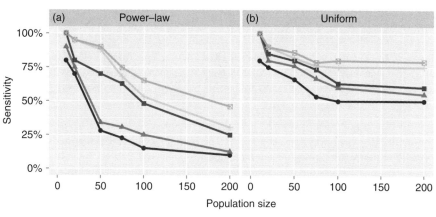

Figure 4.5 Assembly accuracy estimation (see Reference 5). Up to 200 viral genomes were generated from the Gag/Pol 3.4 Kb HIV region. The population diversity is 3–20%. Variant abundances follow power-law (a) and uniform (b) distributions. Consensus error-corrected 2×100 bp paired-end reads were simulated from HIV population. Source: Mangul 2014 (5). Reproduced with permission of Oxford University Press.

an accurate population size estimation. Intuitively, predicting the population size of a large viral population with many rare variants is more difficult than predicting for uniformly distributed or small populations (Figure 4.4). The predicted population size may serve as an indication of insufficient coverage to detect the full viral diversity present in the sample.

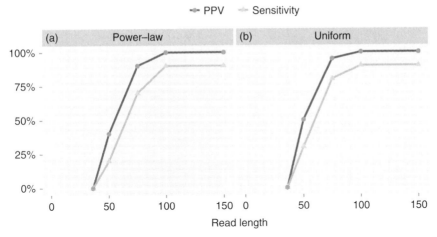

Figure 4.6   Assembly accuracy estimation (see Reference 5). Consensus error-corrected paired-end reads of various lengths were simulated from a mixture of 10 real viral clones from 1.3 kb-long HIV-1 region. Assembly accuracy as measured by sensitivity and PPV when variant abundances follow uniform and power-law distribution. Results are for 50,000 reads, no improvement was observed when increasing the number of reads. Source: Mangul 2014 (5). Reproduced with permission of Oxford University Press.

Deep coverage is a key for accurate estimation of underlying viral diversity. One such platform capable of offering millions of sequencing reads is Illumina HiSeq. The relatively short length of the produced reads is compensated for by sequencing the same fragment from both ends; therefore, producing coupled reads separated by a "gap," known as paired-end read. To our knowledge, VGA is the first method scalable to millions of short paired-end sequencing reads able to produce full-length viral variants spanning the entire viral genome. We explore the influence of sequencing depth on the reconstruction accuracy for varying population structures (uniform and power-law distributions of viral genomes within the population) (5). HIV-1 is known to have greater genetic variability than any other known virus (32). The diversity among viral genomes in an HIV population can vary from 3% to 20% depending on regions (33, 34). Heterogeneous viral samples were prepared by generating viral populations from the Gag/Pol 3.4 Kb HIV region. We simulated variant abundances adhering to either uniform or power-law distribution. Not surprisingly, our simulations suggest that increased sequencing depth has a direct positive effect on the discovery of rare variants and improves the overall assembly accuracy. Figure 4.5 shows the effect of coverage and population size on assembly for reads of length 100 bp. Throughout all experiments, VGA maintained a PPV value of 100%.

In addition to point mutations, genetic recombination facilitates rapid evolution and production of diverse HIV genomes. Indeed, co-infected cells may produce recombinant viral progeny at levels lower than mutation rates in an intrapatient environment (35). Hence, simulated data sets must account for both possible phenomena when determining the quality of assembly. In Reference 5, we utilize a simulation model able to integrate both point mutations and recombination in the generated

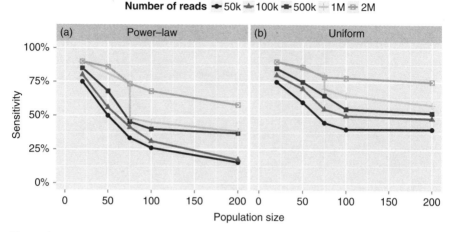

Figure 4.7 Assembly accuracy estimation (see Reference 5). Up to 200 recombinant viral genomes were generated from the from 1.3 kb-long HIV-1 region. Variant abundances follow power-law (a) and uniform (b) distributions. Consensus error-corrected 2×100 bp paired-end reads were simulated from HIV population. Source: Mangul 2014 (5). Reproduced with permission of Oxford University Press.

viral population depending on the amount of diversity required. A mixture of 10 real intrahost viral variants from 1.3 kb-long HIV-1 form the basis population. In addition to point mutations, our simulation model implicitly produces recombinant genomes by first constructing the genotype (i.e., sequence of SNVs) for the population. A random walk is performed over this genotype as specified number of times. Any crossover that occurs represents a new recombination between the "left" and "right" original genomes. Recombinations are implicitly produced, and no control is imposed over the number and length of the recombination. This model produces highly recombinant data on average, posing challenges for assembly and can be used to assess assembly quality. Simulation model incorporates mutation into the process by selecting a position and nucleotide-swap uniformly at random. Simulation results (Figure 4.7) suggest that our method can accurately assemble viral population in the presence of recombinations and point mutations, maintaining PPV of 100%.

Finally, we evaluate population quantification accuracy, that is, the accuracy of our method in prediction abundances of the assembled variants. Taking the results from VGA on 10 real HIV clones with 50,000 reads and 2×100 bp, the JSD was 2.93e-05 for the power law and 0.001 for the uniform-based populations. This already small measure only decreases as the size of the input grows.

4.5 PERFORMANCE OF EXISTING VIRAL ASSEMBLERS ON SIMULATED CONSENSUS ERROR-CORRECTED READS

We have evaluated the performance of ShoRAH (7) and QuasiRecomb (9) for simulated consensus error-corrected read data (5).

ShoRAH disregards pairing information of reads, but it is scalable enough to handle up to 1M reads. ShoRAH fails to produce full-length viral genome but reliably spans 98% of the consensus genome. It reasonably estimates the number of different viral genome, but even the most accurate ShoRAH-assembled viral genome differs from the closest true 1.3 kb-long viral genome in five nucleotides.

QuasiRecomb is designed to handle paired-end read data and manages to produce full-length viral genomes. Unfortunately, it can reliably process no more than 100K reads. Also the number of assembled distinct viral genomes is 10-200 times more than the number of true distinct viral genomes. The most accurate QuasiRecomb-assembled viral genome still differs from the closest true 1.3 kb-long viral genome in four nucleotides.

Unfortunately, we could not compare our method with QColors (23) assembly algorithm, which employs a similar conflict graph to represent viral population. A CSP solver is used by QColors for coloring the graph, which may limit its scalability to high-throughput data sets consisting of millions of sequencing reads. Currently, QColors is not publicly available[1].

4.6 PERFORMANCE OF VGA ON REAL HIV DATA

To further test the ability of VGA to accurately assemble a diverse natural occurring population and predict variant abundance levels in Reference 5, we used an Illumina HiSeq HIV data set, which consisted of 15M 2 × 100 bp paired-end reads with attached barcodes. Next, the high-fidelity sequencing protocol able to eliminate sequencing errors was applied resulting in 3M consensus error-corrected reads (further referred to as reads). The reads were then used to build *de novo* population consensus using Vicuna 1.3 (15). When run on our real data, Vicuna produced four contigs of average length 1195 bp. Each contig was then run through BLAST to check for overlaps. Once overlaps were found, the contigs were assembled into a final consensus of length 4337 bp.

4.6.1 Validation of *de novo* Consensus

A *de novo* assembled consensus was compared against reference-based consensus in Reference 5. To produce reference-based consensus, we iteratively map reads onto the HIV reference (Gag/Pol 3.4 kb HIV region) using InDelFixer. InDelFixer iteratively changes the reference genome based on the mapping of the current iteration. Also, we used InDelFixer in single iteration mode to map reads onto the constructed *de novo* consensus. *De novo* consensus is longer (4337 bp) than the reference-based consensus (3440 bp) and contained two regions with extremely peaked coverage compared to the surrounding regions. Both regions were considered to be the result of technical artifacts and removed from further consideration. After removing both regions, the length

---

[1] Upon querying for information on obtaining QColors, the authors were informed that the original software was tightly coupled for the analyses done in its original manuscript and is not currently available for general use.

of new *de novo* consensus becomes 3452 bp. We also filtered reads that belonged to regions with extreme coverage. Finally, we compared the number of reads mapped to the reference-based consensus versus the *de novo* consensus. A larger amount of reads are mapped onto the assembled consensus, thereby highlighting the advantage of *de novo* procedure for consensus construction over a reference-based.

From the *de novo* consensus, VGA assembled 32 full-length viral genomes that differ from each other in 2145 SNVs. Among known HIV sequences, Gag/Pol is the closest to the *de novo* consensus. Each of the 32 full-length viral genomes do not contain stop codons inside two known coding regions of Gag/Pol of length 1520 bp and 1820 bp, respectively. Alternatively, when VGA is applied to all 15M original uncorrected reads, 57 distinct viral genomes are assembled among which 36 contain stop codons in the two coding regions. This shows that a regular sequencing protocol is unsuitable for viral genome reconstruction.

4.7 COMPARISON OF ALIGNMENT ON ERROR-CORRECTED READS

When the high-fidelity sequencing protocol was applied on real HIV data, 3M error-corrected reads were obtained. To investigate the effect of the aligner on the mapping statistics, we mapped the reads to Gag/Pol sequence using such alignment tools as Mosaik v2.2 (30), Burrows–Wheeler Alignment (BWA) v0.7.12 (36), and Bowtie2 v2.2.5 (37) with default parameters. Mosaik is a tool for mapping second- and third-generation sequencing reads to a reference genome. Mosaik can align reads generated by all the major sequencing technologies. It employs a hash clustering strategy coupled with the Smith–Waterman algorithm, which allows to capture mismatches as well as short insertions and deletions. BWA tool is based on backward search with Burrows–Wheeler transform to efficiently align low-divergent sequences against a large reference sequence such as the human genome. Bowtie2 is the fastest among the three tools. It was developed for aligning sequencing reads to long reference sequences. Bowtie2 indexes the genome with an FM index (38) (based on the Burrows–Wheeler transform) to keep its memory footprint small: for the human genome, its memory footprint is typically around 3.2 GB of RAM. Bowtie2 supports gapped, local, and paired-end alignment modes.

Overall, all mapping tools were consistent in placing the reads onto the viral genome. However, they differ in number of reads mapped. BWA could map totally 56% of the reads where 48% of all reads were mapped with both ends and 8% with only one end mapped. Mosaik mapped 46% of all reads, out of which 36% were both ends mapped and 10% were one end mapped. Finally, Bowtie2 mapped 42% of the reads where 32% were both ends mapped and 10% were one end mapped. BWA produced more overlapping alignments (when both ends have common regions) – there was 11% overlapped paired-end reads while Mosaik resulted in 7.5% and Bowtie2 resulted in 7%. On the other hand, just several overlapped reads disagree between each other in Mosaik and Bowtie2 mappings, while about 5% of the overlapped reads in BWA had mismatches in common regions. BWA also produced the most soft clipping alignments. If we consider a soft clipping region as 10*nt*, then 6% of all reads were mapped as soft clipped in BWA, 0.3% in Mosaik, and no one read in Bowtie2.

4.8 EVALUATING OF ERROR CORRECTION TOOLS BASED ON HIGH-FIDELITY SEQUENCING READS

Another application of the high-fidelity sequencing protocol is the evaluation of error correction methods for NGS reads. The protocol attaches individual barcode to every fragment of the genome. After sequencing, the reads belonging to the original fragment are clustered based on the barcodes. The consensus read of the cluster is considered as erroneous read only if the most popular nucleotide in any position appears in more than 90% of reads in the cluster. Otherwise, the whole barcode cluster is disregard. Thus, after sequencing step, we have original reads with sequencing errors. After de-multiplexing step, we have erroneous reads belonging to genome fragments. We propose to correct original reads with some error correction tool and compare the result with the erroneous reads obtained after barcode correction. Specifically, we compare every erroneous read belonging to a cluster with the reads assigned to the same cluster and corrected by some tool.

In our work, we evaluate BLESS (39) and Fiona (40) error correction tools. We applied them to 15M paired-end original reads. Before error correction, we deleted all barcode parts from the reads so that the length of the reads became 87 bps. After error correction, both approaches changed the length of no more than 5% of all reads. We disregarded those reads for simplicity. Then we assigned every corrected read with length 87 bps to its barcode cluster for which we previously identified erroneous consensus read. We counted the percentage of original reads and reads after error correction identical to corresponding erroneous reads. As a result, both tools overcorrected original reads. Thus, before error correction in about 90% of clusters, all assigned original reads were actually equivalent to the corresponding erroneous consensus. After BLESS error correction, all reads were equivalent to the corresponding erroneous read only in 78% of clusters. Fiona corrected the reverse and forward ends differently. In 78.5% of clusters, the forward ends of all corrected reads were equal to the forward end of the corresponding erroneous reads. Although in 99.1% of clusters, the reverse ends were different from the reverse end of the corresponding consensuses.

ACKNOWLEDGMENT

The authors thank UCLA Clinical Microarray Core for performing the high-throughput sequencing experiment.

Fundings to S.M. and E.E. are supported by National Science Foundation grants 0513612, 0731455, 0729049, 0916676, 1065276, 1302448, and 1320589 and by National Institutes of Health grants K25-HL080079, U01-DA024417, P01-HL30568, P01-HL28481, R01-GM083198, R01-MH101782, and R01-ES022282. S.M. was supported in part by Institute for Quantitative & Computational Biosciences Fellowship, UCLA. N.C.W. was supported by Molecular Biology Whitcome Predoctoral Fellowship, UCLA. A.Z. was partially supported by Agriculture and Food Research Initiative Competitive Grant No. 201167016-30331 from the USDA National

Institute of Food and Agriculture and NSF award IIS-0916401. N.M. was partially supported by Second Century Initiative, Georgia State University.

Conflict of Interest: none declared.

REFERENCES

1. Lauring AS, Andino R. Quasispecies theory and the behavior of RNA viruses. PLoS Pathog 2010;6(7):e1001005.
2. Tsibris AMN, Korber B, Arnaout R, Russ C, Lo C-C, Leitner T, Gaschen B, Theiler J, Paredes R, Su Z et al. Quantitative deep sequencing reveals dynamic HIV-1 escape and large population shifts during CCR5 antagonist therapy in vivo. PLoS ONE 2009;4(5):e5683.
3. Henn MR, Boutwell CL, Charlebois P, Lennon NJ, Power KA, Macalalad AR, Berlin AM, Malboeuf CM, Ryan EM, Gnerre S et al. Whole genome deep sequencing of HIV-1 reveals the impact of early minor variants upon immune recognition during acute infection. PLoS Pathog 2012;8(3):e1002529.
4. Metzker ML. Sequencing technologies-the next generation. Nat Rev Genet 2009;11(1):31–46.
5. Mangul S, Wu NC, Mancuso N, Zelikovsky A, Sun R, Eskin E. Accurate viral population assembly from ultra-deep sequencing data. Bioinformatics 2014;30(12):i329–i337.
6. Astrovskaya I, Tork B, Mangul S, Westbrooks K, Măndoiu I, Balfe P, Zelikovsky A. Inferring viral quasispecies spectra from 454 pyrosequencing reads. BMC bioinformatics 2011;12 Suppl 6:S1.
7. Zagordi O, Bhattacharya A, Eriksson N, Beerenwinkel N. ShoRAH: estimating the genetic diversity of a mixed sample from next-generation sequencing data. BMC Bioinformatics 2011;12(1):119.
8. Prosperi MCF, Salemi M. QuRe: software for viral quasispecies reconstruction from next-generation sequencing data. Bioinformatics 2012;28(1):132–133.
9. Zagordi O, Töpfer A, Prabhakaran S, Roth V, Halperin E, Beerenwinkel N. Probabilistic inference of viral quasispecies subject to recombination. In: Chor B., editor. *Proceedings of the 16th Annual international conference on Research in Computational Molecular Biology (RECOMB'12)*. Berlin, Heidelberg: Springer-Verlag; 2012. p 342–354.
10. Mancuso N, Tork B, Skums P, Ganova-Raeva L, Măndoiu I, Zelikovsky A. Reconstructing viral quasispecies from NGS amplicon reads. In Silico Biol 2011;11(5):237–249.
11. Kinde I, Wu J, Papadopoulos N, Kinzler KW, Vogelstein B. Detection and quantification of rare mutations with massively parallel sequencing. Proc Natl Acad Sci U S A 2011;108(23):9530–9535.
12. Wang C, Mitsuya Y, Gharizadeh B, Ronaghi M, Shafer RW. Characterization of mutation spectra with ultra-deep pyrosequencing: application to HIV-1 drug resistance. Genome Res 2007;17(8):1195–1201.
13. Liu J, Miller MD, Danovich RM, Vandergrift N, Cai F, Hicks CB, Hazuda DJ, Gao F. Analysis of low-frequency mutations associated with drug resistance to raltegravir before antiretroviral treatment. Antimicrob Agents Chemother 2011;55(3):1114–1119.
14. Palmer S, Boltz V, Maldarelli F, Kearney M, Halvas EK, Rock D, Falloon J, Davey RT Jr., Dewar RL, Metcalf JA et al. Selection and persistence of non-nucleoside reverse transcriptase inhibitor-resistant HIV-1 in patients starting and stopping non-nucleoside therapy. Aids 2006;20(5):701–710.

15. Yang X, Charlebois P, Gnerre S, Coole MG, Lennon NJ, Levin JZ, Qu J, Ryan EM, Zody MC, Henn MR. De novo assembly of highly diverse viral populations. BMC Genomics 2012;13(1):475.

16. Gnerre S, MacCallum I, Przybylski D, Ribeiro FJ, Burton JN, Walker BJ, Sharpe T, Hall G, Shea TP, Sykes S et al. High-quality draft assemblies of mammalian genomes from massively parallel sequence data. Proc Natl Acad Sci U S A 2011;108(4):1513–1518.

17. Zerbino DR, Birney E. Velvet: algorithms for de novo short read assembly using de Bruijn graphs. Genome Res 2008;18(5):821–829.

18. Luo R, Liu B, Xie Y, Li Z, Huang W, Yuan J, He G, Chen Y, Pan Q, Liu Y et al. SOAPdenovo2: an empirically improved memory-efficient short-read de novo assembler. GigaScience 2012;1(1):18.

19. Yang W-Y, Hormozdiari F, Wang Z, He D, Pasaniuc B, Eskin E. Leveraging Multi-SNP reads from sequencing data for haplotype inference. Bioinformatics 2013;29(18):2245–2252.

20. Bansal V, Bafna V. HapCUT: an efficient and accurate algorithm for the haplotype assembly problem. Bioinformatics 2008;24(16):i153–i159.

21. Beerenwinkel N, Töpfer A. J Virol. Available at https://github.com/cbg-ethz/InDelFixer.

22. Hormozdiari F, Alkan C, Eichler EE, Sahinalp SC. Combinatorial algorithms for structural variation detection in high-throughput sequenced genomes. Genome Res 2009;19(7):1270–1278.

23. Huang A, Kantor R, DeLong A, Schreier L, Istrail S. QColors: an algorithm for conservative viral quasispecies reconstruction from short and non-contiguous next generation sequencing reads. In Silico Biol 2011;11(5):193–201.

24. Kubale M. *Graph Colorings*. Volume 352. Providence (RI): American Mathematical Society; 2004.

25. Johnson DS, Trick MA. In: Johnson D.J., Trick M.A., editors. *Cliques, Coloring, and Satisfiability: Second DIMACS implementation Challenge, Workshop, October 11-13, 1993*. Volume 26. Boston (MA): American Mathematical Society; 1996.

26. Duitama J, McEwen GK, Huebsch T, Palczewski S, Schulz S, Verstrepen K, Suk E-K, Hoehe MR. Fosmid-based whole genome haplotyping of a hapmap trio child: evaluation of single individual haplotyping techniques. Nucleic Acids Res 2012;40(5):2041–2053.

27. Mitzenmacher M, Upfal E. *Probability and Computing: Randomized Algorithms and probabilistic Analysis*. New York: Cambridge University Press; 2005.

28. Eriksson N, Pachter L, Mitsuya Y, Rhee S-Y, Wang C, Gharizadeh B, Ronaghi M, Shafer RW, Beerenwinkel N. Viral population estimation using pyrosequencing. PLoS Comput Biol 2008;4(5):e1000074.

29. Angly FE, Willner D, Rohwer F, Hugenholtz P, Tyson GW. Grinder: a versatile amplicon and shotgun sequence simulator. Nucleic Acids Res 2012;40(12):e94.

30. Lee W-P, Stromberg MP, Ward A, Stewart C, Garrison EP, Marth GT. MOSAIK: a hash-based algorithm for accurate next-generation sequencing short-read mapping. PLoS ONE 2014;9(3):e90581.

31. Zagordi O, Geyrhofer L, Roth V, Beerenwinkel N. Deep sequencing of a genetically heterogeneous sample: local haplotype reconstruction and read error correction. J Comput Biol 2010;17(3):417–428.

32. Ndung'u T, Weiss RA. On HIV diversity. AIDS 2012;26(10):1255–1260.

33. Martins LP, Chenciner N, Wain-hobson S. Complex intrapatient sequence variation in the V1 and V2 hypervariable regions of the HIV-1 gp120 envelope sequence. Virology 1992;191(2):837–845.

34. Yoshimura FK, Diem K, Learn GH, Riddell S, Corey L. Intrapatient sequence variation of the gag gene of human immunodeficiency virus type 1 plasma virions. J Virol 1996;70(12):8879–8887.

35. Neher RA, Leitner T. Recombination rate and selection strength in HIV intra-patient evolution. PLoS Comput Biol 2010;6(1):e1000660.

36. Li H, Durbin R. Fast and accurate short read alignment with burrows- wheeler transform. Bioinformatics 2009;25(14):1754–1760.

37. Langmead B, Salzberg SL. Fast gapped-read alignment with Bowtie 2. Nat Methods 2012;9(4):357–359.

38. Ferragina P, Manzini G. Opportunistic data structures with applications. Proceedings of the 41st Annual Symposium on Foundations of Computer Science (FOCS '00). Washington (DC): IEEE Computer Society; 2000. p 390.

39. Heo Y, Wu X-L, Chen D, Ma J, Hwu W-M. BLESS: bloom filter-based error correction solution for high-throughput sequencing reads. Bioinformatics 2014;30(10):1354–1362. DOI: 10.1093/bioinformatics/btu030.

40. Schulz MH, Weese D, Holtgrewe M, Dimitrova V, Niu S, Reinert K, Richard H. Fiona: a parallel and automatic strategy for read error correction. Bioinformatics 2014;30(17):i356–i363.

PART II

GENOMICS AND EPIGENOMICS

5

SCAFFOLDING ALGORITHMS

IGOR MANDRIC

Department of Computer Science, Georgia State University, Atlanta, GA, USA

JAMES LINDSAY AND ION I. MĂNDOIU

Department of Computer Science and Engineering, University of Connecticut, Storrs, CT, USA

ALEXANDER ZELIKOVSKY

Department of Computer Science, Georgia State University, Atlanta, GA, USA

5.1 SCAFFOLDING

Due to rapid advances in *high-throughput sequencing* (HTS) technologies, the interest in the problem of *de novo* genome assembly has been renewed. These technologies, also referred to as next-generation sequencing (NGS), can produce millions of short paired-end reads covering whole genome; thus, the throughput is a magnitude higher than the classic *Sanger* sequencing. It is worth noticing that the cost of producing reads keeps a decreasing trend, making NGS a very attractive tool for a high range of applications. For example, Illumina HiSeq platform is able to produce billions of read pairs in a single run at a cost of cents per megabase. Although the number of reads (*shotgun* reads) is significant due to their short length, genome assembly still represents a challenging problem.

Current assemblers (Velvet (1), ALLPATHS-LG (2)) output a set of contiguous DNA chunks, usually referred to as *contigs*. Contig length can vary from hundreds to hundreds of thousands base pairs. As genomes usually constitute millions of base pairs, the resulting assemblies are highly fragmented.

Computational Methods for Next Generation Sequencing Data Analysis, First Edition.
Edited by Ion I. Măndoiu and Alexander Zelikovsky.
© 2016 John Wiley & Sons, Inc. Published 2016 by John Wiley & Sons, Inc.
Companion website: www.wiley.com/go/Mandoiu/NextGenerationSequencing

The advantage of NGS read pairs is the possibility to use them for joining contigs into some larger DNA chunks. Two reads coming from a read pair, which are mapped on to two different contigs, due to constant insert size of the library (it can be deemed as constant, although the mean insert size and its standard deviation are used), infer a certain connectivity information between the contigs. Thus, such a read pair suggests a certain relative order, relative orientation, and an estimation of gap length between the two contigs. A set of contigs joined into a chain, where relative order and orientation, as well as the estimation of the gap length between neighboring contigs is provided, is called *scaffold*.

Software programs, usually referred to as *scaffolders*, construct scaffolds based on contigs output by an assembler and the connectivity information provided by the NGS read pairs. Due to misassemblies in contigs, repeats, and chimeric reads, the information about relative ordering and orientation of two contigs connected with a set of read pairs can be contradictory and not reliable. Thus, choosing a wrong subset of read pairs as an evidence for connection between two contigs can result in inferring a wrong relative ordering and/or orientation as well as the gap estimation between them. Edges that comply with the true orientation of contigs and the distance between them are usually called concordant, otherwise discordant edges.

As usual, the scaffolding problem is formulated as an optimization problem on a graph $G = (V, E)$ with the vertex set V usually being the set of contigs and the edge set E usually corresponding to the read pair links connecting contigs. Graph G is called the *scaffolding graph*. Different tools use different optimization criteria for the scaffolding problem. For example, OPERA (3) maximizes the number of concordant edges in the scaffolding graph; SCARPA (4) minimizes the number of discordant edges; SILP2 (5) uses an integer linear program based approach.

In Reference 3, it is proven that the scaffolding problem is an NP-hard problem. Thus, all state-of-the-art scaffolders use different heuristic approaches. OPERA tackles the problem using an elegant dynamic programming algorithm, SCARPA, MIP (6) and SILP2 solve separately the orientation and the ordering problems, the last two using decomposition of the scaffolding graph into biconnected components. SSPACE (7) uses a simple but powerful greedy heuristics.

In the next section, we briefly describe some of the state-of-the-art scaffolding tools: SSPACE, OPERA, SOPRA, MIP, and SCARPA. Section 1.4 provides a deep insight into two recent scaffolding tools: SILP2 and ScaffMatch. In Section 1.5, we introduce performance metrics and evaluate scaffolding tools based on these metrics.

5.2 STATE-OF-THE-ART SCAFFOLDING TOOLS

5.2.1 SSPACE

SSPACE (SSAKE-based Scaffolding of Pre-Assembled Contigs after Extension) is a stand-alone scaffolder of pre-assembled contigs (7). SSPACE is based on the short-read assembler SSAKE (8).

The first step of the algorithm is reading and filtering the input data. The reads are mapped on to contigs, and if a connection between two contigs is supported by a number of read pairs less than a user-specified threshold, such read pairs are filtered out. The unmapped reads are cut into pieces (by default 5 bp) and stored in a hash table.

An optional contig extension step is followed. The contigs are extended greedily nucleotide by nucleotide using the pieces of the unmapped reads stored in the hash table.

After the preprocessing is done, the reads are mapped as single-end reads to contig edges (the length of the contig edges is determined as sum of insert length and insert standard deviation) and are paired if the gap constraints are satisfied.

In the final step, SSPACE greedily builds scaffolds using contigs as seeds. It starts from the largest contig and iteratively extends the scaffold based on the number of supporting read pairs. If a contig has multiple alternatives, then the ratio between two best alternatives is computed. If the ratio is less than 0.7 (default value), the scaffold is extended with the best alternative, otherwise the extension is stopped. The process is repeated until the scaffold cannot be extended in either direction. Once the scaffold is constructed, SSPACE starts building a new scaffold from the remaining contigs using exactly the same procedure. Thus, the scaffolds are greedily built until no extension is possible.

SSPACE allows scaffolding using multiple libraries. In this case, it hierarchically builds scaffolds starting with the library having the smallest insert length.

5.2.2 OPERA

OPERA implements an elegant dynamic programming scaffolding algorithm that solves the scaffolding problem of maximizing the number of concordant edges in the scaffold.

In the scaffolding graph $G = (V, E)$, the vertex set V corresponds to the set of contigs and the set of edges E corresponds to the set of connections between contigs with weight equal to the bundle size (see bundling step).

The idea of the algorithm relies on some important definitions. A *partial scaffold* S' is a scaffold on a subset of contigs, the *dangling set* $D(S')$ is the set of edges of G from S' to $V\setminus S'$, the *active region* $A(S')$ is the shortest suffix of the partial scaffold S' for which all dangling edges are connected with a contig in $A(S')$. A partial scaffold S' is called *valid* if all edges in the induced graph are concordant. By Lemma 1 (3), two partial scaffolds are equivalent if their corresponding dangling sets and active regions are equivalent (equal).

For the case of an "optimal" scaffold, that is, a scaffold that does not contain discordant edges, OPERA introduces the following algorithm. It starts from an empty scaffold $S = \emptyset$ and extends it into one contig at a time. In every iteration, in every equivalence class of partial scaffolds, it considers those partial scaffolds that do not contain any discordant edges. The algorithm is proved to be polynomial for the class of bounded-width graphs, that is, scaffolding graphs without discordant edges.

However, if the scaffolding graph G contains discordant edges, the problem of deciding whether G contains a scaffold S with at most p discordant edges is NP-complete. Treating p as a constant, a slight modification of the algorithm discussed earlier can deliver scaffold with at most p discordant edges in polynomial time. In order to improve the running time, OPERA contracts sets consisting of "border" contigs (i.e., contigs, whose length is greater than the library fragment length) together with their adjacent contigs into one node of the scaffolding graph G. This procedure reduces the size of G.

OPERA solves the problem of gap size estimation using a maximum likelihood approach. Repeats are handled by filtering out contigs with read coverage 1.5 times higher than the genomic average.

A more detailed description of the algorithm and the software can be found in Reference 3.

5.2.3 SOPRA

SOPRA (Scaffolding algorithm for paired reads via statistical optimization) (9) solves the scaffolding problem by using methods inspired from statistical physics.

The SOPRA algorithm consists of several stages. In the first stage, the problem of contig orientation is solved. For a pair of contigs i and j, a value J_{ij} is introduced, which is equal to the signed number of contigs supporting a certain relative orientation (in terms of (10), bundle size). The sign of J_{ij} depends on the orientations of the contigs i and j: if they have the same orientation, then the sign of J_{ij} is positive; otherwise, it is negative. SOPRA searches such a configuration of the contigs that minimizes the sum

$$- \sum_{(i,j) \in E(G)} J_{ij} S_i S_j,$$

where $S_i \in \{-1, +1\}$ stands for the contig orientation.

After the relative orientation is determined, the next stage is to solve the problem of the relative contig positioning. This is achieved by maximizing a joint probability distribution associated with adjacent contigs. SOPRA provides a nice physical interpretation for the statistical optimization problem it solves at this stage.

Once the relative orientation and positioning is obtained, SOPRA filters out some dubious (erroneous) contigs based on the so-called density profiles, that is, number of times contigs cover a region of a scaffold. It further decides whether a pair of contigs is to be joined or to be separated by a gap.

5.2.4 MIP

MIP is a stand-alone scaffolding tool based on Mixed-Integer Programming. The main feature of MIP is partitioning of the scaffolding graph into smaller biconnected components and solving the scaffolding problem separately for each component. The solution of each independent biconnected component is independent of the other components. Unlike SOPRA, MIP formulates an optimization problem involving both relative orientation and ordering of the contigs.

The final solution is obtained by combining the solutions of all the biconnected components of the scaffolding graph.

5.2.5 SCARPA

SCARPA formulates the problem of contig orientation as finding a minimal odd cycle transversal in the scaffolding graph $G = (V, E)$ built as follows: each contig c is represented by two nodes c^+ and c^- in V (5' and 3' end of the contig) and the edge set E consists of two types of edges, namely edges connecting the two nodes of each contig and the edges between two nodes of different contigs corresponding to read pairs mapping to the corresponding strands.

In addition to removing contigs such that no odd cycles remain in G, SCARPA allows discarding nonconsistent edges. For each edge in E, two auxiliary nodes are introduced and in case there is a tie between removing a node in V and removing an auxiliary node, the last option is preferred, thus resulting in disregarding a suspicious link between contigs rather than removing a contig from the node set V.

Although the orientation stage removes odd-length cycles, the directed graph obtained may still have cycles. Contig ordering is obtained after all directed cycles are eliminated by solving the feedback arc set problem.

In the last stage, SCARPA solves a linear program for determining a topological ordering such that the distance between the contigs agrees best with the size of the gaps.

5.3 RECENT SCAFFOLDING TOOLS

In this section, we provide a detailed description for two recent scaffolding algorithms, namely SILP2 and ScaffMatch.

5.3.1 SILP2

Given a set of contigs C and a set of read pairs R, the scaffolding problem asks for the most likely orientation of the contigs along with a partition of the contigs into ordered sets connected by read pairs of R. The main steps of the SILP2 algorithm are as follows (see Figure 5.1 for a high-level flowchart). We first map the read onto contigs using Bowtie2, disregarding pairing information in the mapping process. Alignments are processed to extract read pairs for which both reads have unique alignments, and the alignments are onto distinct contigs. A scaffolding graph is then constructed with nodes corresponding to contigs and edges corresponding to extracted read pairs. The scaffolding graph is partitioned into 3-connected components using the SPQR tree data structure (11, 12) implemented in OGDF (13). The maximum likelihood contig orientation is formulated as an ILP that is efficiently solved by applying nonserial dynamic programming based the SPQR tree data structure. Next, scaffold chains are extracted from the ILP solution by using bipartite matching and breaking remaining cycles. Finally, maximum likelihood estimates for the gap lengths are obtained using quadratic programming. In the following, we detail the key steps of the algorithm, including scaffolding graph construction, the maximum likelihood models used for contig orientation, and mapped read pair probability estimation, then we briefly overview the orientation, the ILP formulation, and the improved NSDP algorithm for efficiently solving the ILP.

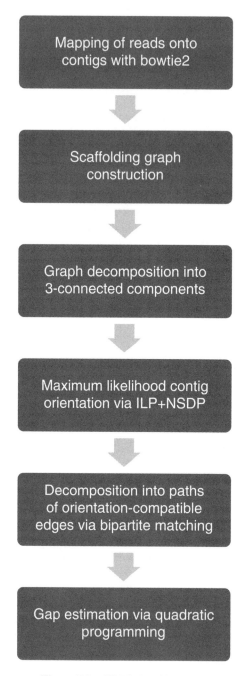

Figure 5.1 SILP2 algorithm flow.

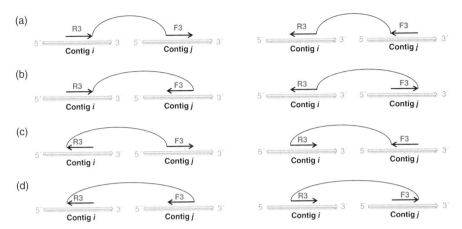

Figure 5.2   Four states A, B, C, and D.

5.3.1.1 Scaffolding Graph   The scaffolding problem is modeled with a *scaffolding graph* $G = (V, E)$, where each node $i \in V$ represents a contig and each edge $(i, j) \in E$ represents all read pairs whose two individual reads are mapped to the contigs i and j, respectively. Each read in a pair is aligned either to the forward or reverse strand of the corresponding contig sequence, and this results in four possible configurations for a read pair (denoted A, B, C, or D, see Figure 5.2) that can be modeled as a bidirected edge (3, 6, 14). Orientation of contigs and the bidirected orientation of edges should agree (be concordant) with each other and should not result in any directed cycles for linear genomes (e.g., eukaryotes).

5.3.1.2 Maximum Likelihood Scaffold Graph Orientation   As an intermediate step toward solving the scaffolding problem, we consider the problem of determining an orientation of the scaffolding graph, which includes choosing one of the two possible orientations for each node (contig) $i \in V$ as well as choosing for each edge $(i, j) \in E$ one of the four bidirections that is concordant with the orientations of i and j. A common way to reduce an inference problem to an optimization problem is to seek a feasible solution with a maximum likelihood. Let each observation, that is, aligned read pair $r \in R$, have a probability p_r of being correct. Any feasible contig orientation $O = O(C)$ either agrees or disagrees with the read pair r. Let R_O be the set of read pairs agreeing with O. Assuming independence of observations, the likelihood of an orientation O can be written as

$$\prod_{r \in R_O} p_r \prod_{r \in R - R_O} (1 - p_r) = \prod_{r \in R}(1 - p_k) \prod_{r \in R_O} \left(\frac{p_r}{1 - p_r} \right)$$

and hence its log-likelihood is $\sum_{r \in R} \ln (1 - p_r) + \sum_{r \in R_O} \ln \left(\frac{p_r}{1 - p_r} \right)$. Since the first sum does not depend on the orientation O, maximizing the log-likelihood is equivalent to maximizing

$$\sum_{r \in R_O} \ln \left(\frac{p_r}{1 - p_r} \right) \tag{5.1}$$

over all contig orientations O.

5.3.1.3 Mapping Probability Estimation If p_r's are assumed to be the same for all read pairs, then the objective (5.1) reduces to maximization of the number of read pairs that agree with the contig orientation O. We consider the following factors that reduce the probability p_r that read pair r is aligned correctly:

1. *Overlap with Repeats.* As noted earlier, only pairs for which both reads map uniquely to the set of contigs are used for scaffolding. Still, a read that fully or partially overlaps a genomic repeat may be uniquely mapped to the incorrect location in case repeat copies are collapsed. We preprocess contigs to annotate repeats from known repeat families and by recording the location of multimapped reads. An estimate of the repeat-based mapping probability p_r^{rep} is found by taking the percentage of bases of r aligned to nonrepetitive portions of the contigs.

2. *Contig Coverage Dissimilarity.* Although sequencing coverage can have significant departures from uniformity due to biases introduced in library preparation and sequencing, the average coverage of adjacent contigs is expected to be similarly affected by such biases (all read alignments, including randomly allocated nonunique alignments, are used for estimating computing average contig coverages). If the two reads of r map on to contigs i, respectively j, the coverage-based mapping probability of r, p_r^{cov}, is defined as

$$p_r^{cov} = 1 - \frac{|coverage_i - coverage_j|}{coverage_i + coverage_j}.$$

Note that factors such as repeat content of the sequenced genome and sequencing depth will determine how informative repeat-based and coverage-based mapping probabilities are. Depending on these factors, either p_r^{rep}, p_r^{cov}, or their product may provide the most accurate estimate for p_r. Mismatches and indels in read alignments, which can be caused by sequencing errors or polymorphisms in the sequenced sample, can easily be incorporated in the estimation of mapping probabilities.

5.3.1.4 Integer Linear Program Our integer linear program maximizes the log-likelihood of scaffold orientation using the following Boolean variables:

– a binary variable S_i for each contig i, with S_i equal to 0 if the contig's orientation remains the same and $S_i = 1$ if the contig's orientation is flipped with respect to the default orientation in the final scaffold.

– a binary variable S_{ij} for each edge $(i, j) \in E$, which equals 0 if none or both ith and jth contigs are flipped and equals 1 if only one of them is flipped.

– binary variables A_{ij} (respectively, B_{ij}, C_{ij}, and D_{ij}), which are set to 1 if and only if an edge in state A (respectively, B, C, or D) is used to connect contigs i and j (see Figure 5.2). For any contig pair i and j, at most one of these variables can be one.

Let A_{ij}^r (respectively, B_{ij}^r, C_{ij}^r, or D_{ij}^r) denote the set of read pairs supporting state A (respectively, B, C, or D) between the ith and jth contig. Define the constant A_{ij}^w by

$$A_{ij}^w = \sum_{r \in A_{ij}^r} \ln \left(\frac{p_r}{1 - p_r} \right)$$

with B_{ij}^w, C_{ij}^w, and D_{ij}^w defined analogously.

We now ready to formulate the ILP for maximizing the log-likelihood of a scaffold orientation:

$$\sum_{(i,j) \in E} (A_{ij}^w \cdot A_{ij} + B_{ij}^w \cdot B_{ij} + C_{ij}^w \cdot C_{ij} + D_{ij}^w \cdot D_{ij}), \tag{5.2}$$

where

$$S_{ij} \leq S_i + S_j \qquad S_{ij} \leq 2 - S_i - S_j \tag{5.3}$$

$$S_{ij} \geq S_j - S_i \qquad S_{ij} \geq S_i - S_j \tag{5.4}$$

$$A_{ij} + D_{ij} \leq 1 - S_{ij} \qquad B_{ij} + C_{ij} \leq S_{ij}. \tag{5.5}$$

In this ILP, constraints ((5.3)–(5.5)) enforce agreement between contig orientation variables S_i's and edge orientation variables S_{ij}'s, A_{ij}'s, B_{ij}'s, C_{ij}'s, and D_{ij}'s.

Since eukaryotic genomes are linear, a valid scaffold orientation should not contain any cycles. The constraints (5.5) already forbid 2-cycles. Additionally, 3-cycles are forbidden with the constraints shown in Figure 5.3. Larger cycles generated in the ILP solution are broken heuristically because it is infeasible to forbid all of them using explicit constraints.

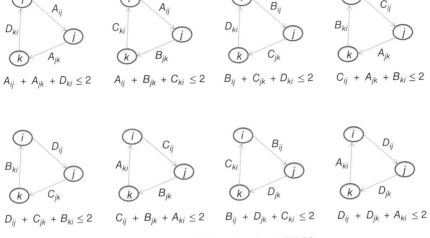

Figure 5.3 Forbidding 3-cycles in SILP2.

5.3.1.5 Nonserial Dynamic Programming For large mammalian genomes, the number of variables and constraints is too large for solving the ILP (5.2)–(5.5) via standard solvers (SILP2 uses CPLEX (15), which is available free of charge for academic institutions). We adopt the nonserial dynamic programming (NSDP) paradigm to overcome this barrier and to optimally solve the problem. NSDP is based on the interaction graph with nodes corresponding to ILP variables and edges corresponding to the ILP constraints—two nodes are adjacent in the interaction graph if their associated variables appear together in the same constraint. Through the NSDP process, variables are removed in the way that adjacent vertices can be merged together (16). The first step in NSDP is identifying weakly connected components of the interaction graph. We find the 2- and 3-connected components of the interaction graph with efficient algorithms, and then we solve each component independently in such a way that the solutions can be merged together to find the global solution.

All constraints ((5.3)–(5.5)) as well as 3-cycle constraints connect S_i's following the edges of the scaffolding graph. Therefore, the S_i-nodes of the interaction graph for our ILP will have the same connectivity structure as the scaffolding graph $G = (V, E)$. As it has been noticed in Reference 3, the scaffolding graph is a bounded-width graph and should be well decomposable in 2- and 3-connected components. The SPQR-tree data structure is employed to determine the decomposition order for 3-connected components the scaffolding graph (11). The solution to each component of the scaffolding graph is found using a bottom-up traversal through which each component is solved two times: for similar and opposite orientations of the common nodes. The objective value of each case is then entered into the objective of the parental component. Having the solution of all components, top-down DFS starting from the same root is performed to apply the chosen solution for each component.

In the following we illustrate the way how the solution is computed in stages and explain how the results of each stage are combined with the previous stage to dynamically solve the problem. Obviously, an isolated, connected component will not influence other components. Moreover, it has been shown in Reference 6 that 2-connected components can be solved independently. As it can be seen in Figure 5.4a, after removing the articulation point (1-cut) to decompose the graph into 2-connected components, each component is solved with the same arbitrary direction assigned to the common node, and then the resulting solutions are collapsed into the parent solution. The pre-assigned direction will never affect the parent solution since all contigs in the scaffold can be flipped at the same time.

Still, 2-connected components can be very large, so we look for 2-cuts in order to decompose the graph into significantly smaller 3-connected components. Figure 5.4b shows that splitting the two 2-cut **nodes i and j decomposes the graph into 3-connected components A and B. The ILP for component A is solved twice to obtain

1. the ILP solution sol_{00} in which the 2-cut nodes i and j are constrained to both have default orientations;
2. the ILP solution sol_{01} in which the 2-cut nodes i and j are constrained to have opposite orientations.

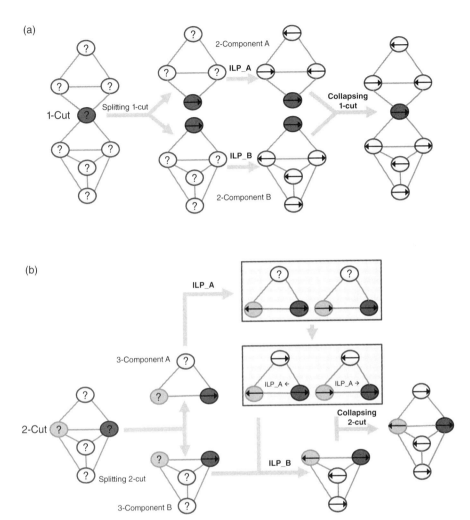

Figure 5.4 Solving the maximum likelihood ILP via graph decomposition. (a) Graph decomposition into 2-connected components: Dark grey (1-cut) node splits the graph into two 2-connected components A and B. The ILP is solved for each component separately. If the direction of the cut node in the ILP solution for B is opposite to the one in the solution for A, then the solution of B is inverted. Then ILP solutions for A and B are collapsed into the parent solution. (b) Graph decomposition into two 3-connected components: dark grey and light grey (2-cut) nodes split the graph into two 3-connected components A and B. The ILP is solved for component A twice—for the same and the opposite directions assigned to two 2-cut nodes. Then these two solutions are used in the objective for the ILP of component B. Finally, ILP solutions for A and B are collapsed into the parent solution.

The two solutions are combined to solve the ILP for component B. The ILP objective for component B should be updated by adding the term of $sol_{00} + (sol_{01} - sol_{00}) \cdot S_{ij}$ or, equivalently, the value sol_{00} should be added to A_{ij}^w and D_{ij}^w and the value sol_{01} should be added to B_{ij}^w and C_{ij}^w. The overall solution is obtained by identifying the common nodes of the components. In the example on Figure 5.4b, the optimal solution happens when 2-cut nodes have opposite directions. The corresponding solution of ILP for the component A should be incorporated in the overall solution. When the scaffolding graph has 3-connected components too large to handle, 3-cuts could also be used for decomposition.

The pseudo-code of the SILP2 NSDP algorithm for processing 3-connected components is given below. SILP2 is different from SILP1 in the else clause—instead of solving ILP for each of four possible combinations of assignments for S_i and S_j as in SILP1, ILP is solved only two times for combinations $S_i = 0$ & $S_j = 0$ and $S_i = 0$ & $S_j = 1$.

Algorithm The Pseudo-Code of the SILP2 NSDP Algorithm for Processing 3-Connected Components.

1:   $STACK \leftarrow$ root of SPQR-tree
2:   $VISITED \leftarrow \emptyset$
3:   **while** $STACK$ is not empty **do**
4:      $p \leftarrow STACK.pop()$
5:      **for each** child q of p **do**
6:          **if** p not in $VISITED$ **then**
7:              $STACK.push()$
8:          **end if**
9:      **end for**
10:      **if** p is root **then**
11:          $sol_{final} = SOLVE(p.skeleton())$
12:      **else**
13:          $(i,j) = getcut(p, parent(p))$
14:          $sol_{00} = SOLVE(p.skeleton(), S_i = 0, S_j = 0)$
15:          $sol_{01} = SOLVE(p.skeleton(), S_i = 0, S_j = 1)$
16:          in $parent(p)$, increase weights A_{ij}^w and D_{ij}^w by sol_{00} and weights B_{ij}^w, C_{ij}^w
     by sol_{01}.
17:      **end if**
18: **end while**

5.3.1.6 *Thinning Heuristic*
Unfortunately the largest tri-connected component may still induce an ILP too large for CPLEX to solve in a reasonable amount of time. In order to address this problem a thinning heuristic is applied to the scaffolding

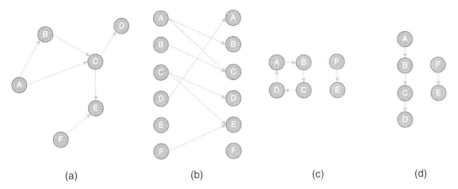

(a)　　　　　　　(b)　　　　　　　(c)　　　　　　　(d)

Figure 5.5 (a) Pairwise ordering obtained from ILP output. (b) The corresponding bipartite graph. (c) The collection of simple paths and cycles in the bipartite matching. (d) The scaffold represented by a collection of paths obtained by deletion of the lightest edges from simple cycles.

graph. This scenario can be detected by setting a threshold on the maximum number of contigs allowed in a tri-connected component. When a component exceeds the threshold the number of read pairs necessary to induce an edge is increased by one and decomposition recomputed until there is no component above the threshold.

5.3.1.7　Contig Ordering via Bipartite Matching The solution to the ILP does not yield a scaffold, instead only gives the pairwise order and orientation of the contigs (see Figure 5.5a). In order to find scaffolding, we need to remove edges such that remaining graph becomes a collection of paths. Naturally, we want to remove the least probable edges, that is, lightest edges. This problem can be solved efficiently by finding matching in the following bipartite graph $B = (V^1 \cup V^2, E)$ where each vertex in V^1 corresponds to the $3'$ end of a contig, each vertex in V^2 to the $5'$ end of a contig, and each edge corresponds to an edge from ILP solution connecting the $5'$ of its beginning to the $3'$ end of its end (see Figure 5.5b). Each edge in B has the same weight as in the ILP solution. Obviously, any matching in B corresponds to a collection of paths and simple directed cycles (see Figure 5.5c). Then from each cycle the lightest one (See Figure 5.5d) is deleted and we output the resulted set of paths.

5.3.2　ScaffMatch

Below we describe the ScaffMatch (17) problem formulation and algorithmic details are explained in the following main scaffolding steps:

- Preprocessing of read pairs including read mapping, handling repeats and gap estimation for each read pair.

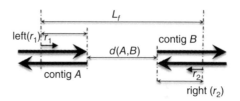

Figure 5.6 Gap estimation d is calculated in conformity with the formula: $d = L_f - (L_A - left(r_1)) - (L_B - right(r_2))$, where L_f is the fragment length, L_A, L_B are lengths of contigs A and B, $left(r_1)$ is the left mapping position of the read r_1, $right(r_2)$ is the right mapping position of the read r_2.

- Scaffolding graph construction with vertices corresponding to contig strands and edges corresponding to read pairs.
- Matching scaffold finding near-maximum weight paths in the scaffolding graph and the corresponding orientation and ordering contigs.
- Insertion of skipped contigs into the matching scaffold.

5.3.2.1 Read Preprocessing Each contig has two reverse complement strands and each read from a pair is mapped to one of the strands. We discard reads aligned to (suspected) repeats. First we filter out read pairs in which at least one read has multiple alignments. Then for each contig we compute its read coverage and filter out contigs with coverage greater by 2.5σ than the average. Multiple reads connecting the same pair of contig strands are bundled according to their gap estimation. Since we want to keep only reads that agree with each other, the reads outside of the largest bundle are discarded. This procedure is illustrated in Figure 5.6.

The next step in the construction of the scaffolding graph $G = (V, E)$ is the so-called edge bundling step (10). Multiple edges that support the same distance, relative ordering and relative orientation are *bundled*.

Consider two contigs A, $B \in V(G)$ and let $E_{AB} = \{e_i\}_{i=1,n}$ be the set of edges (paired-end reads) supporting the same relative ordering and orientation of A and B. The following steps are performed until no edges in E_{AB} remain unused:

1. Choose $e_k \in E_{AB}$ with the median gap estimation $L_{e_k} = L_{median}(E_{AB})$;
2. Determine edges (paired-end reads) $e_i \in E_{AB}$, whose gap estimation L_{e_i} is within $3\sigma(E_{AB})$ of $L_{median}(E_{AB})$, bundle them into one single edge e';
3. For the bundled edge e' set the gap estimation and its standard deviation to $L_{e'} = a/b$ and $\sigma(e') = 1/\sqrt{b}$ respectively, where

$$a = \sum_{e_i \in e'} \frac{L_{e_i}}{\sigma(e_i)^2}, b = \sum_{e_i \in e'} \frac{1}{\sigma(e_i)^2};$$

4. Set the weight of the bundled edge $W_{e'}$ to $|e'|$;
5. $E_{AB} := E_{AB} \backslash e'$.

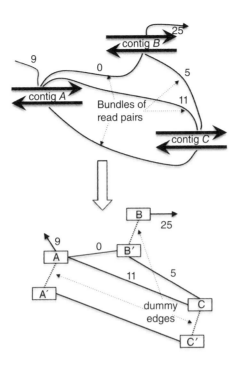

Figure 5.7 Contigs *A*, *B*, and *C* with connecting bundles of read pairs and the corresponding scaffolding graph. Each contig is split into two nodes connected with a dummy edge. Each bundle of read pairs corresponds to an inter-contig edge connecting respective strands with the weight equal to the size of the bundle.

5.3.2.2 Scaffolding Graph ScaffMatch (17) constructs the scaffolding graph in the following way. Each vertex of the scaffolding graph $G = (V, E)$ corresponds to one of the contig strands (SILP2 denotes every contig as a graph node in one of the four states, see Figure 5.2) and each *inter-contig* edge corresponds to a bundle of read pairs connecting two strands of different contigs (see Figure 5.7). The weight of an inter-contig edge is equal to the size of the corresponding bundle. Also for each contig we have a *dummy* edge connecting its two strands.

5.3.2.3 Matching Scaffolding Ideally, we expect that the scaffolding graph would consist of a set of paths each corresponding to a different chromosome (see Figure 5.8a). Unfortunately, repeats introduce noisy edges connecting unrelated contigs even from different chromosomes. Additionally, the paths corresponding to chromosomes may skip short contigs (especially contigs which are shorter than the insert length). Therefore, any set of paths passing through all dummy edges in the scaffolding graph G corresponds to a plausible scaffold (see Figure 5.8b). The most likely scaffold (see Figure 5.8c) would be supported by the largest number of read pairs. Therefore, we can formulate the following

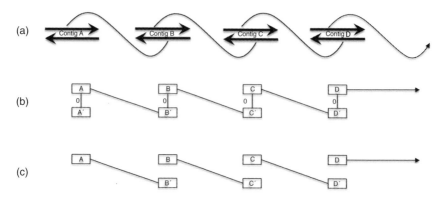

Figure 5.8 (a) A scaffold *A–B–C–D*: the connection of each pair of adjacent contigs is supported by bundles of read pairs. (b) A path *A′–A–B′–B–C′–C–D′–D* in the scaffolding graph corresponding to the scaffold *A–B–C–D*. (c) The matching of the scaffolding graph corresponding to the bunches of read pairs supporting adjacent contigs.

5.3.2.4   The Scaffolding Problem   Given a scaffolding graph *G*, find a set of paths passing through all dummy edges with maximum total weight of all inter-contig edges. By setting the weight of each dummy edge to a large number (e.g., number of all read pairs), we reduce the scaffolding problem to the following

5.3.2.5   Maximum-Weight Acyclic 2-Matching (MWA2M) Problem   Given a weighted graph $G = (V, E, w)$, find a maximum weight subset of edges $M \subseteq E$ such that each vertex $v \in V$ is incident to at most two edges in *M*.

The MW2AM (17) of an *n*-vertex graph *G* with all edge weights 1 has the weight $n - 1$ if and only if *G* has a Hamiltonian path. Therefore, the MWA2M problem is NP-complete since it includes the Hamiltonian path problem. A similar well-known problem, the Maximum Weight 2-Matching (MW2M), allows chosen edges to form cycles. In contrast to the MWA2M problem, the MW2M problem can be efficiently solved (23).

In Reference 17 two heuristics for solving the MWA2M problem are proposed. Below we describe the heuristics.

5.3.2.6   Maximum-Weight Matching Heuristic for the MWA2M Problem   We propose to (sub)optimally solve the MWA2M problem with the following iterative heuristic. It starts with finding the maximum-weight matching *M* among the inter-contig edges. All the dummy edges also form a matching *D*. If the union of these two matchings $M \cup D$ does not contain cycles, then the heuristic reaches the optimal collection of paths. Otherwise, a negative weight −1 is assigned to the least weight inter-contig edge in each cycle. The above steps of finding the intercontig matching *M* and destroying cycles in $M \cup D$ are repeated until the union $M \cup D$ becomes a collection of paths. The output of this heuristic will be called the Matching Scaffold.

In general, the deletion of least-weight edges may significantly reduce (as much as twice in the worst case) the total weight of the collection of paths. Fortunately, the erroneous heavy inter-contig edges are very rare in real data. Our experiments show that for each scaffolding example there is no more than a single cycle in the initial union $M \cup D$ of the maximum-weight matching M and the dummy edges solution and after the second iteration $M \cup D$ does not contain any cycles at all.

5.3.2.7 Greedy Heuristic for the MWA2M Problem Although maximum weight matching can be computed efficiently (see, e.g., the well-known blossom algorithm (24)) even for larger genomes, the runtime can be dramatically decreased using Greedy Heuristic. We provide an option that allows ScaffMatch to run with the Greedy Heuristic reducing the runtime complexity from $O(n^3)$ to $O(n \cdot \log n)$ as we use max heap in our implementation. Our experiments show that the Greedy Matching performs very well in practice but sacrificing not much in quality to the Maximum-Weight Matching heuristic.

5.3.2.8 Contig Ordering and Orientation The Matching Scaffold is represented by a collection of disjoint chains of contig strands. The sequence of edges along each chain alternates: two strands of the same contig are connected with a dummy edge, two strands different contigs are connected with an intercontig edge. When traversing the strands along the paths in the matching scaffold, the order of traversing ends of dummy edges gives us the orientation of the corresponding contigs and the order of traversing intercontig edges gives us the relative order of contigs.

5.3.2.9 Insertion of Skipped Contigs The matching scaffold can skip short contigs whose length is less than the read pair insert size. For example, let the true scaffold contain a triple of consecutive contigs A, B, and C such that $l_A > l_{ins}, l_C > l_{ins}$, but $l_B \ll l_{ins}$. Then instead of picking both edges AB and BC, the matching scaffold may choose one single edge AC since the edge weight between short contigs depends almost linearly on the length of the contigs. Thus, even though the contig B must follow A in the final scaffold, the weight of the edge between A and B is much smaller than the weight of the edge between A and C, which "jumps" over B.

In the following, we describe the insertion of skipped contigs into the matching scaffold (see the following Algorithm). For each skipped contig, we identify the most bundle-supported slot in the matching scaffold satisfying the gap estimations. As a result, several skipped contigs may be assigned to the same slot. In this case, their relative order and orientation are decided based on the gap estimations (see Figure 5.9).

5.3.2.10 Software Implementation The algorithm is implemented as a stand-alone software tool called ScaffMatch. A UNIX shell script is separately provided for mapping reads to contigs. As a short read aligner, Bowtie2 is used. The scaffolder takes as input a fasta file containing the contigs and two SAM files produced by mapping the two read files to the contigs. A small set of mandatory parameters is kept: the mean insert size, the standard deviation, and the orientation (forward-reverse, reverse-forward, or SOLiD-style) of the paired-end reads. The program outputs a fasta file with scaffolds. ScaffMatch uses NetworkX python library for graph computations (25).

Algorithm Insertion of Skipped Contigs

1: $\mathcal{SLOTS} \leftarrow \{\}$
2: $SKIPPED \leftarrow$ the set of skipped contigs
3: $M = \{m_1, m_2, ..., m_n\} \leftarrow$ the Matching Scaffold
4: $G = (V, E, w) \leftarrow$ the Scaffolding Graph
5: $l \leftarrow$ insert length
6: **for all** $\mathcal{X} = (X', X) \in SKIPPED$ **do**
7:     **for all** $m \in M$ **do**
8:         **if** $\exists$ contigs $\mathcal{A} = (A, A')$, $\mathcal{B} = (B, B') \in m$
9:         s.t. $(A, X') \in E$ and $(B', X) \in E$ &
10:         $gap(\mathcal{A}, \mathcal{X}) + l(\mathcal{X}) + gap(\mathcal{X}, \mathcal{B}) \leq l$ **then**
11:             **for** each edge $e = (C_i, C_{i+1}) \in m(\mathcal{A}, \mathcal{B})$ **do**
12:                 **if** $gap(\mathcal{A}, C_i) \leq gap(\mathcal{A}, \mathcal{X})$ **and**
13:                     $gap(C_{i+1}, \mathcal{B}) \leq gap(\mathcal{X}, \mathcal{B})$ **then**
14:                       $\mathcal{SLOTS}[e, \mathcal{X}]$ += $w(A, X') + w(X, B')$
15:                 **end if**
16:             **end for**
17:         **end if**
18:     **end for**
19: **end for**
20: **for all** $\mathcal{X} \in SKIPPED$ **do**
21:     $e \leftarrow \max\{\mathcal{SLOTS}[e, \mathcal{X}] \mid e \in E\}$
22:     **insert** $\mathcal{X}$ into e
23: **end for**

5.4 SCAFFOLDING SOFTWARE EVALUATION

5.4.1 Data Sets

We compare the scaffolding tools on the collection of scaffolding data sets used in Reference 26 including four data sets from the GAGE (27) project (*Staphylococcus aureus*, *Rhodobacter sphaeroides*, and *Homo sapiens (chr14)*,) and one additional data set *Plasmodium falciparum* following (26).

Table 5.1 gives the parameters of all used scaffolding data sets.

5.4.2 Quality Metrics

The main metric that is used for evaluation of scaffolder tools is N50 (28). However, this metric may not be representative enough as mentioned in Reference 26 where a comprehensive evaluation of scaffolders was performed. There state-of-the-art tools were compared based on multiple criteria, such as the number of correct junctions between two adjacent contigs, the number of junctions with incorrect relative order, relative orientation, gap estimation, and their combinations (e.g., incorrect relative order + incorrect gap estimation). The scores assigned to the scaffolders, however, can be misleading. For example, MIP on *S. aureus* (using Bowtie2) got a high score

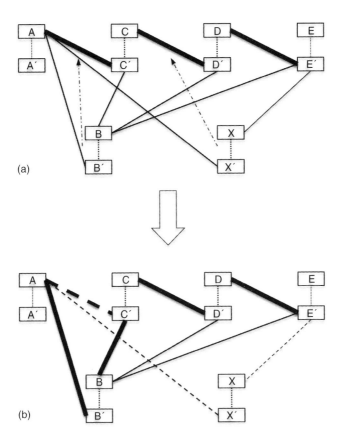

Figure 5.9 Insertion procedure: (a) The matching scaffold *A–C–D–E* is obtained with the maximum weight matching; the contig *B* is connected with edges to all four contigs of the matching, the contig *X* is connected to *A* and *C*; *B* should be placed between *A* and *C* according to the consensus of connecting edges and *X* should be placed between *C* and *D*. (b) Since there is a sufficient distance between contigs *A* and *C*, *B* is placed there, that is, the edge (A, C') from the matching is replaced with (A, B') and (B, C') (the sum of weights of (A, B') and (B, C') is less than the weight of (A, C')); since there is no sufficient room for *X* between contigs *C* and *D*, the edges (A, X') and (X, E') are removed. The resulted scaffold is *A–B–C–D*.

despite the fact that it joined no contigs. Thus, we introduce an F-score-based metric in order to compare the performance of stand-alone scaffolding tools.

We evaluate scaffolders on how well the joins predicted by the tool correspond to the real joins between contigs in the reference genome. Let *P* be the number of potential contigs (equal to the number of contigs minus the number of chromosomes) that can be joined in scaffold, let TP be the number of correct contig joins in the output of the scaffolder (true positives) and FP be the number of erroneous joins (false positives). We compute the following quality metrics:

$$TPR = \frac{TP}{P},$$

$$PPV = \frac{TP}{TP + FP},$$

$$F_{score} = \frac{2 \cdot TPR \cdot PPV}{TPR + PPV},$$

where TPR is sensitivity and PPV is positive predictive value.

5.4.3 Evaluation and Comparison

We compare ScaffMatch and SILP2 with well-established scaffolders SSPACE, OPERA, SOPRA, MIP, SCARPA, as well as a recently published scaffolder, BESST (22). Almost all the scaffolders were run with their default settings using Bowtie2 read aligner. OPERA, SSPACE, and BESST were run using Bowtie.

For computing TPR, PPV, and F-score, we used scripts provided in Reference 26. Note that BESST in the last three data sets failed to produce results due to software exceptions. MIP did not give meaningful results for the first data set.

The results on all scaffolding testcases from Table 5.1 are in Tables 5.2–5.9. In the tables, we used the following notations: C, correct links; E, erroneous links; and S,

TABLE 5.1   Scaffolding Data Sets (29, 30)

Data sets	Inserted size	Relative length	# contigs	# reads
S. aureus	3,600	37	170	3,494,070
R. sphaeroides	3,700	101	577	2,050,868
P. falciparum				
(short insert size)	645	76	9,318	52,542,302
(long insert size)	2,705	75	9,318	12,010,344
H. sapiens (chr14)				
(short insert size)	2,700	101	19,936	22,669,408
(long insert size)	35,000	80	19,936	240,5064

TABLE 5.2   Performance of Different Algorithms on the *S. aureus* Data Set

Scaffolder	C	E	S	N50	Corr. N50	TPR	PPV	F-score
ScaffMatch	139	14	23	1,476,925	351,546	**0.832**	0.908	**0.869**
SSPACE	105	13	13	332,784	261,710	0.629	0.890	0.737
OPERA	112	11	22	1,084,108	**686,577**	0.671	0.911	0.845
SOPRA	40	2	7	112,278	112,083	0.240	0.952	0.383
MIP	0	0	0	46,221	46,221	0	0	0
SCARPA	77	16	10	112,264	112,083	0.461	0.828	0.592
SILP2	121	3	34	645,780	284,980	0.725	0.976	0.832
BESST	112	11	21	**171,6351**	335,064	0.671	0.911	0.772
SGA	83	1	10	309,286	309,153	0.497	**0.988**	0.661
SOAPdenovo2	131	12	13	643,384	621,109	0.784	0.916	0.845

TABLE 5.3   Performance of Different Algorithms on the _R. sphaeroides_ Data Set

Scaffolder	C	E	S	N50	Corr. N50	TPR	PPV	F-score
ScaffMatch	482	18	40	**2,547,706**	2,528,248	**0.845**	0.964	**0.901**
SSPACE	357	7	49	109,776	108,410	0.626	0.981	0.764
OPERA	316	1	23	108,172	108,172	0.554	**0.997**	0.713
SOPRA	242	15	24	32,232	30,492	0.425	0.942	0.585
MIP	419	37	16	488,095	487,941	0.735	0.919	0.817
SCARPA	209	5	23	37,667	37,581	0.367	0.977	0.533
SILP2	425	24	87	471,077	422,445	0.746	0.947	0.834
BESST	367	2	15	1,021,151	1,020,921	0.644	0.995	0.782
SGA	232	1	26	42,825	42,722	0.407	0.996	0.578
SOAPdenovo2	468	8	26	2,522,483	2,522,482	0.821	0.983	0.895

TABLE 5.4   Performance of Different Algorithms on the _P. falciparum_ Data Set (short)

Scaffolder	C	E	S	N50	Corr. N50	TPR	PPV	F-score
ScaffMatch	5648	287	37	**8626**	5872	0.607	0.952	0.741
SSPACE	5746	127	12	6011	5845	**0.612**	0.978	**0.757**
OPERA	3706	116	371	5035	4824	0.398	0.967	0.565
SOPRA	4897	174	34	4954	4632	0.526	0.966	0.681
MIP	5544	359	15	6158	5485	0.596	0.939	0.730
SCARPA	4830	221	38	4912	4628	0.519	0.956	0.673
SILP2	5496	498	48	3109	2601	0.591	0.917	0.719
BESST	2632	462	84	7471	3931	0.283	0.851	0.425
SGA	4940	46	100	5324	5104	0.531	**0.991**	0.691
SOAPdenovo2	5540	84	47	6234	**5981**	0.596	0.985	0.742

TABLE 5.5   Performance of Different Algorithms on the _P. falciparum_ SData set (Long)

Scaffolder	C	E	S	N50	Corr. N50	TPR	PPV	F-score
ScaffMatch	6970	260	1751	41,564	25,380	0.749	0.964	0.843
SSPACE	4610	21	1235	17,796	15,553	0.496	0.995	0.662
OPERA	6257	97	1339	44,667	40,170	0.673	0.985	0.799
SOPRA	7247	181	656	49,671	44,158	0.779	0.976	**0.866**
MIP	7754	707	731	88,297	78,672	**0.834**	0.916	0.873
SCARPA	4882	117	714	14,037	9,708	0.525	0.977	0.683
SILP2	5996	266	2839	45,407	29,399	0.645	0.957	0.771
BESST	1307	46	327	4,133	2,813	0.141	0.966	0.245
SGA	2902	2	652	4,438	4,096	0.312	**0.999**	0.476
SOAPdenovo2	7659	351	803	**167,570**	**83,851**	0.635	0.869	0.734

TABLE 5.6   Performance of Different Algorithms on the Combined *P. falciparum* Data Set (Short + Long)

Scaffolder	C	E	S	N50	Corr. N50	TPR	PPV	F-score
ScaffMatch	8223	425	654	**78,627**	**47,662**	**0.884**	0.951	**0.916**
SSPACE	5889	123	76	6,383	5,982	0.633	0.980	0.769
OPERA	6434	177	1171	42,450	38,409	0.692	0.973	0.809
SOPRA	7018	60	171	16,366	15,511	0.754	**0.992**	0.857
MIP	8082	513	75	56,672	38,704	0.869	0.940	0.903
SCARPA	7336	370	251	36,945	23,951	0.789	0.952	0.863
BESST	3929	541	384	25,300	7,621	0.422	0.879	0.571
SGA	4910	44	419	6,606	6,134	0.528	0.991	0.689
SOAPdenovo2	5977	228	254	12,076	10,629	0.643	0.963	0.771

TABLE 5.7   Performance of Different Algorithms on the *H. sapiens* (chr 14) Data Set (Short)

Scaffolder	C	E	S	N50	Corr. N50	TPR	PPV	F-score
ScaffMatch	12,411	252	3480	131,135	80,329	0.622	0.980	0.761
SSPACE	9,566	43	2754	78,552	77,361	0.487	0.986	0.652
OPERA	12,291	112	2991	214,972	207,047	0.616	0.991	0.760
SOPRA	14,761	381	1441	100,768	96,436	0.740	0.975	0.841
MIP	13,899	954	2735	244,064	**235,731**	0.697	0.936	0.799
SCARPA	9,938	162	1829	58,330	55,760	0.498	0.984	0.661
SILP2	10,548	124	4918	126,689	77,421	0.529	0.988	0.689
BESST	7,970	355	2165	146,749	80,218	0.400	0.957	0.564
SGA	9,761	6	3214	134,574	133,192	0.490	**0.999**	0.657
SOAPdenovo2	15,740	390	2378	**282,437**	234,561	**0.790**	0.976	**0.873**

TABLE 5.8   Performance of Different Algorithms on the *H. sapiens* (chr 14) Data Set (Long)

Scaffolder	C	E	S	N50	Corr. N50	TPR	PPV	F-score
ScaffMatch	5938	443	5198	148,412	42,523	**0.298**	0.933	**0.452**
SSPACE	2750	23	2539	77,832	30,449	0.138	**0.992**	0.242
OPERA	3687	677	3226	73,477	20,677	0.185	0.845	0.303
SOPRA	2938	166	2622	79,517	34,750	0.147	0.947	0.255
MIP	5898	1092	4861	**272,440**	49,800	0.296	0.844	0.438
SCARPA	1603	31	1466	43,969	17,786	0.080	0.981	0.149
SILP2	3899	65	3732	74,094	38,810	0.196	0.984	0.326
BESST	123	13	98	13,815	8,828	0.006	0.904	0.012
SGA	0	0	0	12,211	12,211	0	0	0
SOAPdenovo2	4516	294	3301	220,644	**86,679**	0.227	0.939	0.365

TABLE 5.9 Performance of Different Algorithms on the *H. sapiens* (chr 14) Data Set (Short + Long)

Scaffolder	C	E	S	N50	Corr. N50	TPR	PPV	F-score
ScaffMatch	12,658	348	3874	802,755	195,239	0.635	0.973	0.769
SSPACE	9,249	36	2677	66,271	65,222	0.464	0.996	0.633
OPERA	12,853	58	3409	**1,692,782**	**1,062,031**	0.645	0.996	0.783
SOPRA	10,418	238	3322	112,239	75,046	0.523	0.978	0.681
MIP	8,534	696	3213	44,372	31,148	0.428	0.925	0.585
SCARPA	10,712	161	2376	134,364	106,654	0.537	0.985	0.695
BESST	8,287	286	2347	295,976	114,434	0.416	0.967	0.581
SGA	9,764	3	3214	134,574	133,192	0.490	**0.999**	0.657
SOAPdenovo2	15,748	382	2575	561,198	447,849	**0.790**	0.976	**0.873**

skipped contigs. The entries in the bold font are the best among all 10 scaffolders with respect to the corresponding quality metric. ScaffMatch is the top performer for the first two testcases in both N50 and *F*-score. It also has the best N50 for *P. falciparum* and the best *F*-score for *H. sapiens* (long insert size). MIP has the best N50 for two remaining data sets while SSPACE and SOPRA have each the best *F*-score for one of the remaining data sets.

REFERENCES

1. Zerbino DR, Birney E. Velvet: algorithms for de novo short read assembly using de Bruijn graphs. Genome Res 2008;18(5):821–829.
2. Gnerre S, MacCallum I, Przybylski D, Ribeiro FJ, Burton JN, Walker BJ, Sharpe T, Hall G, Shea TP, Sykes S et al. High-quality draft assemblies of mammalian genomes from massively parallel sequence data. Proc Natl Acad Sci U S A 2011;108(4):1513–1518.
3. Gao S, Nagarajan N, Sung WK. Opera: reconstructing optimal genomic scaffolds with high-throughput paired-end sequences. In Proceedings of the 15th Annual International Conference on Research in Computational Molecular Biology; 2011. p 437–451.
4. Donmez N, Brudno M. SCARPA: scaffolding reads with practical algorithms. Bioinformatics 2013;29(4):428–434.
5. Lindsay J, Salooti H, Măndoiu I, Zelikovsky A. ILP-based maximum likelihood genome scaffolding. BMC Bioinformatics 2014;15(9):1–12.
6. Salmela L, Mäkinen V, Välimäki N, Ylinen J, Ukkonen E. Fast scaffolding with small independent mixed integer programs. Bioinformatics (Oxford, England) 2011;27(23):3259–3265.
7. Boetzer M, Henkel CV, Jansen HJ, Butler D, Pirovano W. Scaffolding pre-assembled contigs using SSPACE. Bioinformatics 2011;27(4):578–579.

8. Warren RL, Sutton GG, Jones SJ, Holt RA. Assembling millions of short DNA sequences using SSAKE. Bioinformatics 2007;23(4):500–501.

9. Dayarian A, Michael T, Sengupta A. SOPRA: Scaffolding algorithm for paired reads via statistical optimization. BMC Bioinformatics 2010;11(1):345+.

10. Huson DH, Reinert K, Myers EW. The greedy path-merging algorithm for contig scaffolding. J ACM 2002;49(5):603–615.

11. Hopcroft JE, Tarjan RE. Dividing a graph into triconnected components. SIAM J Comput 1973;2(3):135–158.

12. Di Battista G, Tamassia R. On-line graph algorithms with SPQR-trees. In: *Automata, Languages and Programming*. Berlin: Springer-Verlag; 1990. p 598–611.

13. Chimani M, Gutwenger C, Jünger M, Klau GW, Klein K, Mutzel P. The open graph drawing framework (OGDF). In: Tamassia R, editor. *Handbook of Graph Drawing and Visualization*. CRC Press; 2011. p 543–569.

14. Lindsay J, Salooti H, Zelikovsky A, Măndoiu I. Scalable genome scaffolding using integer linear programming. Proceedings of the ACM Conference on Bioinformatics, Computational Biology and Biomedicine, BCB '12. New York: ACM; 2012. p 377–383.

15. CPLEX II. V12. 1: User's manual for CPLEX. Int Bus Mach Corporation 2009;46(53):157.

16. Shcherbina O. Nonserial dynamic programming and tree decomposition in discrete optimization. In: Waldmann K-H, Stocker UM, editors. *Operations Research Proceedings 2006*. Berlin Heidelberg: Springer-Verlag; 2006. p 155–160.

17. Mandric I, Zelikovsky A. ScaffMatch: scaffolding algorithm based on maximum weight matching. Bioinformatics 2015;31(16):2632–2638.

18. Langmead B. Aligning short sequencing reads with Bowtie. Curr Protoc Bioinf 2010; Chapter 11:11-7-1–11-7-14.

19. Langmead B, Salzberg SL. Fast gapped-read alignment with Bowtie 2. Nat Methods 2012;9(4):357–359.

20. Li H, Durbin R. Fast and accurate short read alignment with Burrows–Wheeler transform. Bioinformatics 2009;25(14):1754–176.

21. Sahlin K, Street N, Lundeberg J, Arvestad L. Improved gap size estimation for scaffolding algorithms. Bioinformatics 2012;28(17):2215–2222.

22. Sahlin K, Vezzi F, Nystedt B, Lundeberg J, Arvestad L. BESST-Efficient scaffolding of large fragmented assemblies. BMC Bioinformatics 2014;15:281.

23. Pulleyblank W. *Dual Integrality in B-Matching Problems*. Heidelberg: Springer-Verlag; 1980.

24. Edmonds J. Paths, trees, and flowers. Can J Math 1965;17(3):449–467.

25. Hagberg A, Schult D, Swart P. Networkx: Python software for the analysis of networks. Technical report, Mathematical Modeling and Analysis. Los Alamos National Laboratory; 2005. Available at: http://networkx.lanl.gov.

26. Hunt M, Newbold C, Berriman M, Otto TD. A comprehensive evaluation of assembly scaffolding tools. Genome Biol 2014;15(3):R42.

27. Salzberg SL, Phillippy AM, Zimin A, Puiu D, Magoc T, Koren S, Treangen TJ, Schatz MC, Delcher AL, Roberts M, Marçais G, Pop M, Yorke JA. GAGE: a critical evaluation of genome assemblies and assembly algorithms. Genome Res 2012;22(3):557–567.

28. Vezzi F, Narzisi G, Mishra B. Feature-by-feature–evaluating de novo sequence assembly. PLoS ONE 2012;7(2):e31002.

29. Simpson JT, Durbin R. Efficient de novo assembly of large genomes using compressed data structures. Genome Res 2012;22(3):549–556.

30. Luo R, Liu B, Xie Y, Li Z, Huang W, Yuan J, He G, Chen Y, Pan Q, Liu Y et al. SOAPdenovo2: an empirically improved memory-efficient short-read de novo assembler. GigaScience 2012;1:18.

6

GENOMIC VARIANTS DETECTION AND GENOTYPING

JORGE DUITAMA

Agrobiodiversity Research Area, International Center for Tropical Agriculture (CIAT), Cali, Colombia

6.1 INTRODUCTION

Complete and accurate identification of genomic variation among species, populations, organisms, and even tissues is one of the most important steps toward understanding the genetic component of phenotypic variation. Recent developments in genomics have served as building blocks to produce major advances in several fields including evolution and population genetics (1, 2); medical research (3); crop breeding (4); and industrial production of food, organic goods, and even biofuels (5). Since it was discovered that DNA molecules encode most of the genomic information of an organism, continuous efforts have been done to understand the composition of DNA, which is physically organized as a packed long chain of four possible nucleotides: adenine, cytosine, guanine, and thymine. This suggests an initial encoding of the information carried by a DNA molecule as a sequence built from an alphabet in which each nucleotide is represented by a different character ($\{'A', 'C', 'G', 'T'\}$). The process of determining the sequences representing the molecules in a DNA sample is called sequencing (6). Once DNA sequences of two or more samples are determined, they can be compared through sequence alignment algorithms in order to identify genomic variation.

The most widely studied form of genomic variations are single-nucleotide polymorphisms (SNPs). As its name indicates, an SNP is a DNA stretch in which two or

Computational Methods for Next Generation Sequencing Data Analysis, First Edition.
Edited by Ion I. Măndoiu and Alexander Zelikovsky.
© 2016 John Wiley & Sons, Inc. Published 2016 by John Wiley & Sons, Inc.
Companion website: www.wiley.com/go/Mandoiu/NextGenerationSequencing

more sequences only differ in one nucleotide. The base pairs that differ are usually called SNP alleles or just alleles, which in case of maximum variability correspond to the four nucleotides. Once an SNP is identified, the process of determining the alleles present in a sample for a given SNP is called SNP genotyping. SNPs have been thoroughly studied in different organisms because they are well distributed across the genome and because automated methods have been developed to perform cost-effective and high-quality SNP genotyping across several samples (7). These properties make SNPs amenable for evolution and population genetic studies (2) to perform genomic mapping of traits through QTL analysis or genome-wide association studies (GWAS) (4) and to perform marker-assisted selection (8). However, because SNPs only produce minor effects in the DNA Structure, and function (with the exception of SNPs in regulatory or coding regions), they are not normally causative for phenotypic variation and their association with traits is normally due to high linkage disequilibrium (LD) with the causative variants (4). Hence, complete understanding of genomic variation requires the identification of other types of variants with larger structural footprints. These types of variants include insertions and deletions (usually summarized as indels), copy number variants (CNVs), inversions, and translocations (9). Figure 6.1 shows a schematic of the different types of variation that can be identified between the DNA sequences of two different samples.

Indels can be defined as DNA subsequences present only in a subset of the analyzed samples. The determination of insertion or deletion status depends on choosing a particular sequence as reference to determine if the sequence is inserted to the reference or deleted from the reference. Indels are also well distributed across the genome, but they are less frequent and not as easy to genotype as SNPs. Large indels can span whole genes and hence can be causative for significant phenotypic variation. Small indels are more likely to be neutral, again with the exception of indels spanning regulatory or coding regions. CNVs are (usually large) repetitive DNA sequences in

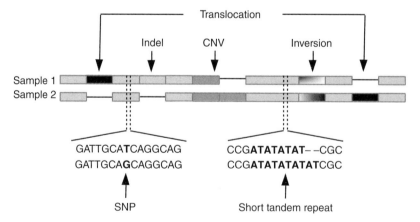

Figure 6.1 Types of genomic variation between the chromosomes of two samples. Structural variants include indels, copy number variants (CNVs), inversions, and translocations. Examples of (a) single - nucleotide polymorphism (SNP) and (b) tandem repeat are shown in the zoomed-in regions.

which the number of copies of the sequence varies among samples. Copies of a CNV are sometimes located in tandem, but they can also be sparsed across the genome. Several recent studies revealed the importance of CNVs for human genetics (10) and medical research (3, 11, 12). Cases in which the repetitive DNA sequence is small and repeats occur in tandem are usually not called CNVs but tandem repeats. Tandem repeats have been important in the history of biotechnology because they are highly polymorphic compared to SNPs, they are feasible to genotype using lab techniques, and it has been shown that they can be causative for major diseases (13). Finally, as their name indicates, inversions and translocations are regions that are, respectively, in opposite strands between samples or that are located in different genomic locations. As in the case of indels, determining the right strand or the right location of the variable DNA stretch requires fixing a particular sequence as the reference. Note that a translocation implies two indel events, and a CNV usually implies several indel events.

Recent development and availability of high-throughput sequencing (HTS) technologies produced a significant breakthrough on the sequencing and genotyping capacity, which jumped from isolated sequencing of genes or genotyping of a few predefined markers to almost complete sequencing of entire genomes and comprehensive genotyping of millions of SNPs and thousands of other types of variants (6). However, this could be possible only after the development of bioinformatic methods to perform *de novo* assembly, variants discovery, and genotyping based on huge amounts of small reads produced by HTS technologies (14). Different strategies for *de novo* assembly combining reads sequenced with different technologies (mainly Illumina and 454) and complementary sources of contiguity information such as synteny, genetic maps, or physical maps produced a recent explosion of draft genomes for several species (see References 15 and 16 for two recent examples). Once a reference genome for a species is available, a cost-effective strategy to study variability within populations (or contrasting samples, or even tissues) consists on sequencing each sample at a low-to-moderate coverage (2×–30× depending on the size of the genome, the number of samples, and the purpose of the study), and then align the reads to the reference genome using different bioinformatic tools such as bwa (14) and Bowtie2 (17). This process allows to efficiently build a virtual multiple sequence alignment which can then be used to identify variation, first against the reference, and later among the sequenced samples. This chapter explains some of the bioinformatic algorithms developed to identify SNPs and CNVs based on reads aligned to a reference genome.

6.2 METHODS FOR DETECTION AND GENOTYPING OF SNPs AND SMALL INDELS

6.2.1 Description of the Problem

Once reads are aligned to the reference genome and alignments are sorted by reference coordinates, the reference genome is traversed base by base and sites for which a base call supporting an allele different from the reference allele is observed are further analyzed. In the absence of sequencing errors, the correct genotype for each of these sites would just be the set of different alleles supported by the base calls aligning at these positions. However, every sequencing technology produces some percentage of erroneous base calls, which would become spurious genotype calls if this naive algorithm is used. Hence, the main challenge to identify and genotype SNPs is to separate base call errors from calls supporting real alleles that differ from the reference.

For each genomic location, the problem can be formalized as follows: Given a set of reads R_i of size d spanning a genomic location i, for $r \in R_i$ let $r(i)$ be the base call from the r aligning to the genomic location i, and let $p_{r(i)}$ be the probability of error sequencing $r(i)$. The value of d is usually called depth of coverage, read depth (RD), or just coverage. In the absence of indels, let $A = \{'A','C','G','T'\}$ be the set of the four possible alleles. The coverage of each allele $a \in A$ can be calculated as $d_a = |\{r \in R : r(i) = a\}|$. Assuming that the sequenced sample comes from a diploid individual, 10 possible genotypes can be made picking two of the four nucleotides with replacement. The problem is then to find the most likely genotype $G_i = H_i H_i'$ given R, where H_i and H_i' belong to A.

6.2.2 Bayesian Model

Simple algorithms to find SNPs ignore the base call error probabilities and set minimum thresholds on the allele coverages d_a either to a fixed number or to a fixed percentage of the total coverage d. The most widely used strategy to take into account the information contained in the error probabilities is based on the Bayes theorem, which for this particular problem can be written as

$$P(G_i|R_i) = \frac{P(R_i|G_i)P(G_i)}{P(R_i)}$$

The main difference among Bayesian algorithms for SNP detection lies in the way conditional probabilities are calculated (18). Here, we will describe a simple, yet accurate model introduced in Reference 19. Assuming independence between the reads, the conditional probability of G_i can be expressed as a product of read contributions as follows:

$$P(R_i|G_i) = \prod_{r \in R} P(r(i)|G_i)$$

For a mapped read $r \in R_i$, let $r(i)$ be the base spanning locus i and $p_{r(i)}$ be the probability of error sequencing the base $r(i)$, which we estimated from the quality score $q_{r(i)}$ calculated during primary analysis using the Phred formula $p_{r(i)} = 10^{-q_{r(i)}/10}$ (20). Let H_i and H_i' be the two real alleles in the locus i, or in other words, let $G_i = H_i H_i'$. The observed base $r(i)$ could be read from either H_i or H_i'. If there is an error in this read, we assume that the error can produce any of the other three possible bases to be observed with the same probability, so the probability of observing a base $r(i)$ given than the real base is different is $p_{r(i)}/3$ while the probability of observing $r(i)$ without error is $1 - p_{r(i)}$.

Assuming a heterozygous genotype $G_i = H_i H'_i$, $H_i \neq H'_i$ if the observed allele $r(i)$ is equal to H_i (H'_i), it could be due to two possible events. Either $r(i)$ was sampled without error from the haplotype H (H') or $r(i)$ was sampled from the haplotype H' (H) but an error turned it to be equal to H_i (H'_i). Assuming that both haplotypes are sampled with equal probability, the first event happens with probability $(1 - p_{r(i)})/2$ while the second happens with probability $p_{r(i)}/6$. For homozygous loci, the probability of observing each possible base does not depend on the haplotype from which the reads were sampled. Hence, the following equation summarizes how to calculate the probability of r for each possible genotype:

$$P(r|G_i = H_i H'_i) = \begin{cases} 1 - p_{r(i)} & : H_i = H'_i = r(i) \\ \dfrac{p_{r(i)}}{3} & : H_i \neq r(i) \wedge H'_i \neq r(i) \\ \dfrac{1}{2} - \dfrac{p_{r(i)}}{3} & : \text{otherwise} \end{cases}$$

We complete the model calculating prior probabilities based on the expected heterozygosity rate h in the following way:

$$P(G_i = H_i H'_i) = \begin{cases} \dfrac{1-h}{4} & : H_i = H'_i \\ \dfrac{h}{6} & : H_i \neq H'_i \end{cases}$$

Other strategies to calculate the prior and posterior distributions have been proposed (14, 21). Assuming that sequenced fragments are well distributed across the genome, coverage is reasonable high (over 20×), error rates are low, and base quality scores reflect the true probability of base calling error, the above-described Bayesian model yields over 95% sensitivity and over 99% specificity to call homozygous and heterozygous variants in unique regions of the genome. We now describe different strategies to maintain high accuracy even when some of these conditions do not hold.

6.2.3 Common Issues Affecting Genotype Quality

The first well-known fact affecting genotyping quality is the increased error rate toward the 3′-end of the reads, which is a consequence of the reduction in throughput and quality of the sequencer as the number of cycles increases. Figure 6.2 shows this effect on a resequencing of the rice cultivar Nipponbare (2), which is the cultivar chosen to build the rice reference genome (22). Because Nipponbare is an inbred line, most of the differences between the reads and the reference genome are sequencing errors. The easiest way to tackle this issue is to build a chart such as the one shown in Figure 6.2 and then choose a cutoff to trim the last portion of each read. Some read alignment tools also offer local alignment strategies in which the region to ignore at the ends of each read is chosen dynamically based on the quality of the local alignment ignoring base pairs at both ends, compared with the quality of a global alignment including the whole read.

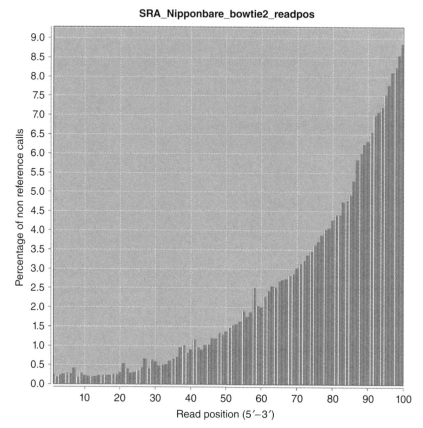

SRA_Nipponbare_bowtie2_readpos

Figure 6.2 Differences between the gold-standard reference assembly of the rice cultivar Nipponbare and an Illumina whole-genome resequencing experiment of the same sample classified by read position from 5' to 3'. Given the inbred nature of rice cultivars, most of the observed differences are due to sequencing errors.

Another well-known factor affecting genotyping quality is the sequencing of multiple copies of the same fragment produced during the polymerase chain reaction (PCR) amplification step (23). This problem creates unexpectedly a high number of reads in the regions of the genome where the multiplied fragment was extracted from. PCR duplicated fragments are likely to produce false-negative homozygous calls in heterozygous sites. A common strategy to reduce the effects of PCR amplification is to identify clusters of alignments of different reads starting and ending at exactly the same genomic location and keeping a small maximum number m of alignments per cluster. This ensures that only up to m copies of a fragment multiplied by PCR amplification will be taken into account for SNP discovery. Smaller values of m increase the specificity at the cost of sensitivity (19).

Miscalibrated base quality scores are another important source of genotyping errors. For an erroneous base call $r(i)$ if the real genotype G does not contain $r(i)$, its conditional probability will be multiplied by $p_{r(i)}$, and if $p_{r(i)} < 10^{-n}$, then this base call will produce a reduction of n orders of magnitude in the conditional probability of the true genotype. The simplest solution to this problem is to set a minimum on the quality score required to consider a base call as real and then just ignore the value of the quality score. Although this strategy solves the issue, it severely limits the power of the Bayesian model. Alternatively, a maximum base quality score can be set as a parameter instead of a minimum base quality score, and the base quality scores larger than this parameter can be equalized to the input value (24). This allows to take into account correctly sequenced bases with low-quality score and alleviates the problem produced by erroneous base calls with high-quality score. Trade-offs between sensitivity and specificity obtained by changing this parameter are shown in Reference 24. A more elaborated solution described in Reference 25 tries to fix possible errors and then produce a new set of recalibrated quality scores. Base calls for all aligned reads are classified according to their (integer) quality score, machine cycle, and dinucleotide context. Then, for each group, the empirical error rate and residual variances and covariances for each of the three classification parameters are estimated. Finally, each quality score is recalibrated adding the observed residual values for the group to which the base call belongs.

Different strategies to handle repetitive regions in the genome produce different sensitivity and specificity trade-offs to discover and to genotype SNPs located in such regions. Most of the currently available alignment tools choose by default a random alignment when there are multiple equally good alignments for one read. Bowtie2 also allows to retain up to k alignments if desired, being k a parameter. As k gets larger, each of the copies of a particular repetitive region retain alignments of reads sequenced from each of the copies of the sample. If secondary alignments are considered for variants detection, the differences among copies of a repetitive element are likely to appear on each copy as heterozygous SNPs. This behavior also happens within regions in which the reference genome has only one copy (novel CNVs). There is currently no standard procedure to accurately separate differences between copies of repetitive elements and heterozygous SNPs within such elements.

6.2.4 Population Variability

The above-described algorithm is able to perform both detection and genotyping of SNPs for an individual sample. When many samples of a population need to be analyzed and compared, the straightforward approach to merge data from different samples can be described in two steps. (1) Discovery: Keep genomic sites with evidence of variation against the reference in at least one individual. (2) Genotyping: Genotype all samples on the sites obtained in step 1. This second step allows to obtain quality scores for homozygous reference genotype calls.

It has been shown that for experimental designs in which individual samples are sequenced at low coverage (such as the design of the 1000 Genomes Project (1)), a significant improvement in sensitivity can be achieved if aligned reads are used instead of individual variants as input for the discovery step (26). However, a specific statistical model needs to be developed to find variants with different minor allele frequencies (MAF) and, moreover, to separate variants with small MAFs from sequencing errors. Several statistical models have been proposed to perform variants discovery from pooled samples (25, 27, 28). Here we describe the statistics proposed by Li (28) to estimate the number of nonreference alleles in the population directly from the data and to subsequently call variants calculating the posterior probability of observing at least one nonreference allele.

The model of (28) assumes independence across sites, errors, and samples. It also assumes that variants are biallelic, which is the case for most of the SNPs within populations of Eukaryotic species. In this case, genotypes can be encoded as integers representing the number of reference alleles in the genotype. For a diploid individual, the homozygous reference genotype is encoded as $g = 2$, the heterozygous genotype is encoded as $g = 1$, and the homozygous nonreference genotype is encoded as $g = 0$. Given sequencing data from n individuals with ploidies $m_1, m_2, \ldots, m_n$, let $\vec{G} = \langle G_1, G_2, \ldots, G_n \rangle$ be the random vector of genotypes for the n individuals. Then, the random variable $K = \sum_{i=1}^{n} G_i$ is the number of reference alleles present in the population. Assuming Hardy–Weinberg equilibrium (HWE), the probability of a genotype configuration given a fixed number of reference alleles is

$$P(\vec{G} = \vec{g} | K = k) = e\left(k, \sum_{i=1}^{n} g_i\right) \prod_{i=1}^{n} \frac{\binom{m_i}{g_i}}{\binom{M}{k}},$$

where e is an indicator function taking the value 1 if the two arguments are equal and 0 otherwise. The conditional probability of the data $\vec{d} = \langle d_1, d_2, \ldots, d_n \rangle$ given $K = k$ is

$$P(\vec{d} | K = k) = \frac{1}{\binom{M}{k}} \sum_{g_1=0}^{m_1} \cdots \sum_{g_n=0}^{m_n} e\left(k, \sum_{i=1}^{n} g_i\right) \prod_{i=1}^{n} \binom{m_i}{g_i} P(d_i | g_i)$$

Efficient calculation of this likelihood can be performed by dynamic programming defining the partial sum:

$$z_{jl} = \sum_{g_1=0}^{m_1} \cdots \sum_{g_j=0}^{m_j} e\left(l, \sum_{i=1}^{j} g_i\right) \prod_{i=1}^{j} \binom{m_i}{g_i} P(d_i | g_i).$$

for $0 \leq l \leq \sum_{g_1=0}^{j} m_i$ ($z_{jl} = 0$ for $l > \sum_{g_1=0}^{j} m_i$). Starting with $z_{00} = 1$, the partial sums and the final value of $P(\vec{d} | K = k)$ are calculated as follows:

$$z_{jl} = \sum_{g_j=0}^{m_j} z_{j-1,l-g_j} \binom{m_j}{g_j} P(d_j | g_j) P(\vec{d} | K = k) = \frac{z_{n,k}}{\binom{M}{k}}$$

The posterior probability $P(K = k | \vec{d})$ can be calculated using the Bayes rule:

$$P(K = k | \vec{d}) = \frac{P(K = k) P(\vec{d} | K = k)}{\sum_{l=1}^{k} P(K = l) P(\vec{d} | K = l)}$$

The probability of having a nonreference allele is then $1 - P(K = M|\vec{d})$ where $M = \sum_{i=1}^{n} m_i$. Two alternatives to calculate the prior distribution of K that estimate the distribution based on the data across different sites are described in Reference 28. The first is an expectation maximization (EM) algorithm with the following iteration equation:

$$P(K = k)^{(t+1)} = \frac{1}{L} \sum_{a=1}^{L} P(K_a = k|\vec{d}, \{P(K = 1)^t \ \ldots \ P(K = M)^t\})$$

The second algorithm calculates a histogram based on the estimations of the likelihoods $P(\vec{d}|K = k)$ across different sites and derives the probabilities based on the histogram.

6.3 METHODS FOR DETECTION AND GENOTYPING OF CNVs

Methods for discovery of CNVs can be generally grouped into two strategies depending on the type of information used as input. Methods based on read depth (RD) start by dividing the genome in nonoverlapping bins and calculating the number of reads aligning within each bin. Assuming that there are no biases in sampling of fragments during library preparation, the read depth per bin should follow a normal distribution centered at the average read depth for the entire genome. A CNV within one sample is called when a group of contiguous bins with total read depth significantly different from the average is observed. CNVs between two samples can also be discovered as contiguous bins in which the difference between read counts, normalized by the total number of reads produced for each sample, is significantly different from zero.

On the other hand, read pair (RP) algorithms use the pairs of reads as main source of information belonging to the same fragment that aligns with different chromosomes, with an unexpected orientation, or with an unexpected insert length. Because these kinds of events can occur due to PCR artifacts or due to unspecific read alignment, RP algorithms cluster read pairs supporting the same event and calculate the probability of existence of each predicted event based on the size of its supporting cluster. Compared with RD approaches, RP algorithms are able to identify more types of genomic variation (e.g., insertions, inversions, and translocations). However, they are not able to call most duplication events, with the exception of some tandem duplications. It has been shown that RD and RP approaches are complementary to identify structural variation (29). Because the next chapter will cover in detail RP approaches, the rest of this chapter will only discuss in detail two of the most effective RD approaches to find, respectively, CNVs within one sample and between two samples.

6.3.1 Mean-Shift Approach for CNVs within a Sample

The first algorithm, called CNVNator, uses mean-shift theory to identify the discrete changes in average read depth produced by regions with copy number variation (30).

The first step of the algorithm is to correct for GC-content biases. Given a bin i, its GC-content is defined as the number of G or C base pairs in the bin divided by the bin length, multiplied by 100, and rounded to the nearest integer. For each possible GC-content $gc \in \{0 \ldots 100\}$, the average read depth $\overline{RD}_{gc}$ is calculated over bins with the same GC content, and then for each bin i the corrected read depth is

$$RD^i_{corrected} = \frac{\overline{RD}_{global}}{\overline{RD}_{gc}} RD^i_{raw}$$

The next step for each bin is to identify if its RD is closer to its left neighbors or to its right neighbors. To accomplish this, each bin is represented as the ordered pair $x_i = \langle i, r_i \rangle$ where r_i is the GC-corrected read depth of the bin i. To represent the signal change over the two-dimensional space, a Gaussian kernel is used with the following density function:

$$F(x_i) = C \sum_{j \neq i} e^{-\frac{(j-i)^2}{2H_b^2}} e^{-\frac{(r_j - r_i)^2}{2H_r^2}},$$

where H_b is the number of bins to look at on each side of i (band width for the genomic location), H_r is the band width for the read depth, and C is a constant normalization factor. The mean-shift vector is the two-dimensional vector of partial derivatives of this function over genomic location and read depth, respectively. Because only the genome needs to be partitioned for CNV discovery, only the partial derivative over genome location needs to be considered:

$$\frac{\delta F}{\delta i}(x_i) = C' \sum_{j \neq i} (j - i) \exp\left(-\frac{(j-i)^2}{2H_b^2}\right) \exp\left(-\frac{(r_j - r_i)^2}{2H_r^2}\right),$$

where C' is a normalization constant that does not need to be calculated in this case because it is always positive and only the sign of the derivative is required. Boundaries of neighbor segments with different copy numbers are calculated finding neighbor bins shifting directions left to right. Smoothed RD signal is calculated for each segment as the average read depth of the bins within the segment.

To complete this partitioning algorithm, the bandwidths H_r and H_b need to be chosen. The bandwidth r_i is calculated as follows:

$$H_r^i = \begin{cases} \sqrt{\frac{r_i}{\overline{r}}} H^0 & : r_i > \frac{\overline{r}}{4} \\ \dfrac{H^0}{2} & : r_i \leq \frac{\overline{r}}{4} \end{cases},$$

where H^0 is estimated as the average of the best fit of a Gaussian function to the distribution of read depth across the genome and $\overline{r}$ is the average read depth across the genome. The second case is made to prevent H_r^i becoming zero. Choosing the bandwidth H_b is challenging because large values of H_b reduce sensitivity to call small CNVs and small values of H_b cause partitioning of large CNVs. To overcome this issue, Abyzov et al. devised an iterative algorithm to combine results using different bandwidths, which temporarily excludes calculations from small segments with significant changes in average read depth as larger values of H_b are used. A pseudocode of this algorithm (reproduced as is from Reference 30) is included here:

(1) Set H_b to 2
(2) While bandwidth is less than limit
   (a) Exclude frozen segments from calculations
   (b) Do three times
      i. Partition by mean-shift with current bandwidth
      ii. Average RD signal within each segment
      iii. Replace RD values within each segment with their corresponding average
   (c) Add frozen segments to partitioning
   (d) Unfreeze all frozen segments
   (e) Restore original RD values for all points
   (f) For each segment
      i. Calculate average and standard deviation of RD signal
      ii. If (segment mean RD is different from genomic average and segment mean RD is different from neighboring segments) freeze the segment
   (g) Increase bandwidth
(3) Merge segments
(4) Make CNV calls

The read depth of a particular segment is significantly different from the global read depth if the p-value of a one sample t-test is less than 0.05. The read depths of two neighbor segments are significantly different if the p-value of a two-sample t-test is less than 0.01 after correction for testing of multiple hypotheses. For small segments (less than 15 bins), they are also cataloged as different if their difference is more than $2H_r^S$, which is the bandwidth corresponding to the average read depth of the union of the two segments being tested. The bandwidth is increased in powers of two until it reaches 128, which is enough to call CNVs in human samples.

Once a partition of the genome is completed using the algorithm explained earlier, CNVs are segments in which the average read depth deviates by at least $\frac{1}{2p}$ from the average read depth, where p is the normal ploidy of the sequenced organism. Significance is assessed by a one-sample t-test and p-values are corrected for testing of multiple hypotheses.

6.3.2 Identifying CNVs between Samples

In several important applications such as identifying CNVs between wild type and tumor samples, the objective is not to identify regions in which the number of copies differs from the genome average but to identify regions in which two samples differ in their number of copies. Different algorithms have been developed to find CNVs in this particular setting. As an example, we will explain in this section the algorithm called CNV-seq, developed by Xie and Tammi (31).

CNV-Seq also starts dividing the genome in nonoverlapping windows of length l_w. The read depth within a window can be approximated with a Poisson distribution with $\lambda = n_g l_w / l_g$ where n_g is the total number of reads and l_g is the length of the genome. Given two samples x and y for which a total of n_x and n_y reads have been sequenced and an arbitrary window w, let N_x and N_y be random variables representing the number of reads falling in w. The raw read depth ratio R and the normalized read depth ratio R' will also be random variables defined as follows:

$$R = \frac{N_x}{N_y}$$

$$R' = R \frac{n_y}{n_x}$$

Both N_x and N_y are Poisson distributions that can be approximated as Gaussian distributions if the mean number of reads per window is greater than 10, making Z be approximated as a ratio of Gaussian distributions. Assuming that N_x and N_y are independent and using the Geary–Hinkley transformation, the variable Z defined by

$$Z = \frac{\mu_y R - \mu_x}{\sqrt{\mu_y R^2 + \mu_x}}$$

will have a standard normal distribution. The p-value for the null hypothesis of no difference in copy number between the two samples in a particular window with normalized read depth ratio r' is

$$p = \begin{cases} 2(1 - \Phi(Z)) & : r' \geq 1 \\ 2\Phi(Z) & : r' < 1 \end{cases}$$

where $\Phi(Z)$ is the cumulative standard normal distribution.

6.4   PUTTING EVERYTHING TOGETHER

In practice, one of the biggest difficulties that scientists face while analyzing HTS data is that different algorithms to discover and genotype variants are implemented in separate pieces of software which are very heterogeneous in programming languages, input and output formats, usage, and even code quality. Moreover, the vast majority of software packages only offer command line interfaces, which make analysis more difficult for people without training in computer science. We finish this chapter mentioning two alternatives currently available to allow researchers to run different algorithms in an integrated environment.

The most widely known integration platform for bioinformatics software is the Galaxy Web portal (32). Galaxy allows to configure different heterogeneous tools, to build custom pipelines, and to run everything into a user-friendly web interface. The main advantages of Galaxy are the flexibility to modify the pipeline including or replacing software tools at any time. Galaxy also allows to share data and pipelines among collaborators located at distant geographical regions. The main drawback of this approach is that it requires people with web server management skills to install local instances of Galaxy and to configure the tools that will be included in the pipelines.

A recently available alternative for this problem is NGSEP (Next-Generation Sequencing Eclipse Plugin) (24). NGSEP is a desktop application developed using the Eclipse framework and is able to perform integrated detection and genotyping of SNPs, indels (small and large), and CNVs, combining some of the algorithms explained in this chapter. One of the main advantages of NGSEP is the tight integration among algorithms produced by their implementation into a single framework. This allows users to obtain variability, copy number, and functional information into a single VCF file without requiring any custom scripting or programming effort. Because NGSEP can be installed and maintained as a normal desktop application, it does not require specialized technical skills to be operated. Moreover, current advances in hardware speed and capacity to store large volumes of data allow users nowadays to analyze relatively large amounts of sequencing data with the computational resources provided by a normal desktop computer. The main limitation of NGSEP compared to a web portal solution is that integrating new algorithms requires a significant programming effort. For this reason, the main functionalities of NGSEP can be called from a command line interface and also can be integrated in Galaxy to allow users familiar with the this portal to include NGSEP in their pipelines. Further development of accurate, efficient, and user-friendly bioinformatic solutions for analysis of HTS data is as important as the development of the sequencing technologies themselves to allow a wider range of researchers in different fields to use sequencing as a building block for their research projects.

REFERENCES

1. The 1000 Genomes Project Consortium. An integrated map of genetic variation from 1,092 human genomes. Nature 2012;491(7422):56–65.

2. Xu X et al. Resequencing 50 accessions of cultivated and wild rice yields markers for identifying agronomically important genes. Nat Biotechnol 2012;30(1):105–111.

3. Zhang J et al. Whole-genome sequencing identifies genetic alterations in pediatric low-grade gliomas. Nat Genet 2013;45(6):602–612.

4. Huang X et al. Genome-wide association studies of 14 agronomic traits in rice landraces. Nat Genet 2010;42(11):961–967.

5. Hubmann G et al. Quantitative trait analysis of yeast biodiversity yields novel gene tools for metabolic engineering. Metab Eng 2013;17:68–81.

6. Bentley DR et al. Accurate whole human genome sequencing using reversible terminator chemistry. Nature 2008;456(7218):53–59.

7. McNally KL et al. Genomewide SNP variation reveals relationships among landraces and modern varieties of rice. Proc Natl Acad Sci U S A 2009;106(30):12273–12278.

8. Mammadov J, Aggarwal R, Buyyarapu R, Kumpatla S. SNP markers and their impact on plant breeding. Int J Plant Genomics 2012;2012:728398.

9. Alkan C, Coe BP, Eichler EE. Genome structural variation discovery and genotyping. Nat Rev Genet 2011;12(5):363–376.

10. Abyzov A et al. Somatic copy number mosaicism in human skin revealed by induced pluripotent stem cells. Nature 2012;492(7429):438–442.

11. Hollox EJ, Hoh B-P. Human gene copy number variation and infectious disease. Human Genet 2014;133(10):1217–1233.

12. Liu B, Morrison CD, Johnson CS, Trump DL, Conroy JC, Wang J, Liu S. Computational methods for detecting copy number variations in cancer genome using next generation sequencing: principles and challenges. Oncotarget 2013;4(11):1868–1881.

13. Gemayel R, Vinces MD, Legendre M, Verstrepen KJ. Variable tandem repeats accelerate evolution of coding and regulatory sequences. Ann Rev Genet 2010;44:445–477.

14. Li H, Durbin R. Fast and accurate short read alignment with Burrows-Wheeler transform. Bioinformatics (Oxford, England) 2009;25(14):1754–1760.

15. Schmutz J et al. A reference genome for common bean and genome-wide analysis of dual domestications. Nat Genet 2014;46(7):707–713.

16. Wang M et al. The genome sequence of African rice (Oryza glaberrima) and evidence for independent domestication. Nat Genet 2014;46:982–988.

17. Langmead B, Salzberg SL. Fast gapped-read alignment with Bowtie 2. Nat Methods 2012;9(4):357–359.

18. Dalca AV, Brudno M. Genome variation discovery with high-throughput sequencing data. Briefings Bioinf 2010;11(1):3–14.

19. Duitama J, Srivastava PK, Măndoiu II. Towards accurate detection and genotyping of expressed variants from whole transcriptome sequencing data. BMC Genomics 2012;13 Suppl 2:S6.

20. Ewing B, Green P. Base-calling of automated sequencer traces using phred. II. Error probabilities. Genome Res 1998;8(3):186–194.

21. McKenna A et al. The genome analysis toolkit: a MapReduce framework for analyzing next-generation DNA sequencing data. Genome Res 2010;20(9):1297–1303.

22. Kawahara Y et al. Improvement of the Oryza sativa Nipponbare reference genome using next generation sequence and optical map data. Rice 2013;6(1):4.

23. Kozarewa I et al. Amplification-free Illumina sequencing-library preparation facilitates improved mapping and assembly of (G + C)-biased genomes. Nat Methods 2009;6(4):291–295.

24. Duitama J et al. An integrated framework for discovery and genotyping of genomic variants from high-throughput sequencing experiments. Nucleic Acids Res 2014;42(6):e44.

25. DePristo MA et al. A framework for variation discovery and genotyping using next-generation DNA sequencing data. Nature Genet 2011;43(5):491–498.

26. Nielsen R, Paul JS, Albrechtsen A, Song YS. Genotype and SNP calling from next-generation sequencing data. Nat Rev Genet 2011;12(6):443–451.

27. Bansal V. A statistical method for the detection of variants from next-generation resequencing of DNA pools. Bioinformatics (Oxford, England) 2010;26(12):i318–i324.

28. Li H. A statistical framework for SNP calling, mutation discovery, association mapping and population genetical parameter estimation from sequencing data. Bioinformatics (Oxford, England) 2011;27(21):2987–2993.

29. Mills RE et al. Mapping copy number variation by population-scale genome sequencing. Nature 2011;470(7332):59–65.

30. Abyzov A, Urban AE, Snyder M, Gerstein M. CNVnator: an approach to discover, genotype, and characterize typical and atypical CNVs from family and population genome sequencing. Genome Res 2011;21(6):974–984.

31. Xie C, Tammi MT. CNV-seq, a new method to detect copy number variation using high-throughput sequencing. BMC Bioinformatics 2009;10:80.

32. Goecks J, Nekrutenko A, Taylor J. Galaxy: a comprehensive approach for supporting accessible, reproducible, and transparent computational research in the life sciences. Genome Biol 2010;11(8):R86.

7

DISCOVERING AND GENOTYPING TWILIGHT ZONE DELETIONS

TOBIAS MARSCHALL AND ALEXANDER SCHÖNHUTH

Centrum Wiskunde & Informatica, Amsterdam, Netherlands

7.1 INTRODUCTION

Although next-generation sequencing (NGS) experiments have become standard, the exploration of the data still poses challenges. NGS experiments usually aim at providing catalogs of genetic variants to be used in downstream analyses of interest. In population studies, such as the "Genome of the Netherlands" or the "1000 Genomes" initiatives (1, 2), such catalogs aim to reflect the full extent of genetic diversity of populations. In cancer genome studies (see Reference 3 for a global initiative), comprehensive lists of somatic variants are sought that help to understand cancer (sub)types and disease progression and to select appropriate therapy protocols.

In this chapter, we focus on techniques for next-generation resequencing studies, which allow to study the differences between a *donor genome*, a genome to be investigated, and a reference genome. The workflow common to resequencing studies proceeds according to the following steps:

(1) The DNA of the genome of interest is broken into fragments.
(2) The fragments are next-generation sequenced, which yields *reads*. A very popular and helpful technique is to generate *paired-end reads*, fragments both ends of which are sequenced with an internal part, the *internal segment*, which remains unsequenced.

Computational Methods for Next Generation Sequencing Data Analysis, First Edition.
Edited by Ion I. Măndoiu and Alexander Zelikovsky.
© 2016 John Wiley & Sons, Inc. Published 2016 by John Wiley & Sons, Inc.
Companion website: www.wiley.com/go/Mandoiu/NextGenerationSequencing

(3) Reads are mapped onto the reference genome, whose sequence is known in its entirety. Mapping requires *read aligners*, algorithms that allow to align reads with the reference.

(4) One then tries to infer the differences between the genome under study and the reference genome, that is, the genetic variants that affect the genome of interest, from the mapped reads.

See Figures 7.1, 7.2, and 7.4 for schematics on scenarios that can result from mapping paired-end reads onto a reference genome.

The computational exploration of certain classes of genetic variants, such as single-nucleotide polymorphisms (SNPs) and small insertions and deletions (indels) have become standard. See, for example, the GATK (4) website for best-practice workflows. Treating other classes of variants, however, such as translocations, inversions, or nested combinations of simpler variants, is still often nonstandard or requires computationally advanced techniques. See, for example, References 5, 6 for reviews on the discovery of structural variants.

Although difficult in general and still lacking best-practice workflows, the analysis of some classes of structural variants has become routine. A predominant example is

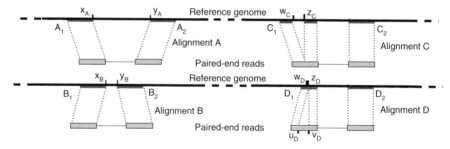

Figure 7.1 (A) Alignment whose interval length indicates a deletion, (B) alignment whose interval length indicates an insertion, (C) alignment where a split (in the left end) indicates a deletion, (D) alignment where a split (in the right end) indicates an insertion.

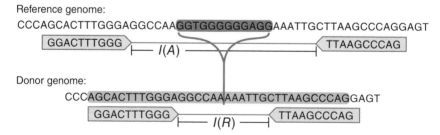

Figure 7.2 Internal segment size-based evidence for a deletion: the piece of sequence "GGTGGGGGAGG" is present in the reference but deleted in the donor genome. The length $I(R)$ of the fragment that is sequenced (in green) is determined during library preparation. When mapped back onto the reference, the internal segment $I(A)$ is longer than $I(R)$ due to the deletion.

the discovery of deletions of more than 200 base pairs (bp), which has been addressed by a large variety of approaches: examples are BreakDancer (7), VariationHunter (8), (MATE-)CLEVER (9, 10), DELLY (11), GASV(-Pro) (12, 13), see also the references therein, and again the above-mentioned reviews (5, 6).

7.1.1   Twilight Zone Deletions

In this chapter, we focus on deletions that are hard to discover because of their length. As mentioned earlier, very short deletions as well as long deletions are no longer posing fundamental difficulties, or even have become part of best-practice workflows. *Mid-sized deletions*, however, which we refer to as *"NGS twilight zone deletions"*, have been posing substantial computational challenges also after 2009. Only most recent advances have made their discovery possible in population-scale practice (9, 10, 14) – earlier approaches had pointed out theoretical ways, but remained slower by orders of magnitude (15). Evidence of this is the fact that catalogs of deletions resulting from projects (1), where References (9, 10, 14) have been in use, finally contain comprehensive amounts of such twilight zone deletions, with excellent validation rates. This is in stark contrast to earlier related projects (in particular the 1000 Genomes Project (2)).

   In this chapter, we review why discovery and genotyping of mid-sized deletions has been difficult and explain the techniques by which this became possible.

   The organization of this chapter is as follows:

- In Section 7.2, we provide the necessary notation.
- In Section 7.3, we give the formal definition of "twilight zone" deletions. We briefly revisit the different approaches suitable for deletion discovery in resequencing studies, and we outline their pitfalls when it comes to discovering mid-sized ("twilight zone") deletions.
- In Section 7.5, we present a novel maximum likelihood approach for genotyping deletions which achieves highly favorable performance rates on twilight zone indels.
- In Section 7.6, we evaluate a comprehensive selection of state-of-the-art tools on NGS reads from a genome containing real variants (Venter's genome (16)), where NGS reads are simulated by means of the Assemblathon (17) read simulator and current NGS technology (Illumina HiSeq and MiSeq reads).
- In Section 7.7, we discuss all results presented and point out challenges that are still open.

7.2   NOTATION

We predominantly focus on paired-end read data, the most widely used data in resequencing studies. Let $\Sigma = \{A, C, G, T, N\}$ be the set of nucleotides, augmented by a character (N), which represents nucleotides that could not be properly read. Throughout this chapter, reads $R = (R_1, R_2) \in (\Sigma^K)^2$ are pairs of strings of length K (where $K = 100$ in Illumina HiSeq, or $K = 250$ in Illumina MiSeq experiments) over Σ. Here, R_1 is the left and R_2 is the right end of R. We refer to single positions in the ends R_i for $i = 1, 2$ by $R_i[t]$ where $t \in [1, K]$. Let $I(R)$ be the length of the *internal segment* between the two sequenced ends R_1, R_2 of a paired-end read $R = (R_1, R_2)$. While sequence and, hence, length of the ends R_1, R_2 are known, neither the sequence of the internal segment nor its length, $I(R)$, is known. This, of course, implies that the length of the entire fragment $2K + I(R)$ is not known either.

We write $\mathcal{G}$ for the reference genome, which we also consider as a sequence over Σ.

7.2.1 Alignments

See Figure 7.1 for the following.

We write $A(R) = (A_1, A_2)$ for an alignment of read $R = (R_1, R_2)$ against the reference. We write x_A, y_A for the rightmost reference position of the alignment of the left read end, and the leftmost reference position of the alignment of the right end. We write $I(A) := y_A - x_A - 1$ for the length of the *alignment interval* of A.

7.2.2 Gaps/Splits

Alignments $A = (A_1, A_2)$ can be gapped, where gaps either indicate insertions or deletions. For the sake of simplicity, we assume that each alignment is affected by at most one gap – note that alignments of NGS fragments containing two gaps are extremely rare. For notational simplicity, we will assume in the examples and explanations to follow that A_1 displays a gap. We write w_A for the reference position that precedes the gap, and z_A for the reference position that immediately follows the gap. In turn, we refer to the position in the read that precedes the gap as u_A and the position in the read that follows the gap as v_A, that is, the reference nucleotide $\mathcal{G}[w_A]$ aligns with $R_1[u_A]$ and $\mathcal{G}[z_A]$ aligns with $R_1[v_A]$. Depending on whether $z_A = w_A + 1$ (see alignment D in Figure 7.1) or $v_A = u_A + 1$ (see alignment C), the gap indicates an insertion in the donor genome (where the inserted sequence is $R[u_A + 1, v_A - 1]$) or a deletion in the donor genome (where the deleted sequence is $\mathcal{G}[w_A + 1, z_A - 1]$).

7.2.3 Deletions

Let D_L and D_R be the reference coordinates of the left and right breakpoint of a deletion D. That is, reference nucleotides from (and including) position D_L till (and including) position D_R, which together form the sequence $\mathcal{G}[D_L, D_R]$, are missing in the donor genome. Let $C(D) := \frac{D_L + D_R}{2}$ be the *centerpoint* of D (which need not be an integer) and $L(D) := D_R - D_L + 1$ be the length of the deletion. We parameterize the deletion $D = (C(D), L(D))$ by its centerpoint $C(D)$ and its length $L(D)$.

7.3 NON-TWILIGHT-ZONE DELETION DISCOVERY

Approaches available for discovering deletions from NGS read data roughly fall into four different categories: internal segment size-based, split-read-based, coverage-based, and assembly-based approaches (see Reference 5 for a detailed review). We solely focus on the first two types of approaches and their hybrids here. Coverage-based approaches can only discover deletions of usually at least 1000 bp in length. Assembly-based approaches face challenges that have not yet been entirely overcome. Note again that very short deletions of length up to 20 bp can be discovered already during the initial mapping stage and therefore pose no unusual computational challenges. While common internal segment size-based approaches work well for deletions of length greater than approximately 150 bp, split-read aligners are able to discover also deletions longer than 20 bp, usually reaching their limits at 30–40 bp. Beyond 40 bp, deletion discovery recall of split-read aligners usually substantially drops.

7.3.1 Internal Segment Size-Based Approaches

The basic idea that underlies internal segment size-based approaches is that the alignment interval length $I(A)$ deviates from $I(R)$, the length of the internal segment of the read that gave rise to A, by the length of a deletion affecting the internal segment of R. That is, $I(A) = I(R) + L(D)$ for a deletion D in the internal segment of R, see Figure 7.2 for an illustration. While $I(A)$ is known, $I(R)$ is not, however. Therefore, one estimates both mean and standard deviation of the empirical distribution of $I(R)$ from uniquely mappable fragments using robust estimators (18, 19). For modern DNA sequencing protocols, the fragment length distributions are approximately Gaussian with low standard deviations. We note already here that well-shaped (Gaussian) distribution is essential for discovering mid-sized deletions.

See Figure 7.3 for such a distribution, derived from the NGS reads of one of the individuals of the GoNL project (1), all of which were sequenced by BGI in 2010.

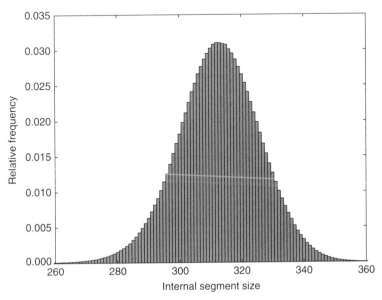

Figure 7.3   Internal segment size distribution for GoNL individual.

After the determination of the internal segment size distribution, the following generic workflow for discovering deletions in NGS data is widely used:

1. Collect all reads whose alignments *statistically significantly deviate* in terms of alignment interval length, the so-called *discordant reads.*
2. Cluster all such reads into groups that support the same deletion.
3. Make predictions from the resulting groups of discordant reads.

The majority of approaches follows this workflow (e.g., References 7, 8, 12, 13, 20, 21). They can be distinguished by their definition of discordant read, their clustering/grouping techniques, and their details in deriving predictions from groups of discordant reads. Note that handling of reads that became multi-mapped due to repetitive sequence often plays a major role (22), see, for example, Reference 8 for a combinatorially principled approach.

The key factor is the definition of a discordant read, as those are supposed to represent fragments whose internal segment is affected by the deleted sequence. Again, the idea is that $I(A)$, the alignment interval length of a discordant read, significantly deviates from the distribution of $I(R)$. Thus, the definition of a discordant read depends on the standard deviation of the distribution of $I(R)$, which in current protocols amounts to about 15. Up to six to seven times this standard deviation is required to obtain sufficiently low, genome-wide false discovery rates. This then translates into the fact that deletions shorter than 100 bp remain undiscoverable.

7.3.2 Split-Read Mapping Approaches

Split-read mapping approaches aim at making direct use of alignment information. As per a usual workflow, a split-read mapper processes only read that standard read aligners fail to align correctly. This is most often due to insertions and deletions that affect the read. Standard read aligners usually face difficulties in aligning such reads properly because correctly placing longer gaps can be computationally (too) expensive.

Therefore, aligning reads affected by longer indels requires extended techniques: "split-read alignments." A generic workflow common to many split-read aligners (e.g., References 14, 23–25) looks as follows:

1. Collect all reads where one end remained unaligned and/or the other end became only partially aligned ("soft-clipped") by the standard aligner in use.
2. In case of entirely unaligned read ends, split the end in parts, or "seeds," and try to align those parts, or "seeds" (see Figure 7.4 for a resulting alignment).
3. For such aligned parts and/or for soft-clipped reads, try to align the remaining part(s) of the read somewhere "nearby."
4. For each such read, collect all possible partial alignments and compute "split alignments," using banded alignment techniques, to connect them. Output the most likely such split alignment(s) as the alignment(s) of the read end.

Figure 7.4 Split-read evidence for deletion.

To date, common split-read aligners can successfully align reads with non-negligible amounts of deletions of length up to 40–50 bp. While aligning reads exhibiting larger deletions is not impossible, the discovery rates of split-read aligners significantly decrease with increasing deletion size. Thereby, recall, that is, the rate of discovered deletions, usually drops below 60–70% when reaching the 30-bp mark, which renders split-read based approaches unable to discover sufficient amounts of "twilight zone deletions", if they follow the workflow from above.

The bottlenecks of split-read aligners are steps 2 and 3 as in the above-mentioned workflow. Split or yet unaligned parts of reads can be small, which can drastically increase the number of locations in the reference genome where these parts can be aligned. The fact that genomes are highly repetitive in general can significantly add to these difficulties – resolving the resulting ambiguity among those multiple mappings is another involved step.

Therefore, one has to limit the size of the regions in which one searches for alignments of split parts. Due to those limitations, although substantially raising the limits of standard aligners, split-read aligners can also quickly reach their limits – every implementational detail can count. Note that the internal segment size distribution also plays a decisive role here, as it is used to appropriately quantify "nearby" in step 3 and as a guide when placing read alignments of split parts in step 2.

7.3.3 Hybrid Approaches

One decisive general advantage of split-read aligning approaches over internal segment size-based approaches is the base pair resolution of deletion breakpoints: if both an internal segment size-based method and a split-read aligner call a deletion, the breakpoints predicted by the split-read aligner are usually much more accurate than those of the internal segment size-based approach.

However, as outlined earlier, split-read aligners usually cannot detect many deletions larger than 30–40 bp. The motivation of the so-called hybrid approaches is to also call breakpoints of large deletions at base pair resolution. A common generic workflow thus is as follows:

1. Run an internal segment size-based approach and collect all deletion calls.
2. Collect all alignments nearby deletion calls collected in step 1.
3. Split-align all unaligned and partially aligned read ends in those regions. Thereby, the split-aligner is "guided" by the deletion calls of the internal segment size-based approach when determining the correct placements of shorter read end parts.
4. Output all variant calls with breakpoints corrected (or removed, if no split alignments could be determined) as per the split alignments determined in step 3.

The results are usually calls for large deletions whose breakpoints come at base pair resolution (see References 10, 11, 26, 27 for most prevalent approaches). In essence, the major bottleneck of hybrid approaches is step 1, that is, they inherit the computational bottlenecks of internal segment size-based approaches in terms of size range limitations.

7.3.4 The "Twilight Zone": Definition

We conclude that both internal segment size-based and hybrid approaches can discover deletions at sufficient power only in the size range of 100 bp (approximately six to seven times the standard deviation of the internal segment size distribution) and larger. Split-read aligners, on the other hand, are able to discover deletions only of size up to 30 bp (approximately two times the standard deviation). We consequently suggest 2 to 6–7 times the standard deviation in terms of base pairs as the *"twilight zone" of NGS deletions*.

7.4 DISCOVERING "TWILIGHT ZONE" DELETIONS: NEW SOLUTIONS

Since 2012, new solutions have been presented for discovering deletions of length 30–100 bp, at both sufficient power and precision (9, 10). The earliest approach that immediately addressed the discovery of mid-sized insertions and deletions was MoDIL (15). While successful by principle, MoDIL is too slow in practice: a single genome, sequenced at the standard coverage of 30×, needs more than 3 days on a large computer cluster, which is no option in a population-scale genome project. In contrast, CLEVER (9) needs only 6–8 hours on a single CPU. CLEVER also significantly outperforms MoDIL in terms of both recall and precision in discovery. We therefore focus on CLEVER and its relatives in the following. We will also mention PINDEL (14), as the possibly most favorable contemporary split-read aligner, which can make considerable contributions in the twilight zone.

7.4.1 CLEVER

The key insight of CLEVER (9) is that exchanging steps 1 and 2 in the workflow outlined in Section 7.3.1 leads to success when aiming at the discovery of insertions and deletions smaller than 100 bp. CLEVER clusters read alignments before discarding nondiscordant reads. The point is that, although single concordant (= nondiscordant) reads make no strong enough statistical signal for a deletion on their own, groups of concordant reads, all of which transmit a rather weak statistical signal for a deletion, can together "bundle up" to form a strong signal for a deletion. CLEVER aims at the discovery of such group signals.

While this sounds easy in principle, it is not in practice. The clear advantage of the previous approaches was that the workflow from Section 7.3.1 allowed to discard all concordant reads, which drastically reduces the number of reads to be processed. Thus, step 2 translated into a clustering problem of decisively smaller scale: instead of clustering billions of reads, only small fractions of discordant reads, which come in amounts that are smaller by orders of magnitude, needed to be grouped and further processed. CLEVER clusters *all read alignments*. To achieve low enough run-times in practice, CLEVER makes use of a highly engineered implementation of a max-clique enumeration technique as underlying clustering algorithm.

CLEVER solves this problem by formally collecting all read alignments into a *read alignment graph*, where each node represents a read alignment and edges indicate that two overlapping alignments are likely to reflect identical alleles. Maximal cliques represent maximal groups of read alignments all of which reflect the same allele (see the right panel of Figure 1 in Reference 9). If there are indel alleles in the donor genome, the max-cliques reflecting such alleles deviate from the internal segment size statistics. If sufficiently many read alignments participate in such a max-clique, they give rise to statistically significant signals even when reflecting only relatively small indels, as revealed by common multiple-sample Z-tests. The statistical model of CLEVER further allows to address that read alignments can be ambiguous due to repetitive sequence and corrects for multiple testing, thereby keeping control of the false discovery rate.

Key to success for enumerating all such max-cliques finally is a bitvector-driven implementation of a max-clique enumeration algorithm that exploits the particular structure of read alignment graphs. See References 9 and also 28 for details and corresponding run-time analyses.

7.4.2   MATE-CLEVER

While CLEVER discovers mid-sized deletions at both high recall and precision (see Section 7.6), the accuracy of the predicted breakpoints suffers from the usual deficits that are common to internal segment size-based approaches. MATE-CLEVER, as a hybrid approach, aims at curing this issue and discovers deletions not only at high recall and precision but also at high accuracy of their breakpoints (see Figure 7.5 for an illustration of MATE-CLEVER).

The workflow of MATE-CLEVER (10) is that of a common hybrid approach (see Section 7.3.3). Thereby, it makes use of CLEVER in the first step. Subsequently, in step 3, it makes use of a novel split-read aligner, LASER (29), which has been particularly trimmed to compute highly accurate split-read alignments reflecting larger gaps also. The outputs of MATE-CLEVER are mid-sized (and long-sized) deletions, discovered by CLEVER, where breakpoints are corrected and therefore highly accurate.

7.4.3   PINDEL

PINDEL (14) is a split-read aligner that specializes in also discovering deletions longer than 20 bp at extremely high precision, with more than 90% of all calls being true positives that are also highly accurate in terms of breakpoint annotations, across all size ranges. In doing so, it achieves clearly the highest recall rates among all (split-read) alignment-based approaches. Note that GSNAP (30, 31) achieves higher

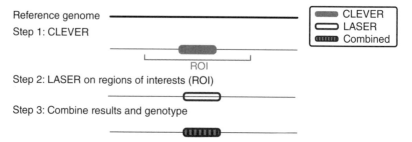

Figure 7.5   MATE-CLEVER. First, the internal segment size-based tool CLEVER discovers deletions (red). The split-read aligner LASER then finds corresponding split-read alignment (blue) in the respective regions. The resulting prediction (red-blue) is that of LASER, as split-read aligner discovers deletion breakpoints at higher accuracy. Please see www.wiley.com\go\Mandoiu\NextGenerationSequencing for a color version of this figure.

recall for deletions of 30–50 bp, which, however, comes at the price of reduced precision and (much) less accurate breakpoint annotations.

Overall, PINDEL follows the workflow common to split-read aligners. Its achievements are due to an accumulation of improvements in the fine details, which in combination yield a superior method. We refer the reader to (14), the original publication, for details.

7.5   GENOTYPING "TWILIGHT ZONE" DELETIONS

7.5.1   A Maximum Likelihood Approach under Read Alignment Uncertainty

Let G_i for $i = 0, 1, 2$ represent the genotypes of an indel, where G_0 indicates absence of the indel, G_1 indicates that the indel is heterozygous, and G_2 indicates that the indel is homozygous. Let A be a read alignment. Let $\mathcal{R}$ be all reads. For $R \in \mathcal{R}$, let $A(R)$ be the alignment of R with the region we would like to genotype. We write A^+ for the event that A is the correct alignment of R, and we write A^- for the event that it is not. Note that $\mathbf{P}(A^-) = 1 - \mathbf{P}(A^+)$. We further formally consider each read $R \in \mathcal{R}$ as the disjoint union of the two events $A^+(R)$ and $A^-(R)$. Let $S \subset \mathcal{R}$ be a subset of reads. In slight abuse of notation, we also consider S as the event where precisely the alignments of reads from S are correct, while all others are not. Hence,

$$\mathbf{P}(S) = \prod_{R \in S} \mathbf{P}(A^+(R)) \cdot \prod_{R \notin S} (1 - \mathbf{P}(A^+(R))) \tag{7.1}$$

is the corresponding probability.

In the following, we consider a maximum likelihood (ML) setting, which in particular reflects that our prior belief in genotypes is the same for all types:

$$\mathbf{P}(G_0) = \mathbf{P}(G_1) = \mathbf{P}(G_2) = \frac{1}{3}. \tag{7.2}$$

Making use of an ML approach allows to attain an efficient computation scheme. We point out below that a full Bayesian approach is infeasible because of being exponential in the number of reads that align with the region of interest – note that we wish to genotype hundreds of thousands of regions of interest, such that run-time considerations are a crucial factor.

We are interested in maximizing

$$\mathbf{P}(G_i \mid \mathcal{R}) \propto \mathbf{P}(G_i, \mathcal{R}) = \sum_{S \subseteq \mathcal{R}} \mathbf{P}(S) \cdot \mathbf{P}(G_i \mid S). \tag{7.3}$$

By taking probabilities $\mathbf{P}(S)$ into account, we would like to appropriately address alignment uncertainty, which can be due to several factors such as multiple mappings and alignment artifacts. Let $K := |\mathcal{R}|$ be the number of reads that align to the region to be genotyped. By Bayes' formula, Equation (7.2) further implies that

$$\mathbf{P}(G_i \mid S) \stackrel{(7.2)}{\propto} \mathbf{P}(S \mid G_i) = \prod_{R \in S} \mathbf{P}(A^+(R) \mid G_i) \cdot \prod_{R \notin S} \mathbf{P}(A^-(R) \mid G_i)$$

$$\stackrel{(7.2)}{\propto} \prod_{R \in S} \mathbf{P}(G_i \mid A^+(R)) \cdot \prod_{R \notin S} \mathbf{P}(G_i \mid A^-(R)), \tag{7.4}$$

where the equality is justified by assuming that reads have been generated independently of each other. Note that the computation (7.4) is not possible in the frame of a fully Bayesian approach because only the assumption (7.2) of constant priors implies the first proportionality. This renders such an undertaking infeasible. From (7.4), we conclude that

$$\mathbf{P}(G_i \mid \mathcal{R}) \propto \sum_{S \subseteq \mathcal{R}} \mathbf{P}(S) \cdot \prod_{R \in S} \mathbf{P}(G_i \mid A^+(R)) \cdot \prod_{R \notin S} \mathbf{P}(G_i \mid A^-(R))$$

$$\stackrel{(7.1)}{=} \sum_{S \subseteq \mathcal{R}} \prod_{R \in S} \mathbf{P}(A^+(R)) \mathbf{P}(G_i \mid A^+(R)) \cdot \prod_{R \notin S} (1 - \mathbf{P}(A^+(R))) \mathbf{P}(G_i \mid A^-(R))$$

$$= \prod_{R \in \mathcal{R}} [\mathbf{P}(A^+(R)) \mathbf{P}(G_i \mid A(R)) + (1 - \mathbf{P}(A^+(R))) \mathbf{P}(G_i \mid A^-(R))], \tag{7.5}$$

where the second row results from expanding the third row. The last term, finally, can be computed in time linear in the number of reads $\mathcal{R}$, which had been our goal.

It remains to compute reasonable probabilities $\mathbf{P}(G_i \mid A^+)$ and $\mathbf{P}(G_i \mid A^-)$ for read alignments A. While

$$\mathbf{P}(G_i \mid A^-) = \mathbf{P}(G_i) \tag{7.6}$$

is obviously reasonable, because the read that underlies A does not stem from the region to be genotyped, computation of terms $\mathbf{P}(G_i \mid A^+)$ requires further reasoning based on the type of evidence that A can provide about G_i.

One has to distinguish the following two cases (see Figure 7.6):

1. *Split-Read Evidence*: A aligns with the region of interest such that one read end stretches across the (potential) variant

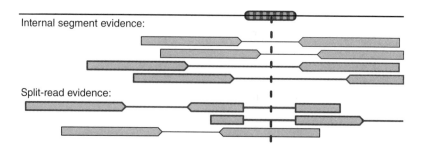

Figure 7.6 Different types of evidence for a heterozygous variant. While the gray alignment rather provide evidence against a deletion, the alignments in red rather provide evidence for it. In case of internal segment evidence with alignments (counting alignments from above) 3–6, and gray alignments with alignments 1, 2, and 7 A reflect the case $\mathcal{N}_{\mu+L,\sigma}(I(A)) > \mathcal{N}_{\mu,\sigma}(I(A))$ in (7.9), whereas the gray alignments reflect the opposite case.

2. *Internal Segment-Based Evidence*: A aligns with the region of interest such that the internal segment of its read pair stretches across the (potential) variant

7.5.1.1 Split-Read Evidence Let us first consider the case of no alignment uncertainty. That is, if read alignments are correct, then they precisely reflect the differences between the donor and the reference.

Let D be the deletion to be genotyped and let A be an alignment where, for example, A_1 stretches across the breakpoints of D. Under the assumption of no alignment uncertainty, we obtain that A stems from a chromosomal copy that is affected by D if and only if w_A and z_A precisely agree with the left and right breakpoints of D. If the split disagrees with the deletion breakpoints or there is no split, the read behind A stems from a chromosomal copy that is not affected by the deletion with probability one.

Let A be an alignment with a split/gap that agrees with D. By the above considerations, the read behind A stems from a chromosomal copy that is affected by D. By Bayes' formula and (7.2), $\mathbf{P}(G_i \mid A^+) \propto \mathbf{P}(A^+ \mid G_i)$. First, $\mathbf{P}(A^+ \mid G_0) = 0$ because the read behind A cannot stem from the region and $\mathbf{P}(A^+ \mid G_1) = \frac{1}{2}(\mathbf{P}(A^+ \mid G_0) + \mathbf{P}(A^+ \mid G_2))$, which reflects that one first randomly selects one of the two chromosomal copies, only one of which is affected by D, and then generates the read from it. The *case of B being an alignment in disagreement with D* is treated analogously, where in this case $\mathbf{P}(B^+ \mid G_2) = 0$. Transforming this into a posterior probability consequently yields

$$\mathbf{P}(G_i \mid A^+) := \begin{cases} \frac{2}{3} & i = 2 \\ \frac{1}{3} & i = 1 \\ 0 & i = 0 \end{cases} \quad \text{and} \quad \mathbf{P}(G_i \mid B^+) := \begin{cases} 0 & i = 2 \\ \frac{1}{3} & i = 1 \\ \frac{2}{3} & i = 0 \end{cases}. \quad (7.7)$$

In general, however, the assumption of no alignment uncertainty does not hold. In fact, split-read alignments can be affected by several sources of errors, the most evident of which are probably repetitive areas, such that both position and length of alignment splits disagree with the positions and the length of the true variants – nevertheless, the split is indeed due to the variant. Therefore, we declare A, where $C(A) := (w_A + z_A)\backslash 2$ is the centerpoint and $L(A) := z_A - w_A$, in case of deletions, (see Figure 7.1) is the length of the split in A, to support the deletion $D = (C(D), L(D))$, with centerpoint $C(D)$ and of length $L(D)$ iff

$$|C(D) - C(A)| \leq 50 \quad \text{and} \quad |L(D) - L(A)| \leq 20. \tag{7.8}$$

While these values may seem large, they are well supported by statistics on the uncertainty of (split-)alignment.

7.5.1.2 *Internal Segment-Based Evidence*

Internal segment-based evidence is provided by evaluating the empirical statistics on fragment length inherent to the library the read stems from. We develop this part here in view of fragment length statistics being approximately Gaussian. However, this can be easily generalized to arbitrary empirical statistics. Let D be the deletion to be genotyped and let $C(D)$ be its centerpoint. Let R be the read that has given rise to alignment A where $x_A < C(D) < y_A$, that is, the alignment interval of A contains the centerpoint of the breakpoints of D. Note that a centerpoint-oriented selection leads to a situation that is balanced in terms of choosing equal amounts of reads that provide evidence for and against the deletion, as outlined in Reference 10. Let μ and σ be mean and standard deviation of the internal segment size distribution of the library R stems from. So, internal segment length, as a random variable X, is distributed as the normal distribution $X \sim \mathcal{N}_{\mu,\sigma}$ for the library under consideration. There are two cases: first, alignments A *whose reads stem from a chromosomal copy that is not affected by D*, and second alignments B *whose reads stem from a chromosomal copy that is affected by D.* We obtain

$$I(A) \sim \mathcal{N}_{\mu,\sigma} \quad \text{and} \quad I(B) \sim \mathcal{N}_{\mu+L,\sigma}, \tag{7.9}$$

where the second case reflects that the alignment interval contains the deletion of length L. Refer to Figure 7.7 for an illustration. We compute that $\mathbf{P}(A^+ \mid G_0) \propto \mathcal{N}_{\mu,\sigma}(I(A))$ and $\mathbf{P}(A^+ \mid G_2) \propto \mathcal{N}_{\mu+L,\sigma}(I(A))$ as appropriate densities for the cases of no variant and a homozygous variant.

Let $Z := \frac{3}{2}(\mathcal{N}_{\mu,\sigma}(I(A)) + \mathcal{N}_{\mu+L,\sigma}(I(A)))$. In analogy to considerations for the split-read case, we arrive at

$$\mathbf{P}(G_i \mid A^+) := \begin{cases} \frac{1}{Z} \cdot \mathcal{N}_{\mu+L,\sigma}(I(A)) & i = 2 \\ \frac{1}{3} & i = 1 \\ \frac{1}{Z} \cdot \mathcal{N}_{\mu,\sigma}(I(A)) & i = 0 \end{cases} \tag{7.10}$$

as an appropriate probability distribution for reads whose alignments span the breakpoints of deletions by their internal segments.

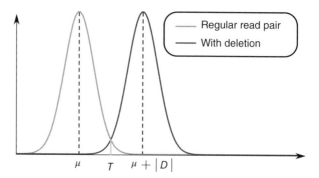

Figure 7.7 Gaussian distribution on interval size for alignments of normal reads (read) and reads indicating a deletion of length $|D|$. Alignments whose intervals are of length T provide no evidence, as both the existence and the nonexistence of the deletion are equally likely.

The procedure described earlier is implemented as part of MATE-CLEVER (10), which can use prior information in form of the Mendelian laws, if the input consists of multiple, ancestry-related genomes. In order to genotype, MATE-CLEVER takes all (split-read) alignments resulting from step 3 of the generic hybrid approach workflow (see Section 7.3.3) and executes the genotyping-related computations as mentioned earlier, by plugging Equations (7.7) and (7.10) into Equation (7.5), thereby inferring the most likeliest genotype.

7.6   RESULTS

7.6.1   Data Set

We downloaded all variant annotations for Craig Venter's genome from the HuRef database (16). We generated a diploid genome using those annotations, as per the procedure described in Reference 10 (to generate a 'father' genome), which results in a genome that is realistic in terms of both amounts of variants and zygosity status of variants. Note that direct usage of the annotations results in a genome with an unrealistic ratio of heterozygous to homozygous deletions (it vastly overrated homozygous indels), which is likely due to the difficulties in determining the zygosity status of insertions and deletions during the original assembly stage. Here, we have resolved these issues such that the ratio of heterozygous and homozygous deletions is realistic.

Subsequently, we simulated reads for each of those copies, using Simseq, the read simulator of the Assemblathon (17), at 15× coverage, which results in 30× coverage overall. We opted to simulate reads according to two prevalent and most recent Illumina protocols: HiSeq and MiSeq, where mean and standard deviation for the size of the internal segment were 112 and 15 (HiSeq) and 250 and 15 (MiSeq), respectively.

TABLE 7.1   List of Used Software Tools

Tool	Read mapper	Internal segment	Split reads	Genotyping	Version	References
Bowtie2	X				2.1.0	32
Breakdancer		X			1.4.4	7
BWA	X				0.7.5a	18
CLEVER		X			v2.0-rc3	9
DELLY		X	X		0.0.11	11
GATK				X	2.8-1-g932cd3a	4
GSNAP	X				2014-01-21	30, 31
MATE-CLEVER		X	X	X	v2.0-rc3	10
PINDEL			X		0.2.4t	14
Socrates			X		–	24
Stampy	X				1.0.23	33
VariationHunter		X			0.3	8

Read mapper: Programs ticked in that column are standard read mappers. *Internal segment*: SV detection methods that use internal segment size information. *Split reads*: SV detection methods based on split-read alignment. *Genotyping*: Methods able to genotype SVs. *Version*: The given version was used in our experiments.

7.6.2   Tools

Table 7.1 gives an overview of the used tools. We aimed at selecting state-of-the-art tools from different categories. From the internal segment size-based methods, we chose Breakdancer, CLEVER, and VariationHunter. Pindel and Socrates represent split-read approaches, where MATE-CLEVER and DELLY are hybrids between internal segment size and split read. Furthermore, we included four standard read mappers in the analysis: Bowtie2, BWA (MEM), GSNAP, and Stampy. Although these tools do not explicitly target deletion discovery, they have some capabilities of mapping reads across deletions. For these tools, we extracted all deletions that were contained in two or more read alignments to compile a set of predictions.

7.6.3   Discovery

See Tables 7.2 and 7.3 for the following. We ran all tools on the two (HiSeq and MiSeq) data sets described in Section 7.6.1. In Tables 7.2 and 7.3, tools are grouped by the class of approach they belong to. The first group are internal segment size-based, the second group are split-read alignment-based, the third group are hybrid, and the fourth group are direct alignment-based approaches.

We evaluate all tools in terms of four different categories.

1. *Strict precision*, which is the fraction of calls where the centerpoint of the breakpoints deviates by not more than 20 bp and the length by not more than 10 bp from that of a true deletion.
2. *Relaxed precision* allows deviations of 100 bp for both centerpoint placement and length. Note that such calls are still statistically highly significant, because

TABLE 7.2   Results for SV Prediction Tools on 30× HiSeq/MiSeq Data for Deletions from 10 to 69 bp

Deletion length	Tool	Precision (strict)	Precision (relaxed)	Recall (hom.)	Recall (het.)
10–29	BreakDancer	0.0 / 0.0	83.7 / 32.8	0.6 / 0.2	0.3 / 0.0
	CLEVER	38.4 / 30.3	80.3 / 76.9	25.1 / 17.7	6.9 / 0.7
	VariationHunter	3.3 / 0.4	89.1 / 53.7	0.7 / 0.4	0.5 / 0.0
	PINDEL	**91.2 / 91.3**	93.0 / 93.4	**89.0 / 92.4**	**80.7 / 83.8**
	SOCRATES	8.0 / 4.5	11.5 / 7.1	1.4 / 0.3	1.2 / 0.3
	DELLY	– / –	– / –	0.0 / 0.0	0.0 / 0.0
	MATE-CLEVER	87.3 / 89.6	**93.2 / 94.1**	23.1 / 15.3	5.8 / 2.7
	Bowtie2	90.5 / 33.2	92.2 / 35.1	61.4 / 82.2	51.5 / 73.5
	BWA MEM	84.8 / 80.0	88.5 / 85.0	79.6 / 86.1	72.4 / 80.8
	GSNAP	69.6 / 68.4	90.0 / 90.9	83.6 / 85.5	75.2 / 73.7
	Stampy	35.0 / 20.8	65.3 / 46.1	83.8 / 86.4	78.3 / 47.3
30–49	BreakDancer	7.5 / 3.0	81.4 / 37.3	18.6 / 5.7	8.1 / 0.1
	CLEVER	26.2 / 19.6	71.2 / 69.2	**90.5 / 80.7**	**61.8 / 15.1**
	VariationHunter	17.1 / 5.6	88.1 / 68.0	34.6 / 12.0	16.2 / 0.5
	PINDEL	77.0 / 83.7	84.1 / 90.2	65.6 / 79.2	54.4 / 67.2
	SOCRATES	44.7 / 32.7	49.1 / 35.6	9.9 / 5.7	8.2 / 3.3
	DELLY	– / –	– / –	0.0 / 0.1	0.1 / 0.0
	MATE-CLEVER	81.4 / **86.1**	89.8 / **93.0**	76.8 / 70.5	51.8 / 38.4
	Bowtie2	87.5 / 71.7	**100.0** / 75.4	2.8 / 63.6	2.0 / 47.5
	BWA MEM	**88.8** / 79.1	92.1 / 86.2	26.0 / **85.7**	21.5 / **75.3**
	GSNAP	43.9 / 56.7	69.5 / 82.9	72.3 / 83.3	57.4 / 61.3
	Stampy	57.4 / 33.8	85.1 / 66.0	56.3 / 84.2	46.2 / 43.5
50–69	BreakDancer	8.4 / 1.3	82.0 / 23.5	40.7 / 11.3	39.6 / 0.4
	CLEVER	30.3 / 21.2	75.3 / 75.3	86.0 / **78.7**	70.2 / 16.6
	VariationHunter	21.2 / 8.9	79.7 / 61.2	**86.7** / 56.0	**70.6** / 2.8
	PINDEL	64.0 / 76.2	72.8 / 83.9	48.0 / 72.3	35.4 / **53.6**
	SOCRATES	29.3 / 27.0	40.4 / 33.3	8.3 / 6.3	6.2 / 4.6
	DELLY	– / –	– / –	1.3 / 0.0	1.0 / 0.0
	MATE-CLEVER	**73.2 / 85.4**	**84.0 / 92.4**	66.0 / 64.0	51.4 / 41.8
	Bowtie2	– / 74.0	– / 74.0	1.0 / 13.7	0.2 / 4.0
	BWA MEM	– / 79.9	– / 86.2	5.3 / 64.0	3.2 / 46.2
	GSNAP	36.7 / 46.2	65.1 / 74.4	20.0 / 25.0	17.6 / 9.2
	Stampy	60.7 / 40.4	78.5 / 73.8	24.7 / 67.7	18.2 / 19.8

the deviations are small relative to genome length and overall numbers of calls, hence are still of great potential interest to the researcher. In essence, these calls just require further refinement.

3. We also evaluate the callsets in terms of *Recall (hom.)* and *Recall (het.)*, which are the fractions of *true homozygous* and *true heterozygous* deletions that were correctly discovered, according to the criteria for relaxed precision.

As becomes immediately clear from Tables 7.2 and 7.3, split-read, hybrid, and direct alignment-based approaches clearly outperform the internal segment

TABLE 7.3   Results for SV Prediction Tools on 30× HiSeq/MiSeq Data for Deletions from 70 to 199 bp

Deletion length	Tool	Precision (strict)	Precision (relaxed)	Recall (hom.)	Recall (het.)
70–99	BreakDancer	5.8 / 1.9	85.3 / 25.8	59.6 / 12.3	50.7 / 0.6
	CLEVER	38.0 / 23.4	88.8 / 78.9	81.3 / **77.9**	60.2 / 16.9
	VariationHunter	15.7 / 12.0	83.1 / 67.4	**83.0** / 77.0	**79.1** / 11.5
	PINDEL	56.1 / 72.1	64.6 / 80.9	38.3 / 65.5	27.2 / 39.0
	SOCRATES	49.1 / 55.8	54.7 / 60.5	7.2 / 8.5	10.9 / 8.9
	DELLY	66.7 / –	66.7 / –	0.4 / 2.1	2.0 / 0.0
	MATE-CLEVER	80.9 / **89.6**	90.1 / **94.6**	57.4 / 63.4	41.5 / **42.1**
	Bowtie2	– / –	– / –	0.9 / 3.0	0.9 / 0.0
	BWA MEM	– / 81.4	– / 86.4	4.3 / 40.4	2.6 / 22.3
	GSNAP	– / –	– / –	13.2 / 12.3	7.7 / 0.0
	Stampy	**87.5** / 34.9	**100.0** / 67.4	20.0 / 54.0	12.3 / 2.0
100–149	BreakDancer	3.1 / 0.2	78.2 / 21.8	48.7 / 19.3	48.6 / 0.4
	CLEVER	31.4 / 21.6	83.6 / 70.8	69.0 / 64.0	48.6 / 14.7
	VariationHunter	5.2 / 5.5	65.3 / 60.6	**76.6** / **72.6**	**66.9** / 8.0
	PINDEL	61.1 / 77.6	65.5 / 85.3	16.8 / 40.1	15.9 / 24.3
	SOCRATES	36.0 / 41.9	44.0 / 48.4	5.6 / 5.6	5.6 / 4.8
	DELLY	40.7 / 50.0	69.5 / 80.4	6.1 / 6.1	8.0 / 5.2
	MATE-CLEVER	**75.5** / 77.9	**85.4** / 87.5	44.2 / 47.7	30.3 / **32.3**
	Bowtie2	– / –	– / –	0.0 / 1.0	0.8 / 0.0
	BWA MEM	– / **100.0**	– / **100.0**	1.5 / 7.1	2.8 / 0.0
	GSNAP	– / –	– / –	8.6 / 5.1	5.6 / 0.0
	Stampy	– / 0.0	– / 0.0	6.6 / 20.8	5.6 / 0.0
150–199	BreakDancer	1.9 / 0.3	41.5 / 18.3	40.7 / 19.8	30.9 / 0.7
	CLEVER	32.0 / 16.3	**86.4** / 67.0	61.5 / 63.7	36.2 / 13.8
	VariationHunter	4.3 / 3.8	35.8 / 31.4	**70.3** / 67.0	**63.8** / 8.6
	PINDEL	55.4 / 69.0	58.5 / 81.0	20.9 / 34.1	11.2 / 18.4
	SOCRATES	20.0 / 31.0	22.9 / 34.5	6.6 / 7.7	2.0 / 2.6
	DELLY	25.0 / 56.4	45.0 / 74.4	9.9 / 12.1	7.2 / 9.9
	MATE-CLEVER	**74.7** / **83.5**	86.1 / **91.3**	37.4 / 46.2	19.7 / **30.9**
	Bowtie2	– / –	– / –	0.0 / 0.0	0.0 / 0.0
	BWA MEM	– / –	– / –	0.0 / 0.0	0.0 / 0.0
	GSNAP	– / –	– / –	0.0 / 0.0	0.0 / 0.0
	Stampy	– / –	– / –	0.0 / 0.0	0.0 / 0.0

size-based approaches in terms of accuracy of breakpoint annotation, as indicated by the much improved strict precision rates. It is also obvious that in the lower part of the twilight zone (see Table 7.2), (split-read) alignment approaches have certain advantages, where we note that among the alignment-based approaches PINDEL excels in terms of precision, while GSNAP excels in terms of recall on HiSeq data. On MiSeq data, BWA-MEM has clear advantages. Approaches from the other classes that achieve high recall in the lower part are CLEVER and MATE-CLEVER, certainly because they are the only such approaches tailored toward discovery of twilight zone deletions.

In the upper part of the twilight zone (see Table 7.3), internal segment size-based and hybrid approaches have clear advantages over (split-read) alignment-based approaches because the recall rates of alignment-based approaches drop to zero or the precision considerably suffers. Among the alignment-based approaches, PINDEL clearly has the best recall rates on longer deletions, at least at sufficiently high precision, and therefore makes a valuable contribution in those size ranges. Note that among the internal segment size-based approaches, VariationHunter puts clear emphasis on recall, which comes at the expense of inaccurate breakpoint annotations, as indicated by very low strict precision.

The only approaches that achieve high recall rates across all size ranges of the twilight zone are CLEVER and MATE-CLEVER. The only real weaknesses are heterozygous deletions of length 30–50 bp, where, however, none of the other tools achieve better recall rates on HiSeq data. In this category, one has to step up from HiSeq to MiSeq in combination with using BWA-MEM. On HiSeq data, heterozygous deletions of 30–50 bp can still be considered a weak spot for *all* approaches, with CLEVER achieving 61% recall, as the best performance rate.

When comparing HiSeq with MiSeq data in general, an immediate observation is that alignment-based approaches tend to achieve higher recall on MiSeq data, but sometimes at the cost of lower precision. Internal segment size-based approaches often incur non-negligible losses on MiSeq data, which is likely due to the longer average internal segment size for the MiSeq data set.

7.6.4 Genotyping

See Tables 7.4 and 7.5 for the following. We evaluated five state-of-the-art protocols for genotyping deletions. We ran the *UnifiedGenotyper (UG)* and the *Haplotype-Caller (HC)* on both BWA-MEM and GSNAP alignments to produce genotyped deletion calls. We also ran MATE-CLEVER in genotyping mode, as outlined in Section 7.5.1.

It becomes evident that, in an overall statement, MATE-CLEVER is the most favorable approach when it comes to genotyping deletions longer than 30 bp, both at sufficiently high recall (see statistic "# of Calls") and good precision (see column "homozyg. *correct*" in Table 7.4 and "heterozyg. *correct*" in Table 7.5). Usually, MATE-CLEVER predicts homozygosity in about 90% and heterozygosity in about 80% of the cases correctly.

While MATE-CLEVER seemingly is the most favorable approach overall, the other approaches have certain partial strengths. Most notably, both BWA-HC and GSNAP-HC achieve better performance rates than MATE-CLEVER on homozygous deletions of 30–49 bp, in terms of both recall and precision. It remains to add, however, that the value of this remains somewhat unclear because these tools do not achieve similar good rates on heterozygous deletions.

When comparing UG with HC, one observes that HC leads to considerable increases in terms of recall over UG, while incurring certain losses in terms of genotyping precision. In conclusion, UG is seemingly the more conservative post-processing method, while HC is a more aggressive variant calling postprocessor for read alignments. Using HC for genotyping deletions longer than 50 bp is seemingly

TABLE 7.4   **Genotyping Performance for Homozygous Calls**

Deletion Length	Tool	True Annot. → ↓ # of Calls ↓	Absent Wrong Call	Heterozyg. Wrong Type	Homozyg. Correct
10–29	BWA/UG	6109 / 7365	7.4 / 9.3	2.9 / 3.9	92.2 / 90.0
	BWA/HC	6327 / 6750	9.2 / 10.0	2.7 / 2.4	90.2 / 89.8
	GSNAP/UG	6392 / 6943	6.8 / 7.1	2.7 / 3.3	92.8 / 92.4
	GSNAP/HC	6380 / 6721	9.3 / 9.9	2.7 / 2.4	90.1 / 89.8
	MATE-CLEVER	1546 / 1222	6.7 / 6.7	7.0 / 8.6	91.8 / 90.8
30–49	BWA/UG	147 / 794	8.2 / 9.9	2.0 / 4.3	91.2 / 88.8
	BWA/HC	567 / 700	10.9 / 10.4	2.6 / 2.6	88.2 / 89.1
	GSNAP/UG	633 / 785	13.4 / 11.0	2.2 / 3.1	86.1 / 88.4
	GSNAP/HC	579 / 686	10.7 / 10.5	3.1 / 2.6	88.1 / 89.1
	MATE-CLEVER	758 / 743	9.8 / 7.9	7.4 / 10.4	86.5 / 85.3
50–69	BWA/UG	0 / 172	– / 12.8	– / 4.1	– / 86.6
	BWA/HC	127 / 207	18.9 / 15.9	2.4 / 4.8	79.5 / 83.1
	GSNAP/UG	31 / 43	6.5 / 9.3	3.2 / 0.0	90.3 / 90.7
	GSNAP/HC	128 / 180	18.0 / 13.9	1.6 / 2.8	81.2 / 85.6
	MATE-CLEVER	236 / 225	14.0 / 9.3	11.0 / 8.9	78.8 / 86.2
70–99	BWA/UG	0 / 55	– / 14.5	– / 5.5	– / 83.6
	BWA/HC	81 / 117	14.8 / 13.7	2.5 / 1.7	85.2 / 86.3
	GSNAP/UG	0 / 0	– / –	– / –	– / –
	GSNAP/HC	80 / 115	17.5 / 12.2	1.2 / 0.9	82.5 / 87.8
	MATE-CLEVER	135 / 154	11.9 / 5.8	3.7 / 5.2	85.2 / 90.3
100–149	BWA/UG	0 / 1	– / 0.0	– / 0.0	– / 100.0
	BWA/HC	28 / 72	14.3 / 13.9	0.0 / 0.0	85.7 / 86.1
	GSNAP/UG	0 / 0	– / –	– / –	– / –
	GSNAP/HC	33 / 73	18.2 / 15.1	0.0 / 0.0	81.8 / 84.9
	MATE-CLEVER	100 / 107	15.0 / 12.1	2.0 / 0.9	84.0 / 86.9
150–199	BWA/UG	0 / 0	– / –	– / –	– / –
	BWA/HC	0 / 11	– / 0.0	– / 9.1	– / 90.9
	GSNAP/UG	0 / 0	– / –	– / –	– / –
	GSNAP/HC	2 / 11	100.0 / 18.2	0.0 / 9.1	0.0 / 72.7
	MATE-CLEVER	39 / 51	7.7 / 7.8	7.7 / 9.8	92.3 / 84.3

not an option, as precision on heterozygous deletions suffers quite substantially, achieving only 50–60%. In this size range, MATE-CLEVER seemingly is the only sound option that is available among the typing pipelines evaluated.

7.7   DISCUSSION

In this chapter, we review the current state of the art about calling deletions of length 30–150 bp from NGS data. As deletions in this length range pose extraordinary computational and statistical challenges, they have been referred to as *"twilight zone deletions"*. Recent approaches [9, 10, MATE-/CLEVER], however, have pointed out novel and successful ways to discover twilight zone deletions at both good recall and high precision. Moreover, it was described in Reference 10 how to reliably

TABLE 7.5 Genotyping Performance for Heterozygous Calls

Deletion length	Tool	True Annot. → ↓ # of Calls ↓	Absent *Wrong call*	Heterozyg. *Correct*	Homozyg. *Wrong type*
10–29	BWA/UG	6,188 / 10,856	8.4 / 10.8	85.5 / 84.9	7.7 / 6.0
	BWA/HC	10,182 / 12,464	11.6 / 10.7	80.5 / 80.6	10.1 / 11.3
	GSNAP/UG	7,748 / 9,848	7.9 / 7.7	86.2 / 86.7	7.5 / 7.1
	GSNAP/HC	10,619 / 12,506	11.6 / 11.3	80.3 / 80.0	10.4 / 11.2
	MATE-CLEVER	1,022 / 333	7.0 / 3.0	61.7 / 86.2	37.1 / 12.9
30–49	BWA / UG	119 / 971	7.6 / 9.6	87.4 / 86.2	9.2 / 6.5
	BWA/HC	961 / 1456	19.6 / 10.4	72.3 / 78.4	11.1 / 15.0
	GSNAP/UG	728 / 1,024	22.9 / 12.3	71.2 / 83.4	8.7 / 6.2
	GSNAP/HC	1,122 / 1,454	22.1 / 11.2	69.5 / 77.8	11.1 / 14.9
	MATE-CLEVER	877 / 597	10.6 / 5.9	80.8 / 89.9	10.9 / 5.4
50–69	BWA/UG	0 / 177	– / 16.4	– / 79.7	– / 5.1
	BWA/HC	262 / 438	32.8 / 14.4	53.4 / 69.4	15.6 / 18.9
	GSNAP/UG	43 / 63	32.6 / 15.9	67.4 / 77.8	0.0 / 6.3
	GSNAP/HC	315 / 410	32.1 / 18.3	54.6 / 66.6	15.6 / 18.5
	MATE-CLEVER	313 / 233	17.6 / 6.0	75.4 / 91.4	7.3 / 3.9
70–99	BWA/UG	0 / 31	– / 16.1	– / 80.6	– / 6.5
	BWA/HC	148 / 190	38.5 / 16.3	52.0 / 68.9	13.5 / 20.5
	GSNAP/UG	0 / 0	– / –	– / –	– / –
	GSNAP/HC	182 / 222	39.0 / 20.7	51.1 / 65.3	12.1 / 18.5
	MATE-CLEVER	169 / 163	8.3 / 4.9	85.2 / 92.0	8.3 / 4.3
100–149	BWA/UG	0 / 0	– / –	– / –	– / –
	BWA/HC	78 / 90	56.4 / 15.6	25.6 / 56.7	19.2 / 35.6
	GSNAP/UG	0 / 0	– / –	– / –	– / –
	GSNAP/HC	100 / 117	59.0 / 17.1	31.0 / 59.8	11.0 / 28.2
	MATE-CLEVER	92 / 101	14.1 / 12.9	80.4 / 86.1	5.4 / 2.0
150–199	BWA/UG	0 / 0	– / –	– / –	– / –
	BWA/HC	18 / 29	88.9 / 31.0	5.6 / 51.7	11.1 / 20.7
	GSNAP/UG	0 / 0	– / –	– / –	– / –
	GSNAP/HC	26 / 20	92.3 / 25.0	7.7 / 50.0	3.8 / 30.0
	MATE-CLEVER	40 / 52	20.0 / 9.6	75.0 / 86.5	5.0 / 3.8

genotype twilight zone deletions, which we re-visit here in detail. In addition to those novel strategies, several well-maintained SV calling tools have constantly undergone improvements. Thereby, they have grown into methods by which one can at least make a good amount of calls in partial areas of the twilight zone (7, 8, 14, 30, 31). As many of those callsets are complementary, combining MATE-/CLEVER with a reasonable selection of other tools, where we favor PINDEL, and on MiSeq data BWA-MEM in particular, should lead to successful twilight zone deletion calling pipelines. In essence, we consider the discovery of twilight zone deletions a resolved issue, at least when operating on carefully prepared sequencing libraries with reasonably small standard deviations.

Here, we sketch the novel, successful strategies and evaluate a large range of tools, some of which have helped considerably shedding more light on the

NGS twilight zone of deletions. In brief, while some advanced internal segment size-based approaches "tackle" the twilight zone from above (among which (7–9)), alignment-based approaches, both regular and split-read oriented (see References 14, 18, 34), tackle the twilight zone from below. A general disadvantage of internal segment size-based approaches is that breakpoints predicted are rather inaccurate (see differences in *relaxed* and *strict* precision statistics). Hybrid approaches address this issue, and therefore enjoy the advantages of both internal segment size-based approaches in terms of being able to also call larger deletions and alignment-based approaches in terms of highly accurate breakpoint predictions.

7.7.1 HiSeq

Among all approaches evaluated, only CLEVER (9) and its hybrid version MATE-CLEVER (10) deliver comprehensive deletion callsets that span the entire size range of the twilight zone (30–150 bp). Among the internal segment size-based callers, VariationHunter delivers good callsets for deletions of length at least 50 bp, and Breakdancer for deletions of length at least 70 bp. The high recall, however, comes at the cost of breakpoint accuracy, which can be considerably improved in both cases. In this aspect, CLEVER takes the lead among the internal segment size-based approaches, as its strict precision rates are clearly superior.

Among the hybrid approaches tested, only MATE-CLEVER makes contributions in the twilight zone. DELLY clearly focuses on longer deletions and achieves very favorable performance rates for deletions longer than 200 bp (data not shown).

Among the (split-)alignment-based approaches, PINDEL is best in terms of an overall assessment, achieving excellent performance rates for calls up to 50 bp and also discovering non-negligible amounts beyond 50 bp. GSNAP and Stampy also deliver substantial amounts of excellent predictions, where, however, the recall of both tools becomes negligible for deletions of 40–50 bp and longer.

7.7.1.1 *Heterozygous deletions 30–50 bp*

A major challenge that has remained when processing HiSeq data is heterozygous deletions of length 30–50 bp. In this case, CLEVER's recall is best (61%). Still, novel solutions are yet to be developed for this class of calls based on HiSeq experiments. The combination of heterozygosity and size range seemingly has (partially) remained a weak point of all classes of deletion discovery approaches, which can not (yet) entirely be overcome when making use of only HiSeq experiments.

7.7.1.2 *Genotyping*

In summary, one can recommend GSNAP-(UG/HC) for deletions of length up to 30 bp and MATE-CLEVER for all deletions longer than 30 bp, with GSNAP-based genotyping pipelines also achieving competitive performance rates for homozygous deletions of 30–50 bp. Since one usually has little information on relative amounts of heterozygous and homozygous deletions, MATE-CLEVER is the superior overall choice for deletions longer than 30 bp, achieving an overall genotyping precision of greater than 90%, as the only tool on 30+ bp deletions.

7.7.2 MiSeq

MiSeq is a recent sequencing technology that allows for reads of length 250 bp and longer. As such, it also holds major promises in terms of spotting longer deletions by (split-)alignment-based approaches. In fact, MiSeq experiments do not suffer from the "blind spot" of 30–50 bp heterozygous deletions, which still applies for HiSeq experiments. In particular BWA-MEM (18) and partially also GSNAP (30) profit from the advance in sequencing technology, achieving recall above 80% (BWA-MEM even 85.7%) at reliable precision rates (BWA-MEM: 86.2/79.1% relaxed/strict precision). If MiSeq technology is available and the throughput is sufficient for the application at hand, clearly these are the methods of choice.

For deletions of 50 bp and longer, CLEVER and MATE-CLEVER are again the methods of choice, achieving highest performance rates throughout, with VariationHunter as the only rival, whose calls, however, are highly inaccurate (Variation Hunter, strict precision <10%).

Usage of MiSeq data leads to losses in performance for internal segment size-based approaches, and hence also for hybrid approaches, which is most likely due to the increased internal segment length. For (split-)alignment-based methods it often leads to a considerable increase in recall, while also leading to losses in precision here.

In summary, usage of MiSeq data in twilight zone deletion discovery has clear advantages when discovering heterozygous deletions of length 30–50 bp. In all other classes of calls, its usage leads to more "aggressive" twilight zone deletion calling when used in combination with (split-)alignment-based approaches, while HiSeq data-based callsets tend to contain less false positives.

7.7.2.1 Genotyping
In the genotyping methods presented, MiSeq data usually leads to losses in recall while leading to enhanced precision for MATE-CLEVER. The explanation is that MATE-CLEVER is a hybrid approach. While MATE-CLEVER achieves best genotyping precision in general, for deletions longer than 30 bp, the HC-based tool combinations lead to more "aggressive" genotypers, with less genotyping precision, and also with a clear increase in calls that are deemed being "typable" for deletions of length 30–50 bp.

In summary, also on MiSeq data, MATE-CLEVER is the only approach that reliably types twilight zone deletion calls across all size ranges. For deletions from the lower ranges of the twilight zone, alignment-based methods in combination with UG and/or HC can be helpful to generate callsets of "higher sensitivity."

7.7.3 Conclusion

7.7.3.1 HiSeq
As a general advice for HiSeq data, one can consider PINDEL the strongest approach for calling deletions of length 10–30 bp. For calls between 30 and 50 bp, a combination of PINDEL, CLEVER, and/or MATE-CLEVER is recommended, which together should yield comprehensive callsets for homozygous deletions, but leave room for improvements for heterozygous deletions. From 50 bp onwards, when focusing on a single tool, MATE-/CLEVER are the methods of choice.

For genotyping, one can recommend usage of alignment-based methods, such as BWA-MEM or GSNAP, in combination with UG and/or HC for deletions of length 10–30 bp (and shorter, data not shown), while MATE-CLEVER is the method of choice for deletions of length longer than 30 bp.

On a side remark, these insights lead to the selection of those tools in the most recent Genome of the Netherlands Project (1), which delivers callsets for deletions in these size ranges, which are decisively more comprehensive than those of other projects, such as the 1000 Genomes Project (2).

7.7.3.2 MiSeq A general advice for MiSeq data is to make use of PINDEL for calls of 10–30 bp, BWA-MEM for calls of 30–50 bp, and MATE-/CLEVER for deletions longer than 50 bp. Still, also MiSeq data analysis technology leaves room for improvements: heterozygous deletions of length 70+ bp can still be considered to be extremely challenging, with no tool operating at a recall rate of 60% or higher, without making drastical sacrifices in terms of precision (note that VariationHunter sometimes achieves relatively high recall, without, however, achieving "operable" precision).

For genotyping, again BWA-MEM and/or GSNAP, postprocessed by UG and/or HC, are the methods of choice for deletions shorter than 30 bp, while MATE-CLEVER is the method of choice for deletions of 30+ bp. For deletions of 30–50 bp, combining GSNAP with HC can make a valuable contribution, beyond using MATE-CLEVER.

7.8 AVAILABILITY

See

http://homepages.cwi.nl/~as/twilightzone/Snakefile

for a snakefile containing all command line instructions that were necessary to generate results.

ACKNOWLEDGMENTS

A. Schönhuth acknowledges funding by the Nederlandse Wetenschappelijke Organisatie (NWO) through Vidi grant 639.072.309.

REFERENCES

1. The Genome of the Netherlands Consortium. Whole-genome sequence variation, population structure and demographic history of the Dutch population. Nat Genet 2014;46(8):818–825.
2. The 1000 Genomes Project Consortium. A map of human genome variation from population-scale sequencing. Nature 2010;467(7319):1061–1073.

3. The International Cancer Genome Consortium. International network of cancer genome projects. Nature 2010;464(7291):993–998.

4. McKenna A, Hanna M, Banks E, Sivachenko A, Cibulskis K, Kernytsky A, Garimella K, Altshuler D, Gabriel S, Daly M, DePristo MA. The Genome Analysis Toolkit: a MapReduce framework for analyzing next-generation DNA sequencing data. Genome Res 2010;20(9):1297–1303.

5. Alkan C, Coe BP, Eichler EE. Genome structural variation discovery and genotyping. Nat Rev Genet 2011;12(5):363–376.

6. Medvedev P, Stanciu M, Brudno M. Computational methods for discovering structural variation with next-generation sequencing. Nat Methods 2009;6(11s):S13–S20.

7. Chen K, Wallis JW, McLellan MD, Larson DE, Kalicki JM et al. BreakDancer: an algorithm for high-resolution mapping of genomic structural variation. Nat Methods 2009;6(9):677–681.

8. Hormozdiari F, Alkan C, Eichler EE, Sahinalp SC. Combinatorial algorithms for structural variation detection in high-throughput sequenced genomes. Genome Res 2009;19(7):1270–1278.

9. Marschall T, Costa IG, Canzar S, Bauer M, Klau GW, Schliep A, Schönhuth A. CLEVER: clique-enumerating variant finder. Bioinformatics 2012;28(22):2875–2882.

10. Marschall T, Hajirasouliha I, Schönhuth A. MATE-CLEVER: mendelian-inheritance-aware discovery and genotyping of midsize and long indels. Bioinformatics 2013;29(24):3143–3150.

11. Rausch T, Zichner T, Schlattl A, Stütz AM, Benes V, Korbel JO. DELLY: structural variant discovery by integrated paired-end and split-read analysis. Bioinformatics 2012;28(18):i333–i339.

12. Sindi S, Helman E, Bashir A, Raphael BJ. A geometric approach for classification and comparison of structural variants. Bioinformatics 2009;25(12):i222–i230.

13. Sindi S, Önal S, Peng LC, Wu H-T, Raphael BJ. An integrative probabilistic model for identification of structural variation in sequencing data. Genome Biol 2012;13:R22.

14. Ye K, Schulz MH, Long Q, Apweiler R, Ning Z. Pindel: a pattern growth approach to detect break points of large deletions and medium sized insertions from paired-end short reads. Bioinformatics 2009;25(21):2865–2871.

15. Lee S, Hormozdiari F, Alkan C, Brudno M. MoDIL: detecting small indels from clone-end sequencing with mixtures of distributions. Nat Methods 2009;6(7):473–474.

16. Levy S, Sutton G, Ng PC, Feuk L, Halpern AL et al. The diploid genome sequence of an individual human. PLoS Biol 2007;5(10):e254.

17. Earl D, Bradnam K, St. John J, Darling A, Lin D et al. Assemblathon 1: a competitive assessment of de novo short read assembly methods. Genome Res 2011;21:2224–2241.

18. Li H, Durbin R. Fast and accurate short read alignment with Burrows-Wheeler transform. Bioinformatics 2009;25(14):1754–1760.

19. Li H, Ruan J, Durbin R. Mapping short DNA sequencing reads and calling variants using mapping quality scores. Genome Res 2008;18(11):1851–1858.

20. Korbel JO, Abyzov A, Mu X, Carriero N, Cayting P et al. PEMer: a computational framework with simulation-based error models for inferring genomic structural variants from massive paired-end sequencing data. Genome Biol 2009;10(2):R23.

21. Quinlan AR, Clark RA, Sokolova S, Leibowitz ML, Zhang Y et al. Genome-wide mapping and assembly of structural variant breakpoints in the mouse genome. Genome Res 2010;20(5):623–635.

22. Treangen TJ, Salzberg SL. Repetitive DNA and next-generation sequencing: computational challenges and solutions. Nat Rev Genet 2012;13:557–567.

23. Emde A-K, Schulz MH, Weese D, Sun R, Vingron M, Kalscheuer VM, Haas SA, Reinert K. Detecting genomic indel variants with exact breakpoints in single- and paired-end sequencing data using SplazerS. Bioinformatics 2012;28(5):619–627.

24. Schroeder J, Hsu A, Boyle SE, MacIntyre G, Cmero M, Tothill RW, Johnstone RW, Shackleton M, Papenfuss AT. Socrates: identification of genomic rearrangements in tumour genomes by re-aligning soft clipped reads. Bioinformatics 2014. DOI: 10.1093/bioinformatics/btt767.

25. Zhang ZD, Du J, Lam H, Abyzov A, Urban AE, Snyder M, Gerstein M. Identification of genomic indels and structural variations using split reads. BMC Genomics 2011;12:375.

26. Jiang Y, Wang Y, Brudno M. Prism: pair-read informed split-read mapping for base-pair level detection of insertion, deletion and structural variants. Bioinformatics 2012;28(20):2576–2583.

27. Zhang J, Wang J, Wu Y. An improved approach for accurate and efficient calling of structural variations with low-coverage sequence data. BMC Bioinf 2012;13 Suppl 6:S6.

28. Toepfer A, Marschall T, Bull RA, Luciani F, Schönhuth A, Beerenwinkel N. Viral quasispecies assembly via maximal clique enumeration. PLoS Comput Biol 2014;10(3):e1003515.

29. Marschall T, Schönhuth A. Sensitive long-indel-aware alignment of sequencing reads. Technical report; 2013. arXiv:1303.3520.

30. Wu TD, Nacu S. Fast and SNP-tolerant detection of complex variants and splicing in short reads. Bioinformatics 2010;26:873–881.

31. Wu TD, Watanabe CK. GMAP: a genomic mapping and alignment program for mRNA and EST sequences. Bioinformatics 2005;21:1859–1875.

32. Langmead B, Salzberg SL. Fast gapped-read alignment with Bowtie 2. Nat Methods 2012;9(4):357–359.

33. Lunter G, Goodson M. Stampy: a statistical algorithm for sensitive and fast mapping of illumina sequence reads. Genome Res 2011;21(6):936–939.

34. Langmead B, Trapnell C, Pop M, Salzberg SL. Ultrafast and memory-efficient alignment of short DNA sequences to the human genome. Genome Biol 2009;10(3):R25.

8

COMPUTATIONAL APPROACHES FOR FINDING LONG INSERTIONS AND DELETIONS WITH NGS DATA

JIN ZHANG

McDonnell Genome Institute, Washington University in St. Louis, MO, USA

CHONG CHU AND YUFENG WU

Department of Computer Science & Engineering, University of Connecticut, Storrs, CT, USA

8.1 BACKGROUND

Structural variation (SV), or genomic structural variation, is the variation in the structure of an organism's genome. When there is a reference genome of a species, a variation is usually defined relative to the reference genome. Among different genomes of one species, there can be many types of SVs, such as deletions, insertions, duplications, copy number variants, inversions, and translocations. While many structural variations do not affect disease susceptibility, some other structural variations can be associated with complex diseases, such as cancer. Great efforts have been taken to map and categorize SVs in large-scale projects, such as the 1000 Genomes Project (1) and The Cancer Genome Atlas (2). Many computational tools of calling SVs have been developed and applied to these projects.

In this chapter, we will cover some aspects of the computational methods that call certain types of SVs (namely long insertions and deletions) using next-generation sequencing (NGS). It is known that insertions and deletions are among the more common SVs in the genome. Moreover, long indels may potentially have significant impact on some phenotypes. NGS has become the standard technology in discovering

Computational Methods for Next Generation Sequencing Data Analysis, First Edition.
Edited by Ion I. Măndoiu and Alexander Zelikovsky.
© 2016 John Wiley & Sons, Inc. Published 2016 by John Wiley & Sons, Inc.
Companion website: www.wiley.com/go/Mandoiu/NextGenerationSequencing

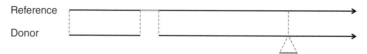

Figure 8.1 Donor genome with one long deletion and insertion. The figure shows that it has been aligned to reference genome.

SVs recently. High-throughput and low-cost NGS technologies make it possible to discover SVs from thousands of samples. On the other hand, there are also challenges for calling SVs using NGS. First, NGS sequence reads are rather short and may contain sequencing errors and artifacts. Second, these sequence reads might be generated not perfectly uniformly and coverage may vary at many regions of the genome. Third, the volume of sequence reads is often very large.

One approach is assembling the sequence reads to obtain the complete donor (or alternative) genome. If the donor genome is completely known, one may simply compare the donor genome with the reference genome (which is available for many species) to find the SVs. Figure 8.1 shows an example of a simulated donor genome which contains one long deletion and one long insertion. The simulated reference genome is 1 million base pairs (bp) long. The donor genome does not have the sequences from 250,001 to 251,000 (the portion marked with light grey), and it has a segment that does not exist in the reference genome (the portion marked with dark grey). So compared to the reference genome, the donor genome has a long deletion and a long insertion.

In practice, however, it is often difficult to assemble a perfect donor whole genome using NGS reads. A main issue for assembly using NGS reads is the short reads length. Also, the distribution of reads is usually uneven, which may lead to low coverage on some regions of the genome. The noise contained in the reads also post challenges for assembly. Moreover, the donor genome itself can be very repetitive.

An alternative strategy for finding SVs from sequence reads is calling SVs directly from the reads without performing assembly. There are many existing computational tools that use NGS data to call SVs. Most of these tools fall in one of the four categories of methods of calling structural variations, namely: (i) assembly (AS), (ii) read depth (RD), (iii) read pair (RP), and (iv) split read (SP). In the following, we will describe the main idea of each of these methods in details. We will introduce some tools as examples to illustrate these approaches, by primarily concerning some techniques in calling long deletions and insertions (especially deletions). Some of these techniques may be applicable to other types of SVs. We note that certain kinds of SVs may need other techniques not described in this chapter.

Each category of methods mentioned earlier has its advantages and disadvantages in calling different SVs. It is unlikely that a tool from one category would be able to call all the SVs from the sequence reads. One category of method may be intrinsically not able to find certain SVs. So instead of using tools of one category, it might be sensible to combine results of tools from different categories in calling SVs, as The 1000 Genomes Project does (3). There are different ways of running multiple tools and generating a list of candidates. A straightforward way is running several tools on the same data with various parameters, and then taking the intersection, union, or some other subsets of the output of these tools as the final result. The overall precision is now the percentage of true positives (TPs) in the final list, and the overall sensitivity is the percentage of TPs out of all the TPs. For example, running each tool with relatively good accuracy (by using more stringent parameters) may lead to good overall accuracy when using the union as the final list. Note that the overall accuracy might be somewhat lower than using a single tool because different tools may share some true SVs, and false positives (FP) of each methods can be different. But by using multiple tools together, the overall sensitivity can be much higher than using a single tool, since there can be many TPs that are picked up only by one or a small number of tools. There are other approaches for running multiple methods, each with their own pros and cons. Overall, combining results from multiple methods is usually an *ad hoc* process.

Since tools from different categories all have strength or weakness, a natural approach is developing a tool that *combines* approaches from several categories. Such tool may explicitly use different types of information about SVs contained in the reads. Indeed, developing such combined approach is the current trend in developing SV calling tools. Note that this combined approach is different from taking the intersection of the outputs of two or more methods that use single signature. The signal of one type (e.g., RP) of an SV can be very small so that it may lead to many FPs. An SV calling method using this single approach may fail to include this SV in the final list to obtain reasonable FDR rate. Even if we do have two very long lists of SV candidates with two different categories, by chance, there could be many FPs that are shared in both lists. Instead of looking at the two types of signals separately, we can examine the two signals together, which can be stronger. Thus, a combined method may have higher accuracy and sensitivity. Also note that the combined approach, if devised properly, can lead to more efficient algorithms. We will introduce one such combined method in Section 8.2.4.

In this chapter, we first describe the main approaches from different categories, including the combined approaches. We then introduce some ongoing projects that SV calling methods have been applied to. Finally, we provide a conclusion and discussion for future research directions.

8.2   METHODS

We now describe the technical ideas of various approaches. Each approach relies on some kind of signal contained in the reads that indicate the occurrence of SVs. We call such signals the *signatures* of SVs.

8.2.1   Signatures of Long Indels in Sequence Reads

In the problem of SV calling, we care more about whether there are some SVs in the donor genome and the coordinates of true SVs. Coordinates are measured according to the positions of the bases of reference genome. For example, the coordinates for the long deletion in Figure 8.1 is *reference* : 250,001–251,000, and the long insertion is between *reference* : 750,000 and *reference* : 750,001. Note that the NGS reads are

Figure 8.2   A paired-end read. Different from Figure 8.1, which uses one line to represent a genome, in this figure we use two lines to represent a segment of a double-stranded genome. A paired-end read is generated from this segment of DNA with each end from 5′ of each strand. The length of the segment is called insert size.

generated from the donor genome that contains SVs, and the reads from the regions of SVs carry information about the SVs. When mapped to the reference genome, the alignments of these reads could have some characteristics different from the reads from regions without SVs. We call these characteristics as signatures. The categories of methods we mentioned in the last section utilize various signatures of SVs. We focus on signatures of long deletions and long insertions in this section. For more signatures on other types of SVs, the reader can refer to reviews in the literature, for example, Reference 4.

To illustrate the concepts of signatures, we generate some small-scale simulated NGS data sets. We only generate one type of NGS reads called paired-end reads. In Figure 8.2, a paired-end read is generated with two ends from two strands of a segment of the donor genome. The length of the segment is called insert size. Since we do not have the sequences of the entire segment from the donor genome, we do not know the insert size of the paired-end read. But when the read is properly mapped, we use the distance of the two ends on the reference genome as the insert size. The underlying assumption is that the donor genome has similar structure as the reference genome within this region. For resequencing, the donor genome is usually very similar to the reference genome. So for most of the paired-end reads, the insert sizes calculated from read mapping are reasonably accurate. Usually, paired-end reads from the same library has similar library insert size. So we assume that the standard deviation of insert sizes is not very large. In our simulation, the simulated reference genome is a random sequence of A, C, G, and T's, with each letter equally likely. The donor genome is generated from the reference genome by removing the bases from 250,001 to 251,000 and inserting a random sequence of length 1000 between bases *reference* : 750,000 and *reference* : 750,001. The setting is illustrated in Figure 8.1.

We generate paired-end reads from the donor genome with coverage 30× (coverage is the average number of reads covering one base). In total, 30× coverage of data is also generated from the reference genome. The reads are put together to simulate that the sample is heterozygous with one copy of genome having the SVs and the other copy of genome being the same as the reference genome. The data set is then 60× coverage. Note that in practice, the two sequences may have different coverage. We use this simplified simulation to illustrate the basic concept. The reads of the data set are generated with the read simulation tool wgism (https://github.com/lh3/wgsim) with insert size, standard deviation, and read length set to 500, 50, and 100 respectively. Error rates are chosen to be the default. The data set is very ideal compared with the real situation, but is suffice to show the signatures. The reads are mapped by BWA (5) using default parameters. We can view the alignments using IGV (6).

Figure 8.3 shows the region of the long deletion and Figure 8.4 shows the region of the long insertion.

From Figure 8.3, we can see that the coverage at reference: 250,001–251,000 (the deletion) is about one half of that at other coordinates (without SVs). This is because only one copy of genome of the sample has the sequence from 250,001–251,000, and we sequenced it with 30×. The regions without SVs are sequenced with 60× coverage, 30× for each copy. So one signature of long deletion when using NGS reads is that the coverage of the deletion site is smaller compared with regions without SVs. Ideally, for organisms with two copies of genomes, if the deletion is heterozygous (only in one copy of genome) then the coverage is about one half. If the deletion is homogeneous (in both copies of genomes), then the coverage is about 0. This is the read depth (RD) signature. For real data sets, the RD signature can suffer from several aspects. Reads may be generated not as even as the simulated data set. Coverage at different locations of the same copy of genome can fluctuate. Coverage from the two copies of genome at the same location can be different. Sequencing errors and mapping errors may add false-positive reads to the region and remove true-positive reads from the region. Also, not every data set has high coverage. Therefore, it may happen that the coverage drop at the location of deletion cannot be reliably detected from the background fluctuation of coverage. Moreover, the number of copies of genomes is not necessarily two. Some organisms have more copies of genomes. Some

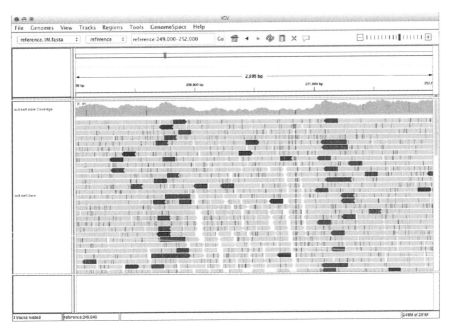

Figure 8.3   IGV view of an example of long deletion. The deletion is from 250,001 to 251,000. We can see that the read coverage in the region of deletion is lower, and there are many discordant paired-end reads with very large insert sizes. Please see www.wiley.com/go/Mandoiu/NextGenerationSequencing for a color version of this figure.

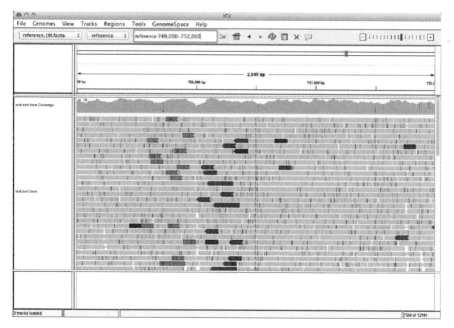

Figure 8.4　IGV view of an example of long insertion. The insertion is between 750,000 and 750,001. Please see www.wiley.com/go/Mandoiu/NextGenerationSequencing for a color version of this figure.

organisms have two copies of genomes for normal tissues, but genomes from some samples, for example, tumor cells, can be highly altered and thus the copy number may not be two. So when we see some region with lower coverage, instead of calling it as deletion immediately, more work has to be done to filter out the false positives. We will give an example of an SV calling tool that uses this signature in the next section.

In Figure 8.3, some reads are marked with colors. The ones with red color are very important for the purpose of calling deletions. They are marked with red because the insert size of these reads is significantly longer than expected. As mentioned earlier, we expect the insert size of paired-end read to be similar to the library insert size. The simulated data sets use the mean insert size of 500, but those reads marked with red apparently have much longer insert sizes. The values are longer than 1000, since they are out of the region of 250,001–251,000. Recall that the insert size of these reads is about 1500, which is the size of the deletion plus the mean insert size. The reason for the longer insert sizes is illustrated in Figure 8.5. The donor genome is shown both as split and consecutive sequences. A paired-end read is generated from the donor genome, which is shown at the bottom of the figure. The read is from a segment of sequence that spans the breakpoint. The length of the segment follows from an assumed normal distribution whose mean is the library insert size. But we only know the true length of the segment in simulation. In practice, we only have the sequences of the reads. So we need to map the read to the reference genome to get the estimate of

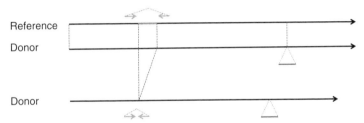

Figure 8.5 Insert size of an encompassing paired-end read is enlarged after mapping because of the deletion. When generated from the donor genome, the insert size of an encompassing paired-end read is normal (bottom). But because it encompasses a deletion, when mapped onto the reference genome, the insert size is enlarged by the size of the deletion (top).

the insert size. The alignment of the read is shown at the top of the figure. The insert size is enlarged because the reference has a segment of sequence (the red portion) that the donor does not have. So the second signature of long deletion is the existence of paired-end reads with insert size significantly larger than library insert size. In other words, a long deletion may lead to reads with significantly larger than library mean. If we find one or more encompassing reads covering the same location with much larger insert sizes than expected, this can be an indication of a long deletion. This is the read pair (RP) signature. Note that the insert sizes of different reads generated in the same library are different. By chance there will be many reads that are having an insert size that is significantly large, since the sheer number of reads from NGS data can be as large as hundreds of millions. So we cannot assert that when there is a paired-end read mapped with a very large insert size, then there is a deletion. Smaller deviation of insert size and higher coverage will help when calling deletions using the RP signature. But in practice, even if the average coverage of an NGS data set is large enough, there is no guarantee that enough encompassing reads are generated and mapped properly for a specific deletion. By chance, there would be many reads that suggest that there might be deletions, especially not very large ones. It is not easy to distinguish between a true positive of this type and the false positives caused by chance.

The RD and RP signatures discussed earlier have a disadvantage of not being able to detect the *exact* breakpoints of deletions. For example, in Figure 8.3, we can see that the read coverage does not change abruptly at the breakpoints even using an ideal data set with a high read coverage of 60×. We can also see that most of the red reads are encompassing the breakpoints of the deletion, but the region that is encompassed is longer than the size of the deletion. We can roughly see where the deletion is, but not able to tell the exact breakpoints. Also there might be reads like the red ones mapped within the region of the deletion. These reads can further blur the positions of the breakpoints. There is another signature that can be used not only to detect the existence of deletions but also to get the exact breakpoints. A spanning read that contains the breakpoints of a deletion consists of two segments from the reference genome, and thus can be mapped as a split read with two separate portions. This is the split-read (SP) signature. Figure 8.6 shows a paired-end read generated from the donor genome, with one end of the pair spanning the deletion junction. The

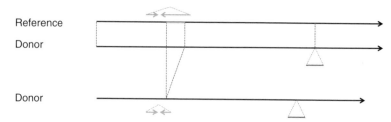

Figure 8.6 Signature of paired-end split read for long deletion. One end of the read is generated spanning the breakpoint of a long deletion on the donor genome (bottom). The sequence of the end does not match a consecutive sequence in the reference genome. It has to be split in two separate segments, with each mapped onto a flanking region of the deletion on the reference genome.

other end of the pair is mapped as a whole read onto the reference genome. It is called the anchor. The spanning end cannot be mapped onto the reference genome as a whole read. It is composed of two flanking segments that are separated on the reference genome. It is called spanning read or split read. Currently, popular NGS sequence reads aligners, such as BWA (5) and Bowtie (7) are able to map the anchor if the read quality is not too bad and the sequence is not too repetitive. The spanning read cannot be mapped as a whole read. Split-read SV calling tools map these reads as split reads first and use this signature to call SVs. For many reads that are not able to be mapped as whole reads, some of them may contain more errors and some other reads may be artifacts. By chance, some of these reads could be mapped as split reads. This may lead to many false positives. Also note that genomes like the human genome can be repetitive, and splitting short NGS reads into even shorter segments may lead to multiple hits. Also different ways of splitting a read can lead to multiple hits. For the true split reads, it can be difficult to pick out the correct split mappings.

A main difference between calling long deletion and long insertion is that the inserted sequence of long insertions is not known. An inserted sequence can be from another location of the genome so that it is in the reference genome. It can also be a new sequence that is not in the reference genome. Either way, the reads generated from the inserted sequence are not mapped to the location of the insertion. Thus, there is no read coverage change around the breakpoints of insertions, for example, Figure 8.4. Also for large insertions, it is not likely that there are encompassing reads because library insert size can be much smaller compared to the sizes of the insertions. Paired-end reads with one end outside the inserted sequences may have this end mapped near the breakpoints of insertions. The other end can be contained in the inserted sequence or can span one breakpoint of the insertion. The two cases are illustrated in Figure 8.7, which only shows one read for each case. With reasonably high coverage, both types of reads may appear at both sides of the breakpoint of a long insertion. The inserted sequence may be assembled using the split ends and the unmapped ends. Note that because of sequencing errors and artifacts, by chance there would be many reads that are mapped having the signature that are the same as the signature of long insertions.

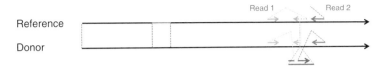

Figure 8.7 Signature of paired-end reads for long insertion. The bottom shows where the sequences of two signature reads come from. The top shows the signatures they have when mapped onto the reference genome.

Next, we describe several tools that only use single signatures and then show some examples of tools with combined signatures.

8.2.2 Methods for Discovering Long Indels without Exact Breakpoints

Starting from this section, we will give several examples of SV calling tools that implemented certain signatures discussed in the previous section. In this section, we will cover an RD method called CNVnator (8), which calls copy number variations (CNV). A long deletion is a loss of copy and a long insertion is a gain of copy. We will also cover BreakDancer (9), an RP method. There are more RD methods, for example, References 10–12 and more RP methods, for example, References 13–15.

8.2.2.1 *CNVnator* CNVnator is one of the tools that use RD to discover CNVs from family and population genome sequencing. The main idea of this approach is based on the observation: RD signal for CNVs is different from that of normal regions, which is also the main observation of all RD-based approaches. One of the techniques CNVnator uses is called mean-shift approach. This approach is borrowed from imagine processing technique, and it presents an elegant way to locate the density maxima without having to estimate the density directly, which is an iterative procedure that shifts each data point to the density maxima along the mean-shift vector. CNVnator combines the established mean-shift approach with additional refinements to extend the range of discovered CNVs, which are obtained from the initial RD. Detailed implementation of CNVnator include three steps:

Firstly, CNVnator calculates the unique RD for each bin and correct GC-bias. It maps reads to genome (which is divided into consecutive nonoverlapping bins), and if a read (or a read pair with concordant mapping distance) can map to more than one position with equal mapping score, it randomly picks one (or a pair) to get unique RD. Then GC-bias is corrected to eliminate the correlation of RD signal with GC content.

The second step is mean-shift-based segment partition and signal merging, which is the key step of the approach. As mentioned earlier, the mean-shift technique is borrowed from imagine processing to detect edges in computer vision. And using mean-shift approach to participate genome, each bin is viewed as a point, and mean-shift vector of each point is calculated with the direction pointing to bins with the most similar RD signal. Segment breakpoints are determined where two neighboring vectors direct oppositely but do not point to each other. Then, neighboring segments are merged if their RD signal difference is less than one-quarter of the genomic average.

The last step is CNVs calling. At first, segments with a mean RD signal deviating by at least a quarter from genomic average RD signal are selected, and regions with p-value by a t-test less than 0.05 is called. Then, call additional ones by performing a one-side test that all values of the RD signal within a segment are smaller than the maximum RD signal within the segment.

Experiment results on data from the 1000 Genomes Project (reads length 36) show that CNVnator is able to achieve high sensitivity (86–96%) with false-discovery rate 3–20%, and high resolution in breakpoint discovery (< 200 bp in 90% of cases with high sequencing coverage). However, CNVnator is sensitive to read length, and for constant coverage, as read length increases, less reads will be generated, which will degrade the sensitivity. And also, longer reads are more difficult to be fully mapped around the breakpoints. Thus, it is harder for CNVnator to call small CNVs. Therefore, RD-based approach is useful for CNVs discovery, but still has its limitations.

8.2.2.2   BreakDancer   It is one of the first RP-based computational approaches. It is designed for the genome-wide detection of five types of structural variants, including deletions, insertions, inversions, and intrachromosomal and interchromosomal translocations. It takes paired-end reads as input. If both single-end and paired-end reads are given, only the paired-end ones will be used for further analysis.

First, BreakDancer tallies the anomalous read pairs. Different types of structural variations have different anomalous patterns on paired-end reads, as explainedearlier. According to these different patterns, this algorithm tries to locate a set of anomalous regions and then identify interconnected clusters. Here, different SV types are considered. Each region should have much more anomalous read pairs than expected by chance. This allows BreakDancer to filter some false variants due to sequencing and mapping errors.

A putative structural variant is then considered to derive from a set of regions, each of which interconnects others via several anomalous read pairs. For example, sometimes BreakDancer uses a preset threshold of at least two anomalous read pairs. Compared to average coverages, the more pairs supporting the region(s), the stronger statistical signals are obtained. BreakDancer applies a statistical test, called confidence score estimation, to quantify the underlying error probabilities of each potential structural variant. This test assumes that the genomic locations of each type of structural variants follow a uniform distribution. Thus, for each pair of adjacent sites, the probability of harboring a structural variant follows mixture binomial distributions with each mixture component representing one of the insert types. On the scale of genome-wide, the probability of multiple structural variants locating at a particular site follows a mixture of Poisson distribution.

In practice, BreakDancer suggests to define three types of inserts: long, medium, and short that are measured according to preset thresholds. Then the indel identification process is to find the regions that contain significantly more long inserts and insertions from regions that contain significantly more short inserts via the Poisson test.

Note that the original BreakDancer contains two different packages: BreakDancer-Max focuses on long indels, while BreakDancerMini is mainly for small indels. However, the BreakDancerMini is no longer updated. Thus, BreakDancer is considered as a method specific to long indels.

8.2.3 Methods for Discovering Long Indels with Exact Breakpoints

To call SVs with exact breakpoints, AS and SR methods could be used. *De novo* assembly methods may need higher read coverage than SR method, but the throughput of the NGS technologies is increasing constantly. The sequences of long insertions are already known when called with AS methods. AS methods usually require more computational resources (especially memory). On the other hand, SR methods could be faster and require less memory if implemented properly. We will not describe any particular short reads assembly method since *de novo* short reads assembly is an interesting problem on its own. We will give more detail about an SV calling tools that uses split-read mapping algorithm, called Pindel (16).

8.2.3.1 Pindel It is first introduced in Reference 16, which was an SR method that calls long deletions and medium-sized insertions. The SR method of Pindel is implemented as a pattern growth substring mapping, which can be used to find minimum and maximum unique substrings of a given pattern P in a reference sequence S. Suppose the reference is "ATCAAGTATGCTTAGC" and the pattern P is "ATGCA." First, the locations of all "A"s are searched and stored. In this case, they are 1,4,5,8,14. Then T is searched at the second base of these locations, since T is the second base of the pattern. Keep searching this way, "ATGC" is unique at location 8, but "ATGCA" is not a substring, so "ATG" is the minimum unique substring and "ATGC" is the maximum unique substring.

To call indels, Pindel uses paired-end reads that only have one end mapped as whole read, and maps the unmapped end as split reads using the above method. Instead of mapping split reads to the whole genome, Pindel uses a threshold of maximum event size, that is, mapping 5′ of a split read only within the threshold of distance from the mapped 3′ of a split read. Smaller threshold makes the procedure faster and take less memory. More importantly, imposing a threshold makes the SV calling more accurate, since less false positive (FP) hits are expected in a smaller region. However, there might be a chance that some extremely large SVs are lost, although those SVs may be rare. Also, a cutoff value on number of reads supporting the same SV is used by Pindel so that the final list is more accurate.

The signatures of SVs can be found by applying the procedure of split-read mapping using pattern growth. We have shown the SR signature of long deletions. For medium insertion, the signature is reads with a portion in the middle not mapped, but the segments at 5′ and 3′ are mapped and stitched together. Since its first release, Pindel has been updated: new features include mapping allowing sequencing errors, calling more types of SVs such as translocations and so on.

There are more methods dedicated to find exact breakpoints. SRiC (17) is a split-read method mainly working on longer single reads such as the Sanger and 454 reads. AGE (18) maps an assembled contig onto a reference genome to detect the exact breakpoints of multiple SVs. There are also methods (e.g., CREST (19)) that do not detect the breakpoints themselves but rely on the exact breakpoints provided by local alignments mapped by other tools.

8.2.4 Combined Approaches

We have mentioned previously that there are two types of combined approaches of SV calling. One is combining the results of different tools that use single signature, and the other is combining signatures to call SVs. In this section, we focus on describing computational tools that use combined signatures. We will use a tool called SVseq (20, 21) as an example of an SV calling tool that combine two signatures, namely, split read and read pair. Only the methods of calling long deletions are covered in detail in this section. The two signatures of calling long deletions with split read and read pair are discussed in details in Section 8.2.1.

Intuitively, calling very long deletions (e.g., 1 Mbps) using split-read method is harder than calling shorter deletions (e.g., 50 bps). This is because the split segments of a split read may hit more locations of the reference genome by chance if we need to search for larger region (e.g., 1 Mbps). At the same time, more FP split reads can be mapped to this region. So the signal of a split read is relatively weak. Also, mapping split reads in a larger region can be slower. However, if given some encompassing reads that are supporting the TP split read alignment, then situation may improve. On the other hand, calling shorter deletions using read pair signature can be harder, especially when read coverage is lower. For example, if the insert size of an encompassing read of a deletion of length 60 bps is 70 bps longer than mean library insert size, then this is not significant since the standard deviation is 50 bps: by chance a large portion of reads in a library would have insert size longer than mean library insert size by the magnitude of one or two standard deviations. However, if given some spanning reads that support a deletion of length 60, then we are more confident about the deletion.

SVseq implements the above ideas. There are two versions of SVseq: SVseq1 (20) and SVseq2 (21). Both versions combine split read and read pair signatures, but the way of combining the signatures are different. Note that SVseq is able to find the exact breakpoints of SVs since it uses the split-read signature.

8.2.4.1 SVseq1
It first maps both split segments of a split read to a reference genome using a read mapping algorithm based on Burrows–Wheeler transform (BWT). Recall that we introduced the pattern-growth approach in Section 8.2.3. The split-read mapping algorithm of SVseq1 is faster than pattern growth. Also because it is used in a combined approach to discover deletions even with very week splitting signals, the mapping method of this step allows more sequencing errors. Thus, SVseq1 also has more FPs in this step, which will be removed by the second step. When mapping a whole read without errors in it, the BWT mapping method described in Reference 22 can be used directly. Starting from one end of a sequence, this algorithm maps one character of the sequence in one step. The range of hits on the BWT of the sequence's mapped portion is being updated until the whole read is mapped. Handling mismatches and small indels has been introduced into BWT mapping by Bowtie (7) and BWA (5), which use a greedy

algorithm and a branch-and-bound algorithm, respectively. SVseq1 adopts a similar branch-and-bound algorithm in allowing sequencing errors, but instead of mapping whole reads, SVseq1 maps the segments of a split reads efficiently. Note that BWA also can use local alignment to map one split segment of a split read near the anchor (the other split segment is not mapped and is soft-clipped). Both split segments have to be mapped to call deletions.

The breakpoint of a split read is not clear until all ways of splitting the read are examined and the pairs of split segments are mapped. A naive way for this purpose is breaking the read into two portions, take each portion as a new whole read and run a read mapping algorithm, such as BWA, $2(n - 2m + 1)$ times, where n is the length of the read and m is the minimum allowed length of a split segment. The algorithm of SVseq1 only runs the BWT mapping algorithm two times for a split read this way. We omit the details of the mapping step. Refer to (20) for more details. When the pairs of split segments of a split read are mapped, the mappings of one pair correspond to the breakpoints of a potential deletion. Note that because split-read mapping can suffer from noise, repeats and artifacts, FP singleton can be very common. It is not easy to separate many of these from the TPs even if other information is available. Deletions revealed by split reads become *candidate* deletions only if they are supported by a set of split reads. When working with population sequence reads, a candidate deletion can be supported by split reads from different individuals. Since singleton split reads may be unreliable, SVseq1 used a cutoff value of C as the minimum number of split reads need to call candidate deletions. C can be set as 2 or 3, even higher values may have higher accuracy but may reduce sensitivity when sequence coverage is low.

Since candidate deletions found in the first step are bound to contain many false positives, they have to be further examined before they can be put in the final list of candidate deletions. The second step of SVseq1 uses paired-end reads encompassing candidate deletions to filter the candidate deletion list got from the first step. A paired-end read encompassing a candidate deletion if its two ends are mapped to different sides of the candidate on the reference genome. The insert size of each encompassing paired-end read is examined to test whether it supports the candidate. When there is a deletion, an encompassing pair is mapped to the reference genome with insert size extended by the length of the deletion. If such discordant insert size and the length of the candidate deletion match well with each other, then it is a strong signal that indicates the candidate is a true deletion. Matching means that the difference between a discordant insert size and the length of the candidate deletion is not significantly different from the library insert size. If the insert size of a pair minus the length of the candidate deletion is within three standard deviations of library insert size, we say the pair supports the candidate deletion. When a candidate deletion is supported by at least one encompassing paired-end read, a candidate deletion is reported in the final list. Note that the insert size of an encompassing read is not necessarily be discordant, especially for deletions that is not very long. In this sense, the signature of read pair used here is not exactly the same as the signature introduced in Section 8.2.1, when only single signature is considered. It is the combined signal of split read and read pair that is strong enough to call a deletion candidate, while the single signatures may be weak in this purpose.

8.2.4.2 SVseq2 It uses the two signatures together in one step. Read pair signature helps split-read mapping by aligning only to smaller regions (called focal regions). Same as SVseq1, because combined strategy is used, both signatures are less stringent. Insert size of an encompassing read does not need to be significantly longer than mean library insert size and split-read mapping can allow more errors. We will discuss in more detail how the focal regions will reduce more FPs. Note that SVseq2 is able to call deletions with only one spanning read and one encompassing read when the read coverage is very low, while SVseq1 and the other methods mentioned before usually need more spanning reads to discover the exact breakpoints. Also note that RD, PR, and AS methods usually work only on higher coverage data sets. When single signatures are combined and used together, we call a combination as a pattern. Different patterns can be combinations of the same signatures combined in different ways. Also note that since encompassing reads are mapped without imposing a maximum event size, there is no limit on the size of the long indels that SVseq2 is able to call.

SVseq2 relies on two types of patterns formed by split reads to detect deletions.

- Type I pattern: the segment facing the anchor end is mapped (e.g., Read 1 in Figure 8.8).
- Type II pattern: the segment away from the anchor is mapped (e.g., Read 4 in Figure 8.9).

For type I pattern, the mapped segment of a split read based on soft-clip mapping faces the anchor. We denote the mapped location of the mapped segment as $[a, b]$ (where $a < b$). To discover a deletion, the soft-clipped segment needs to be mapped to some region $[c, d]$ (where $c < d < a$). We denote the length of the soft-clipped segment as $l_s = d - c + 1$. Because the length of the true deletion is not known, some existing split-read mapping methods (e.g., References 16, 20) have a parameter on the maximum distance to search for the second (i.e., the soft-clipped) segment. Instead of searching in a large region, SVseq2 only searches a *focal* region by the guidance of encompassing pairs. Our goal here is to infer where the soft-clipped segment is likely to start (i.e., the likely range of c). Our first observation is even with low-coverage sequence data, a deletion is still likely to have at least one paired-end read whose two ends are located on different sides of the deletion (i.e., a encompassing pair). Suppose

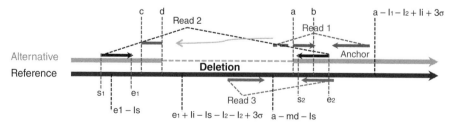

Figure 8.8 Type I pattern of deletion calling. Read 1 is a split read. Read 2 is an encompassing pair. Read 3 is a pair on the other haplotype without the deletion.

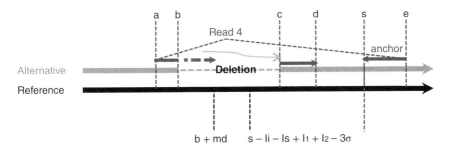

Figure 8.9 Type II pattern of deletion calling. Read 4 itself is an encompassing pair. The left end is split.

there is a read pair whose two ends are mapped to $[s_1, e_1]$ and $[s_2, e_2]$, respectively, on the reference genome (where $s_1 < e_1 < s_2 < e_2$), and this pair is an encompassing pair for the deletion, whose location is determined by the mapping of the soft-clipped segment of the split read. We let l_i be the expected insert size and let σ be the standard deviation of the insert length. Note that l_i measures the outer distance of the pair (i.e., the distance of the two farthest points of the two reads). We denote the length of the two reads of the encompassing pair as l_1 and l_2, respectively. Suppose the minimum deletion size to be detected by SVseq2 is m_d. SVseq2 sets m_d to be 50.

We first show where to find encompassing pairs for a given split read.

Lemma 1. *For type I pattern, $s_2 \geq a$, and with high probability, we have $s_2 \leq a - l_1 - l_2 + l_i + 3\sigma$.*

Proof. If $s_2 < a$, then a is not a breakpoint. This does not agree with our underlying assumption that the mapped segment $[a, b]$ corresponds to a deletion.

To give an upper bound on s_2, note that a is the position of the right breakpoint. The rightmost position of e_1 on the reference is $a - l_{del}$, where l_{del} is the length of the deletion. Now since with high probability, the distance between s_2 and e_1 is at most $l_i - l_1 - l_2 + 3\sigma + l_{del}$ on the reference. So with high probability $s_2 \leq (a - l_{del}) + (l_i - l_1 - l_2 + 3\sigma + l_{del}) = a - l_1 - l_2 + l_i + 3\sigma.$ ∎

Lemma 1 states that where the encompassing pairs are very likely to be located. For a given split read, SVseq2 searches for reads mapped on the reverse strand within this region for encompassing pairs. Now suppose we find one encompassing pair for the given split read. Recall the encompassing pair is mapped onto $[s_1, e_1]$ and $[s_2, e_2]$. The following lemma specifies the range of c (i.e., the starting point of the soft-clipped segment).

Lemma 2. *For type I pattern, $e_1 - l_s \leq c \leq a - m_d - l_s$. Moreover, with high probability, we have $c \leq e_1 + l_i - l_s - l_1 - l_2 + 3\sigma$.*

Proof. Note that the rightmost position a deletion can end is m_d bases to the left of a on the reference because the minimum deletion size is m_d. So $c \leq a - m_d - l_s$.

Since the encompassing pair $([s_1, e_1], [s_2, e_2])$ encompasses the deletion, we know the deletion must occur to the right of $[s_1, e_1]$. The leftmost position of the deletion is thus at least e_1. Since the length of l_s is to be mapped (to the left) from the left end of the deletion, we have $c + l_s \geq e_1$.

We now estimate how large c can be. Note that on the alternative chromosome (the chromosome with the deletion), the left and right breakpoints of the deletion become the same, and the left breakpoint of the deletion must be to the left of the starting position of the right end of the encompassing pair. Thus, on the reference chromosome, with high probability, the left breakpoint is no bigger than $e_1 + l_i - l_1 - l_2 + 3\sigma$. ∎

Lemma 2 states that we only need to search for the second segment of the split read within the region $[e_1 - l_s, \min(e_1 + l_i - l_s - l_1 - l_2 + 3\sigma, a - m_d - l_s)]$. This region is called the "focal" region for the split read being mapped and a encompassing pair. In most current sequence data, the focal region is relatively small. For example, suppose $l_s = 50$ (taken from a read of 100 bps long), $l_1 = l_2 = 100$, $l_i = 200$, and $\sigma = 50$. Then the width of the focal region is not larger than 200. This is much smaller than the focal region that the original split-read mapping would have searched (which can be as long as 1 Mbps). Also, from Lemma 1, the width of the region for encompassing pairs is at most 150.

The processing of split reads with type II pattern is similar in many aspects to that of type I pattern. A main difference between type I and type II patterns is that type II pattern does not need additional encompassing pairs because the paired-end read itself is an encompassing pair. This imposes an additional constraint on the focal region. Refer to (21) for more detailed description.

To search for the occurrence of a soft-clipped segment within an inferred focal region, SVseq2 uses a semi-global alignment algorithm. Briefly, we want to map the entire soft-clipped segment within the focal region. Thus, the gaps outside of the aligned positions for the focal region are without penalty, while we set the gap penalty within the read to 3. The similarity score is 1 for matches and −1 for mismatches. Since the focal region and the read are relatively short (e.g., several hundreds at most for the focal region and less than 100 for paired-end reads), split-read mapping with sequence alignment can be performed relatively fast.

In practice, there may be more than one encompassing pairs for a candidate deletion (corresponding to a split read). When the deletion is heterozygous in a diploid genome, some encompassing pairs may originate from the copy without the deletion while others are from the copy with the deletion. Some other encompassing pairs may be due to mapping errors. One possible scheme is to find a "consensus" focal region by combining information provided by multiple encompassing pairs. SVseq2 simply takes the union of all the focal regions from all the possible encompassing pairs. This is because there may be mapping errors in the encompassing pairs, and thus SVseq2 takes a conservative estimate of the focal region. Our experience shows that the overall focal region is still relatively small and searching for split read can be performed relatively efficiently.

Combined methods have become a hot topic in SV calling recently for the reasons we mentioned in Section 8.2.1. For example, there is another SV calling method that combines RP and SP, called PRISM (23). PRISM first uses an RP method CNVer (24) to identify discordant clusters of RP reads, and then maps split reads to target regions identified by the clusters using a dynamic programming algorithm that allow long jumps between the target regions. LUMPY (25) uses a probabilistic framework to discover SVs by combining signals of RP, SP, and RD together.

8.3 APPLICATIONS

8.3.1 Population SV Calling

The 1000 Genomes Project (1, 3) is the first large-scale project to sequence a large number of people in providing a comprehensive resource on human population genetic variation. The goal of the 1000 Genomes Project is finding most genetic variants that have frequencies of at least 1% in the populations studied. The genetic variants include single-nucleotide polymorphisms (SNV) and structural variations. To achieve the goal of the project, NGS sequencing technologies have been adopted because the cost of NGS data is lower. Even using NGS, it is too expensive to sequence thousands of individuals with high coverage. Instead, the goal has been attained by sequencing many individuals at low coverage. Pilot projects of the 1000 Genomes Project include low-coverage (2–4×) whole-genome sequencing of 180 samples, high-coverage (20–60×) whole-genome sequencing of 2 mother–father–adult child trios, and high-coverage (50×) sequencing of 1000 gene regions in 900 samples. The pilot study discovered 22,025 deletions and 6000 additional SVs, including insertions and tandem duplications, most of the SVs (53%) were mapped onto nucleotide resolution. Dozens of SV calling methods, including some method discussed in this chapter, have been used to make the discoveries. Many methods used are using single signature in calling SVs. These tools have contributed in the analytical framework and SV map that serve as a resource for sequencing-based association studies. The plan for the full project is to sequence about 2500 samples at 4× coverage. In Reference 26, from 1092 genomes of 14 populations, more than 14,000 larger deletions have been discovered using NGS SV calling tools.

Here, we demonstrate with simulation study that SVseq can be used in population SV calling. The simulated data sets consist of several sub data sets. All of them are paired-end reads with length 100 and insert size 500. The data sets are simulated from the sequence of chromosome 15 (100,338,915 bps in length) of human genome build 36. The results of the copy number variation release paper of the 1000 Genomes Project (3) are based on this version of genome. One hundred and thirty-two deletions of the 45 individuals from the CEU population reported by Mills et al. (3) are introduced to the simulation data sets. Since the haplotypes of the deletions are not inferred, for the heterogeneous deletions we arbitrarily place one such deletion to one of the two haplotypes of an individual. Since the deletions are usually far apart from each other, this may not have big effects on the accuracy of the simulation. wgsim (https://github.com/lh3/wgsim) is used to generate paired-end reads from the two copies of genomes of an individual. Single-nucleotide polymorphisms and small

TABLE 8.1 **Simulation Precision and Sensitivity of SVseq1 and SVseq2 for Pooled Population Data Sets**

Coverage	Tool	Findings	True Positives	Precision (%)	Sensitivity (%)
3.2×	SVseq2	114	112	98	85
	SVseq1	111	108	97	82
4.2×	SVseq2	113	112	99	85
	SVseq1	117	109	93	83
6.4×	SVseq2	123	120	98	91
	SVseq1	128	120	94	91

indels in each genome are simulated using the default parameters. All the data sets are generated with base error rate 2%. Three data sets with coverage 3.2×, 4.2×, and 6.4× are used. BWA, which provides soft-clips, is used with default parameters to map these simulated paired-end reads to the entire human genome build 36.

Pooling low-coverage data sets from the same population is a strategy that may discover more SVs than using individual data sets separately. We run SVseq1 and SVseq2 on the data sets. Sensitivity and accuracy of the methods when applied on different data sets can be computed exactly since the correct answer is known for the simulated data sets. The results are shown in Table 8.1. We can see that while SVseq1 is having a satisfied accuracy and sensitivity on all the data sets, SVseq2 performs even better. Coverage definitely affects the sensitivity. When coverage is higher (6.4×), the sensitivity of SVseq2 and SVseq1 is the same. But when coverage is lower, the sensitivity of SVseq2 is getting higher than SVseq1. When the coverage or the frequency of a deletion is very low, SVseq2 may have a better chance of detecting it since SVseq2 is able to call deletions with singleton split reads without loss of accuracy. Note also that SVseq2 is more accurate for all the data sets, this is because the split-read mapping is targeted in focal regions and thus less FPs.

8.3.2 Cancer Genomics

It is another area where SV calling can be very useful. Elucidating the contribution of the genetic variations to complex traits (or disease susceptibilities) is a fundamental problem in human genetics. Several researches report that structural variations are considered playing important roles in the development and the evolution of different human cancers. In principle, most of the tools mentioned in this chapter could be used to discover SVs from tumor samples. However, SVs also show complicated patterns in cancer genomes. For example, we cannot always assume diploid genomes in cancer cells. Several cancer-genome projects, such as TCGA and ICGC, are also proposed and designed based on NGS, which are around the identification and analysis of the germline and the somatic variations (2, 27). In these projects, two separate sequencing samples are generated from each cancer patient, one of which sequences a sample of tumor tissue while the other sample is from normal tissue and functions as a control. In the following analysis, two sequencing alignments are compared to identify plausible differences, many of which may represent genuine germline and somatic variations.

There are many challenges in calling SVs in tumor samples. It is likely to have more computational tools to be developed for cancer SV calling.

8.4 CONCLUSIONS AND FUTURE DIRECTIONS

Finding long insertions and deletions, along with other types of structural variations, from high-throughput sequence reads is likely to be useful in the foreseeable future. Due to the inherent noise in the sequence reads and the complexity of genomic structure, accurate detection of long insertions and deletions is still a challenging problem. We note that the latest methods for finding long insertion and deletions tend to use multiple sources of information rather than a single one (20, 21, 23). Combining multiple sources of information may potentially be more tolerant to sequencing errors and lead to more accurate calling of structural variations.

Looking forward, there are several issues where future computational methods may need to address. First, there are certain types of structural variations that fewer methods are available. While computational calling of deletions performs reasonably well, calling insertions tend to be harder. Even when one may call the size of insertions, reliably calling the inserted novel sequences may be even more difficult (28). Current methods for detecting long insertions and deletions usually perform less well on shorter indels (say 50 bp or shorter). A main reason is that discordant insert size becomes less reliable for such shorter indels. There are also other types of structural variations (e.g., inversions, translocations) that are less studied and analyzed. Second, one may want to find structural variations under more complex context. For example, one may want to call structural variations using both whole genome sequencing and whole transcriptome sequencing data to identify functionally relevant SVs. Moreover, more SV calling tools targeted to some specific application domains (e.g., cancer genomics) may be needed in order to address the specific challenges. Also, with the ongoing population sequencing projects, more population-scale applications of structural variations may be needed. See Reference 29 for a recent work on calling genotypes of insertions and deletions from sequence reads. We expect more work like this in the near future. At last, technological developments (e.g., longer read length but higher error rates in newer sequencing technologies (30)) may lead to newer computational challenges.

ACKNOWLEDGMENT

Work supported in part by the National Science Foundation under awards IIS-0953563 and IIS-1447711. Tests were run on workstations supported by National Science Foundation under award IIS-0916948.

REFERENCES

1. The 1000 Genomes Project Consortium. A map of human genome variation from population-scale sequencing. Nature 2010;467(3):1061.

2. http://cancergenome.nih.gov. Available at Accessed 2016 Mar 17.

3. Mills R et al. Mapping copy number variation by population-scale genome sequencing. Nature 2011;470:59.

4. Medvedev P, Stanciu M, Brudno M. Computational methods for discovering structural variation with next generation sequencing. Nat Methods 2009;6:S13.

5. Li H, Durbin R. Fast and accurate short read alignment with Burrows-Wheeler transform. Bioinformatics 2009;25:1754.

6. Thorvaldsdóttir H, Robinson J, Mesirov J. Integrative Genomics Viewer (IGV): high-performance genomics data visualization and exploration. Brief Bioinform 2013;14(2):178.

7. Langmead B, Trapnell C, Pop M, Salzberg S. Ultrafast and memory-efficient alignment of short DNA sequences to the human genome. Genome Biol 2009;10:R25.

8. Abyzov A, Urban A, Snyder M, Gerstein M. CNVnator: an approach to discover, genotype, and characterize typical and atypical CNVs from family and population genome sequencing. Genome Res 2011;21(6):974.

9. Chen K et al. BreakDancer: an algorithm for high-resolution mapping of genomic structural variation. Nat Methods 2009;6:677.

10. Chiang D et al. High- resolution mapping of copy-number alterations with massively parallel sequencing. Nat Methods 2009;6(1):99.

11. Alkan C et al. Personalized copy number and segmental duplication maps using next-generation sequencing. Nat Genet 2009;41(10):1061.

12. Sudmant P et al. Diversity of human copy number variation and multicopy genes. Science 2010;33(6004):641.

13. Korbel J et al. PEMer: a computational framework with simulation-based error models for inferring genomic structural variants from massive paired-end sequencing data. Genome Biol 2009;10(2):R23.

14. Sindi S, Helman E, Bashir A, Raphael B. A geometric approach for classification and comparison of structural variants. Bioinformatics 2009;25(12):i222.

15. Quinlan A et al. Genome-wide mapping and assembly of structural variant breakpoints in the mouse genome. Genome Res 2010;20(5):623.

16. Ye K, Schulz M, Long Q, Apweiler R, Ning Z. Pindel: a pattern growth approach to detect break points of large deletions and medium sized insertions from paired-end short reads. Bioinformatics 2009;25:2865.

17. Zhang Z, Du J, Lam H, Abyzov A, Urban A, Snyder M, Gerstein M. Identification of genomic indels and structural variations using split reads. BMC Genomics 2011;12:375.

18. Abyzov A, Gerstein M. AGE: defining breakpoints of genomic structural variants at single-nucleotide resolution, through optimal alignments with gap excision. Bioinformatics 2011;27(5):595.

19. Wang J et al. CREST maps somatic structural variation in cancer genomes with base-pair resolution. Nat Methods 2011;8:652s.

20. Zhang J, Wu Y. SVseq: an approach for detecting exact breakpoints of deletions with low-coverage sequence data. Bioinformatics 2011;27(23):3228.

21. Zhang J, Wang J, Wu Y. An improved approach for accurate and efficient calling of structural variations with low-coverage sequence data. BMC Bioinformatics 2012;13:S6.

22. Ferragina P, Manzini G. Opportunistic data structures with applications. Proceedings of the 41st Annual Symposium on Foundations of Computer Science; 2000. p 390.

23. Jiang Y, Wang Y, Brudno M. PRISM: pair-read informed split-read mapping for base-pair level detection of insertion, deletion and structural variants. Bioinformatics 2012;28(20):2576.

24. Medvedev P, Fiume M, Dzamba M, Smith T, Brudno M. Detecting copy number variation with mated short reads. Genome Res 2010;20(11):1613.

25. Layer R, Hall I, Quinlan A. PLUMPY: a probabilistic framework for structural variant discovery. Genome Biol 2014;15:R84.

26. The 1000 Genomes Project Consortium. An integrated map of genetic variation from 1,092 human genomes. Nature 2012;491:56.

27. http://http://icgc.org. Available at Accessed 2016 Mar 17.

28. Hajirasouliha I et al. Detection and characterization of novel sequence insertions using paired-end next-generation sequencing. Bioinformatics 2010;26(10):1277.

29. Chu C, Zhang J, Wu Y. GINDEL: accurate genotype calling of insertions and deletions from low coverage population sequence reads. PLoS ONE 2014;9:e113324.

30. http://www.pacificbiosciences.com. Available at Accessed 2016 Mar 17.

9

COMPUTATIONAL APPROACHES IN NEXT-GENERATION SEQUENCING DATA ANALYSIS FOR GENOME-WIDE DNA METHYLATION STUDIES

JEONG-HYEON CHOI

Cancer Center and Department of Biostatistics and Epidemiology, Medical College of Georgia, Georgia Regents University, Augusta, GA, USA

HUIDONG SHI

Cancer Center and Department of Biochemistry, Medical College of Georgia, Georgia Regents University, Augusta, GA, USA

9.1 INTRODUCTION

DNA methylation is one of the most important epigenetic mechanisms that ensures the maintenance and inheritance of gene-expression programs in mammalian cells (1). DNA methylation primarily involves the methylation of the 5th carbon on cytosine residues (5-mC) in a CpG dinucleotide. Three highly conserved DNA methyltransferase (DNMT) proteins (DNMT1, DNMT3A, and DNMT3B), which complex together in vivo, establish and maintain the DNA methylation landscape within the human genome (2). DNMT1 has a high affinity toward hemi-methylated DNA and is the key maintenance methyltransferase in mammals. DNMT3A and 3B are *de novo* methyltransferases that set up the pattern of methylation in early embryo development. Genome-wide DNA methylation studies have shown that DNA methylation in

Computational Methods for Next Generation Sequencing Data Analysis, First Edition.
Edited by Ion I. Măndoiu and Alexander Zelikovsky.
© 2016 John Wiley & Sons, Inc. Published 2016 by John Wiley & Sons, Inc.
Companion website: www.wiley.com/go/Mandoiu/NextGenerationSequencing

CpG-rich regions called CpG islands (CGIs) located near transcriptional start sites (TSS) are negatively associated with gene expression, while DNA methylation in body of genes are positively associated with gene expression (3). Epigenetic alterations are at least as common as, if not more frequent than, mutational events in the development of cancer (4). As compared to normal cells, cancer cells exhibit global hypomethylation primarily in repeat elements and pericentromeric regions and simultaneously local hypermethylation in normally protected CGIs (5, 6). Hypermethylation within promoters serves to turn off critical genes that could otherwise suppress tumorigenesis (6). Aberrant DNA methylation changes can also cause a number of human diseases such as developmental diseases (ICF syndrome, Prader–Willi, and Angelman syndromes, etc), aging-related diseases (i.e., Alzheimer's disease), and complex trait diseases (i.e., heart disease, diabetes and autoimmune diseases) (7–11). Therefore, it is very important to understand how methylation patterns are established and maintained during normal development and under pathological conditions.

Recently, it was discovered that 5-hydroxymethylcytosine (5-hmC), an intermediate to demethylation of 5-mC, was found to be abundant in human and mouse brains, as well as in embryonic stem cells (12, 13). Ten-eleven translocation (Tet) family of DNA hydroxylase including Tet1, 2, and 3 can catalyze the oxidation of 5-mC to 5-hmC (12, 13). Subsequently, 5-hmC can be further oxidized by Tet enzymes to 5-formylcytosine (5-fC) and 5-carboxylcytosine (5-caC) (14, 15). The oxidation pathway from 5-mC to 5-fC or 5-CaC followed by base excision by thymine-DNA glycosylase (TDG) was proposed as a plausible mechanism of active DNA demethylation (14). It was demonstrated that TET1 is mainly expressed in embryonic stem (ES) cells and maintains the pluripotency of ES cells (16). TET2 regulates cell lineage commitment by modulating the balance between self-renewal and differentiation (17) while TET3 is important in regulating the global DNA demethylation of zygotes (18). Although the biological function of 5-hmC is not fully understood, it has been implicated in transcriptional regulation (19, 20). Loss of 5-hmC has been reported as an epigenetic hallmark of melanoma, with diagnostic and prognostic implications (21). In addition, TET family proteins are frequently altered by genetic mutation or gene translocations in human leukemias.

A variety of methods are available for analyzing gene-specific as well as genome-wide DNA methylation patterns (Figure 9.1). Almost all of the methods utilize three basic approaches: (i) digest unmethylated or methylated DNA with methylation-sensitive restriction enzymes; (ii) use anti-methylcytosine antibodies or methyl-binding domain (MBD) proteins to enrich methylated DNA; and (iii) bisulfite treatment that can convert unmethylated cytosine into uracil while leaving methylated cytosine unchanged. Several excellent reviews of the genome-wide DNA methylation analysis are already available (22–25). These methods have dramatically evolved during the last two decades from using low-throughput PCR and gel-based platforms to the current high-throughput microarray and next-generation sequencing (NGS) technologies.

One of the commonly used methods to study genome-wide DNA methylation patterns is methylated DNA immunoprecipitation sequencing (MeDIP-seq), which utilizes anti-5-mC antibodies to probe for 5-methylcytosines within fragment genomic DNA (26). The MeDIP-seq protocol involves sheering and denaturing the DNA in

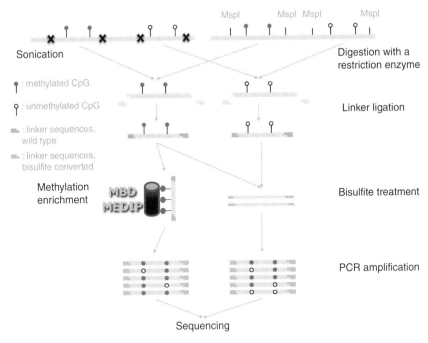

Figure 9.1 Sequencing methods based on enrichment and bisulfite conversion for DNA methylation studies. A genome is fragmented by sonication or a restriction enzyme. After repairing the ends of fragments by linker ligation, fragments that are captured by MBD or MeDIP (left) or bisulfite treated (right) are amplified and sequenced.

order to create short ssDNA fragments, and then immunoprecipitating the methylated DNA fragments from the ssDNA pool using the specified anti-5-mC antibody. The enriched methylated DNA fragment is then ligated with sequencing adaptors for subsequent NGS library preparation. 5-hmC can also be analyzed using this approach by simply replacing with an antibody against 5-hmC (hMeDIP-seq). The hMeDIP-seq has been successfully used to perform genome-wide 5-hmC profiling in both human and mouse genome (27, 28). MBD-seq incorporates the same affinity-based concept used in MeDIP-seq, but use MBD proteins to enrich the methylated DNA. GST or his-tagged MBD2 protein is utilized to capture the methylated DNA (29, 30). The protein–DNA complex can then be extracted using magnetic beads. After enrichment and separation from the protein–DNA heterocomplex, the enriched methylated DNA can then be prepared as a genomic library and sequenced to determine DNA methylation regions. By these affinity purification-based approaches, the global 5-mC and 5-hmC landscapes for the probed DNA samples can be determined with an approximate 100–300 base pair resolution.

Bisulfite sequencing is the golden standard for DNA methylation, which provides a single-base resolution analysis. The whole-genome shotgun bisulfite sequencing (WGBS-seq) that incorporates bisulfite conversion with NGS has been successfully used to analyze the whole methylomes of several human cell types at the single-base resolution (3, 31–36). To perform WGBS-seq, the sheered genomic DNA (average size 200 bp) is first converted into a sequencing library by ligation to adaptors that contains 5-mCs. The adaptor-ligated DNA is then treated with bisulfite to convert all unmethylated cytosines into uracils. After PCR amplification of the bisulfite-treated DNA, the library is sequenced by NGS. The WGBS-seq approach is capable of measuring the methylation statuses of more than 90% of cytosines in human genomes (3). WGBS is instrumental in the identification of non-CG methylation in human ES cells and the identification of large partially methylated domains (up to megabases in length) in cancer genomes. However, the cost of WGBS approach is still quite expensive. In this regard, RRBS (reduced representation bisulfite sequencing) is a cost-effective method for sub-genome scale DNA methylation analysis. For RRBS, the high molecular weight genomic DNA is first digested with a methylation-insensitive restriction enzyme (MspI). A smaller fraction of the digested DNA (40–220 bp) is then extracted. After end-repair and ligation to the methylated adaptors, the library is treated with bisulfite, amplified by PCR, and sequenced by NGS (37). Overall only 8–14% of CpG sites in the human and mouse genome can be analyzed by RRBS. However, MspI restriction sites is biasedly distributed toward to CpG rich regions (CpG island); therefore, more than 80% of CpG island and overall half of promoters in the human and mouse genomes can be surveyed by RRBS. One of the limitations is that the CpG sites analyzed by RRBS highly depend on the MspI restriction sites (CCGG) in the human genome. CpG sites outside of the MspI fragment will not be analyzed by this method. Various targeted bisulfite sequencing approaches such as bisulfite padlock probe (38, 39), solution hybrid selection (18), array capture (40), multiplexed PCR (41) have been developed for analyzing DNA methylation patterns in specific gene of interests. However, most of these methods involve complex protocols and the upfront cost for synthesizing probes and primers is relatively high, and therefore hinder the wide adoption of these approaches (42).

Bisulfite treatment alone cannot differentiate 5-mC from 5-hmC because it converts 5-hmC to cytosine-5-methylsulfonate (CMS) (43). CMS is read as "C" during sequencing. Recently, Booth et al. developed a novel method called "OxBS-seq," which utilizes perruthenate (KRuO4) selectively oxidizing 5-hmC into 5-fC (44). The subsequent bisulfite treatment converts 5-fC to uracil, just like unmethylated cytosine. By subtracting out an OxBS-seq data set from regular bisulfite sequencing results, OxBS-seq can generate high-resolution 5-hmC map in samples (45). TAB-seq, another recently developed method, takes a different approach for single-nucleotide 5-hmC mapping (46). TAB-seq is based on the fact that Tet proteins can oxidize 5-mC to 5-hmC (12, 13), but also 5-hmC to 5-caC, which behaves like unmodified cytosines during bisulfite conversion. TAB-seq first converts 5-hmC to 5-gmC using β-GT in order to prevent the 5-hmC from been oxidized by Tet recombinant enzyme. After β-GT treatment, DNA samples are modified with recombinant mouse Tet1 enzyme during which all 5-mC is converted to 5-caC. The subsequent bisulfite treatment converts unmodified cytosines and 5-caC to uracil and 5-caU, respectively, leaving 5-ghmC unchanged. Once sequencing is complete, the modified 5-hmCs (5-ghmC) are presented as cytosines. By comparing the TAB-seq data set with standard bisulfite sequencing results, precise mapping of both 5-hmC and 5-mC modifications can be achieved. Currently, most of the bisulfite-based NGS analyses are performed using Illumina sequencing technologies, although a few studies also used 454 and SOLiD sequencing platforms. The sequencing cost has come down dramatically over the last 2–3 years. New sequencing platform such as the single-molecule real-time (SMRT) technology developed by Pacific Biosciences already can detect 5-methylcytosine directly without bisulfite conversion (47). Nanopore sequencing can also read the 5-mC signal directly without the need of bisulfite conversion (48, 49). These novel NGS platforms also have many other advantages over the current platforms such as less bias during template preparation, possible longer read length, lower cost, higher speed, and better accuracy.

9.2 ENRICHMENT-BASED APPROACHES

9.2.1 Data Analysis Procedure

9.2.1.1 Sequence Mapping Since enrichment-based sequencing captures genomic DNA fragments with methylated CpGs, the read depth of each location is proportional to the level of methylation. To this end, sequence reads are mapped to a genome using sophisticated programs such as BWA (50), Bowtie (51), and SOAP (52). Mapped reads are then processed in several steps to eliminate sequencing artifacts: PCR and sequencing duplicates. Specifically, all reads mapped to the same location are filtered out except one copy so that each position has at most one mapped read. Since in the current sequencing technologies, the length of sequence reads is shorter than the length of fragments and methylated CpGs could lie beyond the sequenced region of a fragment, mapped reads lengthen toward the 3'-end up to the original (average) length of fragments.

9.2.1.2 Methylation Quantification As in RNA-seq analysis for gene expression, different numbers of sequence reads for samples should be taken into account to measure absolute methylation level and to compare multiple samples. Like ChIP-seq, MeDIP-seq and MBD-seq require input sequencing to take into account genomic differences between the sequenced samples and the reference genome, for example, copy number variants. There are many algorithms and programs to call significantly enriched regions (peaks) for ChIP-seq for transcription factors and histone marks: MACS (53), SICER (54), F-seq (55), and Useq (56). These were used for the analysis of enrichment-based methylome sequencing data. In contrast, Batman (57), MEDIPS (58), and BALM (59) have been developed to identify enriched regions for methylome sequencing because the more methylated CpGs in a fragment, the more captured. In other words, if two loci have the same number of reads but a different number of CpGs, they have different methylation. In Figure 9.2, the two loci p and q have the same number of reads mapped, but a different number of CpGs, that is, a different methylation level. Actually the same number of reads is captured due to the identical

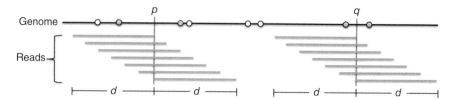

Figure 9.2 The number of mapped reads at p and q is the same, but their methylation level is different because of the different number of CpGs within captured fragments, that is, the region around p shows twice as much methylation as that around q.

number of methylated CpGs, resulting in a methylation at p half of that at q, assuming sequencing and affinity efficiency.

9.2.1.3 Differential Methylation Differential methylation between two samples can be simply defined as log2 fold change of average read depth of a given window or genomic feature. An alternative method is to use the enriched regions (peaks) so that mutually exclusive peaks in two samples are differentially methylated. However, these methods are limited in statistical assessment and sensitivity, respectively. The best solution is to apply statistical models to methylation changes and measure the significance of differentially methylated regions in terms of p-values: SICER (54), ChIPDiff (60), DiffBind (61), DBChIP (62), edgeR (63), and DESeq (64). MEDIPS, a specialized program for MeDIP-seq, used the Student t and Wilcoxon tests, and the latest version uses edgeR (58, 65).

9.2.1.4 Pipelines Two pipelines have been developed from sequence mapping to methylation profiling and differential methylation including quality assessment of sequencing and affinity efficiency. MeDUSA (Methylated DNA Utility for Sequence Analysis) (66) is a full analysis pipeline for MeDIP-seq and performs sequence mapping using BWA, subsequent filtering using SAMtools, quality assessment of sequencing using FastQC, coverage and saturation plots using MEDIPS, identification of DMRs using USeq and R Bioconductor DESeq in the old version and MEDIPS in the new version, and annotation using BEDTools. The output of MeDUSA can be visualized on the UCSC genome browser. Another pipeline, MeQA (MeDIP-seq data Quality assessment and Analysis) (67), uses FastQC for quality assessment of sequencing; BWA for sequence mapping; SAMtools for sorting and filtering; SAMStat for mapping summary and quality check; MEDIPS for saturation, coverage, and CpG enrichment estimation, as well as DMR identification; and CEAS for DMR annotation (Figure 9.3).

9.2.1.5 Tag Density Plot Most programs generate sequence mapping results in the wiggle and BED format, which can be visualized onto the UCSC genome browser (68), gbrowser (69), IGV (70), and IGB (71). Visualization is very useful for interpreting results intuitively and summarizing the overall methylation tendency. A tag density plot is an extended wiggle plot made by drawing the mean read coverage for a group of regions of interest, for example, upregulated genes and even all genes.

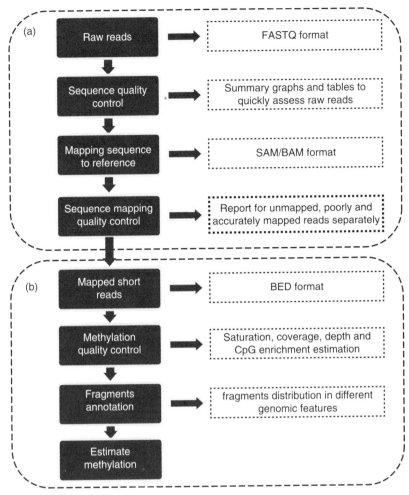

Figure 9.3 Flow diagram of the MeQA analysis pipeline for MeDIP-seq. Source: Huang et al. 2012 (67). Reproduced with permission of Oxford University Press.

Figure 9.4 shows tag density plots for eight samples around RefSeq genes (a) and CpG islands (b).

9.2.1.6 Saturation Test One of the most important questions in capture-based sequencing such as ChIP-seq, MeDIP-seq, and MBD-seq is whether sequencing is sufficient and if not, how much further sequencing is needed. These questions can be answered by saturation testing. MEDIPS developed the idea that if sequencing is sufficient, a set of reads in the sequencing should be highly correlated to an independent set of reads. Therefore, reads are randomly assigned to $2n$ data sets of equal size, $S_1, \ldots, S_n, S_{n+1}, \ldots, S_{2n}$. Let A be $\{S_1, \ldots, S_n\}$ and B be $\{S_{n+1}, \ldots, S_{2n}\}$. In each iteration i, the test increasingly adds data sets S_i and S_{n+i} from A and B to the current data sets A_i and B_i, respectively, counts reads in a window of 50 bp, and

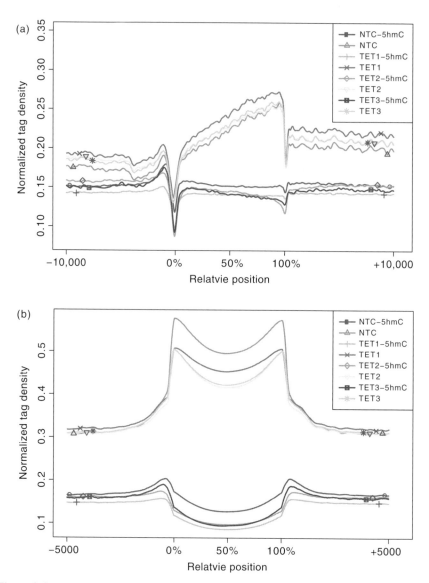

Figure 9.4 Tag density plots around RefSeq genes (a) and CpG islands (b). Each line represents the normalized mean read coverage for a sample. Please see www.wiley.com/go/Mandoiu/NextGenerationSequencing for a color version of this figure.

computes a Pearson correlation coefficient between the read counts in windows for A_i and B_i. In adding a subset, it is expected that the correlation coefficient increases. Since this method can determine the saturation for only half the reads, MEDIPS performs the saturation analysis with artificially doubled reads to estimate the real saturation. However, this leads to overestimation. The real correlation is between the results with half the reads and artificially doubled reads.

Similarly, an alternative test randomly splits mapped reads into n data sets and computes per-base read coverage for incrementally accumulated data sets. Finally, the test performs nonlinear regression using Michaelis–Menten kinetics, that is, $f(x) = M * x/(H + x)$ for pairs of read coverage and number of bases where M represents the maximum number of bases to be gained by infinite sequencing and H represents the number of sequencing run needed to obtain half of the maximum number of bases. Figure 9.5 shows the results of saturation testing by both methods. MEDIPS estimated that the sample is saturated highly between 0.81 and 0.92, while the other method estimated saturation of 73% and predicted that one and two additional sequencing runs will increase by 85% and 89%, respectively. The topmost line represents the maximal number of bases to be sequenced (745,870,270 bp).

9.2.2 Available Approaches

9.2.2.1 Bayesian Tool for Methylation Analysis (Batman) Batman (57) has been developed to address the difficulty of estimating absolute methylation levels due to CpG density effects from DNA immunoprecipitation profiles using oligonucleotide arrays (MeDIP-chip) or NGS (MeDIP-seq), as explained in methylation quantification. To address this issue, Batman defines the coupling factor C_{pw} between a genomic window w and CpG dinucleotide p as the fraction of DNA molecules being sequenced in the window that contains the CpG, so that local CpG density is in proportion to the sum of the coupling factor for a given window. As shown in Figure 9.6, there is an approximately linear relationship between the read depth at a window and the density of methylated CpGs. This observation leads to a model that given a set m of methylation states, the probability distribution S for a complete set of sequencing observations is defined as $f(S\,|m) = \Pi_w G(S_w\,|r\Sigma_p(C_{pw}m_p^2), v^{-1})$ where $G(x\,|\mu, \sigma^2)$ is a rectified Gaussian probability density function, m_p represents the methylation level at position p, and C_{pw} represents the coupling factor between a window w and a CpG p. To calculate $f(m\,|S)$, the posterior distribution of the methylation state parameters given the MBD-seq data, a standard Bayesian inference approach based on nested sampling, a highly robust Monte Carlo technique, has been applied. In other words, for each window of the genome, the approach generates 100 independent samples from $f(m\,|S)$, fits beta distributions to these samples, chooses the most likely methylation state in 100 bp windows, and then chooses the modes of the most likely beta distributions as the final methylation calls.

To run Batman, we need to (i) filter out low-confident alignments from BAM files, (ii) generate mapped reads in GFF2 files, (iii) create a database on MySQL, (iv) generate and load a coupling factor using EstimateCouplingProfileApplication and LoadCouplingProfileApplication, (v) generate a Batman array file from a GFF2 file using ReadsToPseudoArrayApplication, (vi) load the array file to the database using LoadProbes and LoadRatsGFF, (vii) calibrate read counts using the coupling factor with CalibrateApplication, and (viii) generate methylation states using SampleMethStatesApplication. Batman does not provide a user-friendly interface and requires knowledge of a database management system. It is quite slow and recommends running jobs for each chromosome. It assumes that MeDIP experiments do

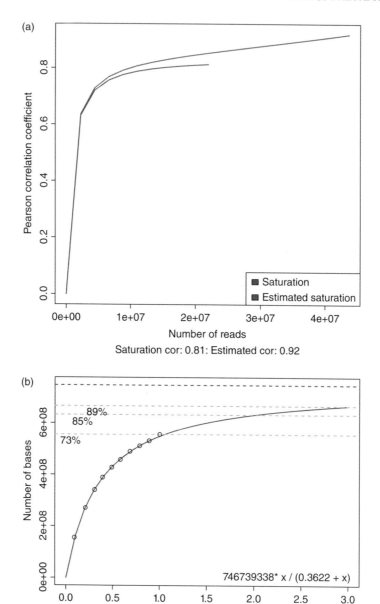

Figure 9.5   Results of saturation test by MEDIPS (a) and Michaelis–Menten kinetics (b).

not produce fragment biases and that affinity efficiency depends only on methylated CpGs. It cannot analyze the methylation of non-CpG sites, that is, CpHs in stem cells and CpHpGs in plant cells, and does not take into account copy number variants, which are especially very important in cancer. The source files are on https://github.com/dasmoth/batman.

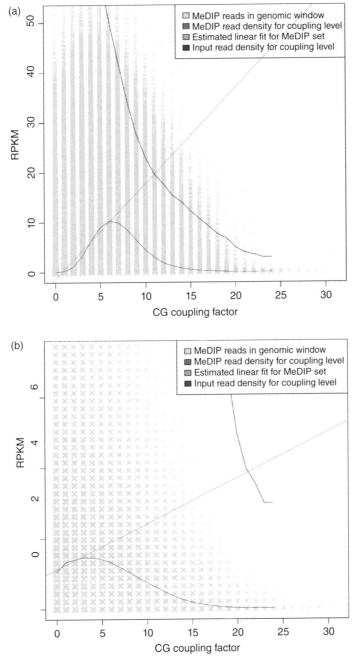

Figure 9.6 Scatter plot of CpG Coupling factor and RPKM (read depth) as marked in X. The dark and medium grey lines represent MeDIP-seq and input read density for coupling level, respectively. The light grey line represents estimated linear fit of the proposed method.

9.2.2.2 MEDIPS Similar to Batman, MEDIPS (65, 72) defines coupling factors
as the local CpG density determined by counting the number of CpGs for a given
position. Let p be a given position, and d the size of library (fragment) chosen for
the sequencing (Figure 9.6). Then the objective is to count the number of CpGs in
$[p - d, p + d]$. Several ways have been proposed: *count, linear, exp, log*, and *custom*.
While *count* simply counts the number of CpGs, the others weight CpGs based on
the distance to p using linear, exponential, or logarithmic function. *Custom* allows
user-defined weights, for example, Batman's coupling factors. Coupling factors
are precomputed for a genome. Since mammalian genomes are big, fixed-size (say
50 bp) bins are used for efficiency, where p is the center of a bin. The authors of
MEDIPS have tested the different weighting functions and library sizes with bins of
50 bp by measuring the correlation of the methylation to local CpG density using
bisulfite sequencing in the human epigenome project (73). For each of about 3000
genomic loci of length 50–500 bp in the bisulfite sequencing, they computed the
mean methylation scores and the mean coupling factors, and measured a Pearson
correlation coefficient. The best parameter setting is to use $l = 700$ and *count*; this
gives the best negative correlation as expected in mammalian cells: regions with low
CpG density are highly methylated while those with high CpG density are mainly
unmethylated (74).

To make the read count in bins comparable among samples, it is normalized in
RPM (Reads Per Millions), that is, the read count when a million reads are sequenced:
$rpm_i = c_i * 10^6/n$ where c_i represents the read count at position i and n represents
the total number of reads mapped. Now we have an RPM value and coupling factor for a given position or bin. The goal is to adjust the RPM value by the coupling
factor, that is, CpG density. To this end, MEDIPS visually inspected a calibration
plot where the x- and y-axes represent coupling factors and read counts, respectively. As shown in Figure 9.6, while the X marks show no dependency, the dark
grey line called a calibration curve shows a linear relationship for the low range of
coupling factor (0–5 in the figure). The high range of coupling factor has low read
count because regions with high CpG density are mainly unmethylated in mammalian
cells. Pelizzoal et al. (75) confirmed the dependency for the high range of coupling
factor in an experiment of artificially fully methylated samples using MeDIP-ChIP
data. Based on this observation, MEDIPS fits a linear regression model to a calibration curve and calls the methylation score adjusted with CpG density. Given
a coupling vector with n genomic bins, $\mathbf{cf} = (cf_1, \ldots, cf_n)$, the algorithm finds the
first local maximum of coupling factors, which is denoted as cf_m, and then performs a linear regression using the least square approach to $\mathbf{cf}^m, = (cf_1, \ldots, cf_m,)$
and $\mathbf{rpm}^m = (rpm_1, \ldots, rpm_m,) : \mathbf{rpm}^m = \alpha * \mathbf{cf}^m + \beta + \epsilon$ where α and β are the
theoretical slope and intercept and ϵ is an error vector. Once the parameters are deter-
mined, adjusted RPM values are calculated using $rms_i = \log 2(rpm_i/(\alpha * cf_i + \beta))$,
which are called RMS (Relative Methylation Score). Figure 9.7 shows large changes
between RPM and RMS for two samples. To compare any specific region of interest
(ROI), especially for different lengths, MEDIPS calculates an absolute methylation
score of an ROI by the mean of rms values divided by the mean of coupling factors.

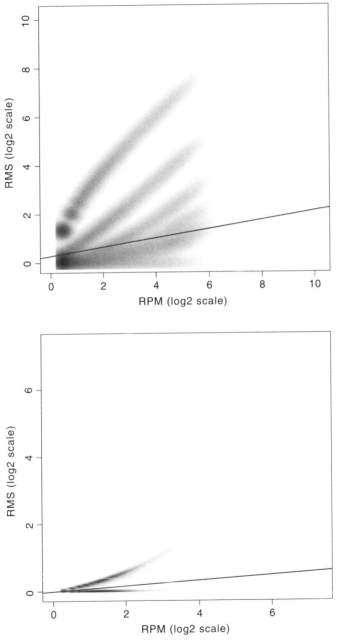

Figure 9.7   Scatter plot of RPM (Reads Per Million reads) and RMS (Relative Methylation Score).

To identify DMRs (Differentially Methylated Regions) between treatment and control samples along with an input sample, the initial version of MEDIPS used Student t and Wilcoxon tests for each ROI or genomic window of arbitrary length with at least 5 bins. First, it calculated the mean rpm and rms values of all samples for each ROI, the ratio of rms of the control to the treatment, and the ratios of rpms of the control and treatment to the input. Let rpm_i^S and rms_i^S be the rpm and rms values at bin i, respectively, of a sample S, then the means of rpm and rms values for an ROI_j are defined as $\text{Mrpm}_j^S = (\Sigma i\,\text{rpm}_i^S)/n_j$ and $\text{Mrms}_j^S = (\Sigma i\,\text{rms}_i^S)/n_j$, where n_j is the number of bins in the ROI. The ratio of rms values of the control to the treatment is $\text{Rrms}_j = \text{Mrms}_j^C/\text{Mrms}_j^T$ and the ratios of rpm values of the control and treatment to the input is $\text{Rrpm}_j^C = \text{Mrpms}_j^C/\text{Mrpm}_j^I$ and $\text{Rrpm}_j^T = \text{Mrpms}_j^T/\text{Mrpm}_j^I$, respectively. Second, for an ROI or genomic window, the Student t and Wilcoxon tests are performed using $\{\text{rpm}_i^T\}$ and $\{\text{rms}_i^C\}$ and the p-values are computed: t.pval_j and w.pval_j, respectively. Finally, for the DMR candidate, several filtering schemes are applied to ROIs with $\text{Mrms}_j^C = \text{Mrms}_j^T = 0$; $\text{t.pval}_j > p$ and $\text{w.pval}_j > p$; $l < \text{Rrms}_j < h$; $\text{Mrpm}_j^C < t$ or $\text{Mrpm}_j^T < t$; and $\text{Rrpm}_j^C < h$ or $\text{Rrpm}_j^T < h$ where p, l, and h are the threshold of the p-value, and lower and upper ratios, respectively, and t is a global background rpm value estimated by the input.

The latest version 1.12.0 of MEDIPS improved DMR call by using the model of edgeR (63), which allows a low number of replicates and uses a linear model for count data, similar to RNA-seq using negative binomial distribution. This version also accounts for copy number variants using DNAcopy (76), which is very important to heterogeneous data, for example, cancer cells, by computing the log ratio of the mean rpm of all samples in a group to the rpm of the input for each window of a fixed length.

9.2.2.3 Bi-Asymmetric-Laplace Model (BALM)

BALM (59) has been developed to accurately estimate the methylation level of each CpG from MBD-seq using a bi-asymmetric-Laplace model, which has been used in recapitulating the tags' bimodal distribution over target sites in a ChIP-seq experiment (53). The authors observed that tag density from the peaks to both directions decreases exponentially as in ChIP-seq, and they found that BALM fits better than Gaussian and Student t distributions, that is, BALM yields a lower value of goodness. Figure 9.8 shows the performance of those models measured by the goodness of fit for four samples. BALM uses a tag-shifting model used in W-ChIPeaks (77) to measure the enrichment in a sliding window of 50 bp across a genome. Significantly enriched windows are decided by Fisher's exact tests with input data and a predefined threshold weighted by the genome region amplification index. For each significantly enriched region, consisting of continuous windows with significant enrichment, a target site is defined as the mean of shifted tag positions in the region. Then BALM performs the fixed number of iterations for subsequent analysis. It measures tag distribution and estimates parameters of a bi-asymmetric Laplace model using maximum likelihood estimators. Since an enriched region could have multiple target sites, it builds a mixture model, estimates the parameters of the model using expectation-maximization algorithms, and determines the number of target sites using the Bayesian information criterion (Figure 9.9).

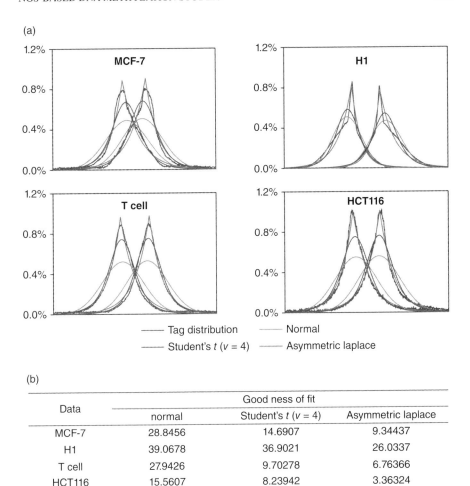

Figure 9.8 Several fitting models with performance measured by the goodness of fit. Source: Lan et al. 2011 (59). CC BY. Please see www.wiley.com/go/Mandoiu/NextGeneration Sequencing for a color version of this figure.

9.3 BISULFITE TREATMENT-BASED APPROACHES

9.3.1 Data Analysis Procedure

9.3.1.1 Mapping In contrast to capture-based sequencing, BS-seq analysis is more complicated than regular sequencing data because bisulfite treatment converts unmethylated cytosines to uracils and following PCR amplification generates two more strands—the reverse complements of Watson and Crick strands. This causes difficulty in mapping sequence reads to a reference genome because thymines in

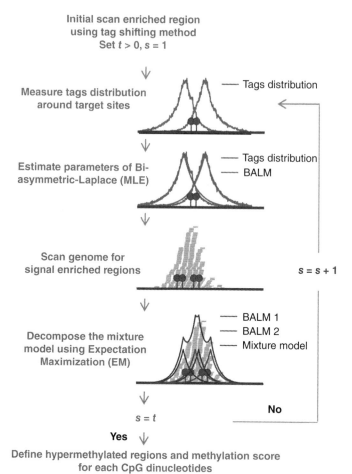

Initial scan enriched region
using tag shifting method
Set $t > 0$, $s = 1$

Measure tags distribution
around target sites
— Tags distribution

Estimate parameters of Bi-
asymmetric-Laplace (MLE)
— Tags distribution
— BALM

Scan genome for
signal enriched regions
$s = s + 1$

Decompose the mixture
model using Expectation
Maximization (EM)
— BALM 1
— BALM 2
— Mixture model

$s = t$
No

Yes

Define hypermethylated regions and methylation score
for each CpG dinucleotides

Figure 9.9 Flow diagram of BALM. Source: Lan et al. 2011 (59). CC BY. Please see www.wiley.com/go/Mandoiu/NextGenerationSequencing for a color version of this figure.

reads represent both original thymines and cytosines converted by bisulfite treatment, resulting in an asymmetric mismatch between read and reference sequences. BSMAP (78), BS-Seeker (79), Bismark (80), BRAT (81, 82), RMAP-BS (83), MethylCoder (84), SOCS-B (85), and B-SOLANA (86) have been developed to overcome this problem. See the previous chapter for details.

9.3.1.2 Methylation Quantification The methylation level of CpG sites is defined as a beta score. For a CpG site with m methylated (CpG) and u unmethylated (TpG or CpA) reads, the beta score is $m/(m + u)$. For a high-quality methylation profile, several filtering schemes are applied to reads if the base quality score of a cytosine is low, for example, 10; if the mapping quality score of a read is low, for example, 10; or if the read depth of a CpG is lower, for example, 6. In Figure 9.10, since the third CpG has Ts in all reads, its beta value is 0, while the beta value of the fourth is $3/3 = 1$.

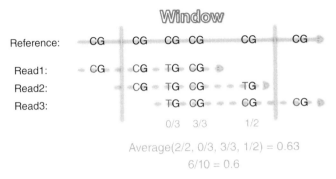

Figure 9.10 Example showing how to compute methylation score for CpGs or windows.

Interestingly, the last CpG has 0.5, which results from differential methylation of a cell population or alleles.

Since adjacent CpGs have similar methylation to enable working together on regulating cell activity, a sliding window approach is frequently used for methylation profiling. Most studies have used 100 or 200 bp windows for hot spots of methylation change and often 1000 or 5000 bp for global methylation. There are two ways to compute the beta score of a sliding window. In Figure 9.10, a window has 4 CpGs. Its beta score is defined as the average of beta scores of the CpGs or the fraction of the number of methylated CpGs to the total number of CpGs in reads: 0.63 or 0.6.

9.3.1.3 Differential Methylation One of the most important goals in DNA methylation studies is to identify DMRs (Differentially Methylated Regions) among samples or groups. Differential methylation is quantified as the difference or fold change between the beta scores of two samples. Since beta values are in the range of 0–1, the difference, hereinafter called the delta score, is preferred more. Therefore, hyper- and hypomethylation have positive and negative delta scores, respectively. Statistical tests produce more relevant results than descriptive analysis for differential methylation. If there is no replicate for each sample or only one sample for each group, Fisher's exact test is performed for each CpG to a 2 by 2 contingency table for the number of methylated and unmethylated reads for two samples. If each sample has replicates or each group has two or more samples, then statistical tests such as the Student t test, Wilcoxon test, Rao–Scott test, and Kolmogorov–Smirnov test can identify DMRs for each CpG. Since statistical tests generate a p-value of each CpG and a genome has many CpGs, the p-values should be adjusted for multiple tests using Bonferroni, or false discovery rate. Analysis of variance (ANOVA) can be used to compare three or more samples of groups simultaneously. Several algorithms have been developed for better or alternative DMR identification. DMEAS (87) measures methylation entropy that represents the complexity of methylation patterns in a region using information content and examines whether a heterogeneous methylation pattern is allele-specific or cell-specific. CpG_MPs (88) searches hot spots of differential methylation using hotspot extension and adapted Shannon entropy. BSmooth (89) employs smoothing

to estimate the methylation level in a genomic region for a single sample and to identify DMRs based on a statistic that appropriately summarizes consistent differences
by accounting for biological variability.

9.3.1.4 Advanced Technologies Using Bisulfite Treatment Although sequencing
cost has decreased dramatically, sequencing whole genomes is still expensive, especially for large-scale projects. RRBS and TBS-seq are subgenomic sequencing protocols. It is important to check the reliability of sequence reads. RRBS uses a restriction
enzyme, so sequence reads must start with the cutting site. TBS-seq defines the capture regions, so the reads should be mapped to the target regions.

Bisulfite treatment cannot distinguish hydroxymethylation (5-hmC) from methylation (5-mC) and does not convert either. In other words, it converts unmethylated
cytosines (neither hydroxymethylated nor methylated) to thymines. OxBS-seq and
TAB-seq have been developed to sequence hydroxymethylation. OxBS-seq converts
both unmethylated and hydroxymethylated cytosines to thymines, while TAB-seq
converts both unmethylated and methylated cytosines (Table 9.1). To profile hydroxymethylation, OxBS-seq requires BS-seq as well and determines methylation status.

9.3.1.5 Pipelines For an end-to-end analysis, AMP-methylKit and BSpipe have
been developed. AMP maps RRBS and ERRBS reads using Bismark and calls
per-base methylation scores (90). methylKit performs summarizing methylation
over predefined regions or tiling windows, and measuring and visualizing similarity
among samples, principal component analysis (PCA), unsupervised clustering, and
DMR detection and annotation (91). As a comprehensive pipeline, BSpipe has the
same functions as AMP and methylKit and also supports different mappers, various
visualization methods, and external programs such as QDMR and BSmooth (92).
BSpipe also provides a user-friendly environment with GenePattern, a web-based
workflow system for microarray and NGS data analyses.

9.3.2 Available Approaches

9.3.2.1 DNA Methylation Entropy Analysis Software (DMEAS) ME (Methylation Entropy) (93) has been proposed to distinguish among distribution of
methylation pattern. As shown in Figure 9.11, the methylation of regions *A* and
B is the same in beta scores, that is, 0.5. However, the methylation patterns are
contrary to each other. In region *A*, half of the reads have fully methylated CpGs
but the other half are fully unmethylated CpGs, while in region *B* the methylation patterns of reads are completely different but all reads have the same beta

**TABLE 9.1   Cytosines in Sequence Reads Based on Different
Sequencing Protocols**

Base	Sequence	BS-seq	OxBS-seq	TAB-seq
C	C	T	T	T
5-mC	C	C	C	T
5-hmC	C	C	T	C

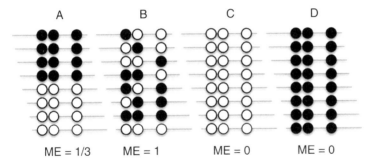

Figure 9.11 Methylation entropy for four examples. While A and B have the same methylation score but different methylation entropy, C and D have the same entropy but different score.

score. This difference is important in the studies of heterogeneous diseases such as cancer and of allele-specific methylation. Methylation entropy is defined as $ME = 1/b \times \Sigma_i(-n_i/N)\log_2(n_i/N)$ where b is the number of CpGs, n_i is the number of occurrence of methylation pattern i among reads, and N is the total number of reads. Therefore, $ME(A) = 1/3 * (-4/8 * \log_2 4/8) \times 2 = -1/3 \times \log_2 1/2 = 1/3$ and $ME(B) = 1/3 * (-1/8 \times \log_2 1/8) \times 8 = -1/3 \times \log_2 1/8 = 1$. Interestingly, regions C and D have the same pattern, that is, the same ME score, although their beta scores are 0 and 1. Therefore, ME can distinguish the consistency of single molecule methylation pattern, that is, the higher the ME, the more random the methylation pattern. This can be used to decide whether intermediate methylation of a cell population is either stochastic or allele-specific.

9.3.2.2 Quantitative Differentially Methylated Region (QDMR)

QDMR (94) has adapted Shannon entropy to identify DMRs across samples. Shannon entropy represents the uncertainty (unpredictability) of a random variable, that is, the complexity of methylation values across samples, and is defined for a site or region as $H_O = -\Sigma_s p_s \log_2 p_s$ where p_s is the probability of occurrence of the methylation value m_s of sample s, that is, $m_s/\Sigma_s m_s$. QDMR requires methylation value in a range of [0,1] or [0,100]. If all samples have the same methylation value, then H_o is maximized (Figure 9.12a) because the probability of each methylation value is equal, and therefore the uncertainty is maximal. If a fair die is rolled, the outcome is maximally uncertain (unpredicted). If a die is highly biased to a side, then the outcome is highly predictable and the entropy is low. Similarly, if a sample's methylation differs from other samples, then the entropy is lower than the maximum. In other words, the higher the difference, the lower the entropy. However, while hypermethylation in a sample results in low entropy (Figure 9.12b and d), hypomethylation does not (Figure 9.12c and e). The ROKU (95) method has been proposed to solve this problem using a one-step Tukey biweight (a weighted mean T), resulting in the processed methylation values $m_s' = |m_s - T|$. See Reference 94 for more details. The processed entropy is defined as $H_P = -\Sigma_s p_s' \log_2 p_s'$ where $p_s' = m_s'/\Sigma_s m_s'$. In Figure 9.12, the green dots represent the processed methylation values, and H_P in plots B (D) and C (E) are the

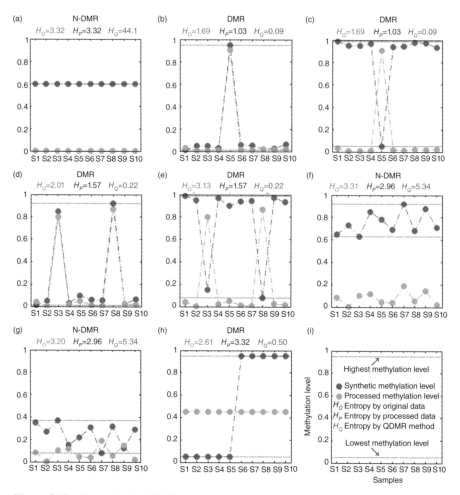

Figure 9.12 Examples for QDMR entropy computation where the red and green points represent the original and processed methylation scores for samples, respectively, and H_O, H_P, and H_Q represent the raw, processed, and QDMR methylation entropy, respectively. Source: Zhang et al. 2011 (94). Reproduced with permission of Oxford University Press. Please see www.wiley.com/go/Mandoiu/NextGenerationSequencing for a color version of this figure.

same as expected because they have the same difference from 0.5. Similarly, plots f and g show that the difference in the range of methylation values does not matter.

However, H_P in plot H is high although this is a DMR. Even in Figure 9.13, H_O and H_P in the plots are the same and lower than a threshold for DMRs, but plot a is only a true DMR. To overcome this issue, QDMR proposed a new entropy H_Q, which is adjusted by a methylation weight w, that is, $H_Q = H_p \times w$ and $w = |\log_2(\max(m_s) - \min(m_s)/(MAX - MIN) + \epsilon|$ where MAX and MIN are the highest and lowest methylation values, respectively, that is, 1 and 0, and ϵ is a small value (say 0.0001) used to avoid zero values in the logarithm. In Figure 9.13,

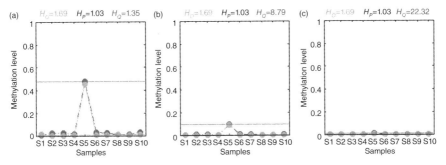

Figure 9.13 Examples with the same H_O and H_P but different H_Q. Source: Zhang et al. 2011 (94). Reproduced with permission of Oxford University Press.

H_Q is different for each plot and Figure 9.12h has lower H_Q than the threshold for DMRs.

It is very important to choose an appropriate threshold for H_Q to identify DMRs. To model experimental variability, QDMR assumed that the distribution of methylation values in a region follows a normal distribution with mean equal to zero and an unknown small standard deviation (SD). First, it chose 0.07 for SD, which allows methylation variation between 43 and 57 in 68% of the samples, between 36 and 64 in 95%, and between 29 and 71 in 99%. Second, it simulated 5000 uniformly methylated regions across samples and then computed entropy H_Q for each region. Third, it determined the entropy value at p-value of 0.05 (one-sided) as a threshold from the distribution of 5000 entropies. Finally, it repeated this process 10 times and used the mean of the 10 thresholds as the final threshold for DMR identification.

9.3.2.3 Identification of CpG Methylation Patterns of Genomic Regions (CpG_MPs)
The tool CpG_MPs (88) searches hot spots of differential methylation using hotspot extension and adapted Shannon entropy. It first measures beta values of CpGs with 5 or more reads and divides a CpG into four categories based on its beta score: *unmethylated* for [0,0.3], *partially unmethylated* for (0.3,0.5], *partially methylated* for (0.5,0.7), and *methylated* for [0.7,1]. Next, it searches for *unmethylated* and *methylated* hot spots that have at least n consecutively *unmethylated* and *methylated* CpGs, respectively. Then, it extends the unmethylated hotspots toward both ends until reaching the second partially methylated or methylated CpG. Similarly, it extends methylated hotspots. Finally, it combines two extended hotspots with the same methylation if the distance is less than 200 bp and computes the mean and standard deviation of the beta values of CpGs for each methylated or unmethylated region (finally merged, extended hotspots) as shown in Figure 9.14a.

To identify DMRs, CMRs (Conservatively Methylated Regions), and CUMRs (Conservatively UnMethylated Regions) among multiple samples, CpG_MPs defined two measures for an OR (Overlapping Region) among methylated or unmethylated regions for two or more samples: $u = (n_m - n_u)/(n_m - n_u)$ and $v = (n_m + n_u)/N$ where n_m and n_u represent the number of samples with methylated and unmethylated regions, respectively, and N represents the number of samples

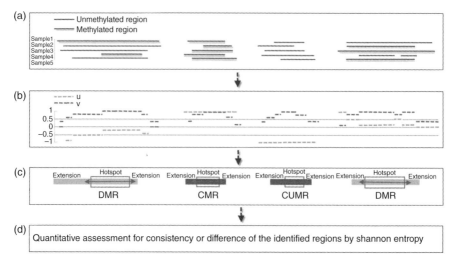

Figure 9.14 Algorithm to identify DMRs, CMRs, and CUMRs among samples. Source: Su et al. 2013 (88). Reproduced with permission of Oxford University Press. Please see www.wiley.com/go/Mandoiu/NextGenerationSequencing for a color version of this figure.

(Figure 9.14b). The measure u determines the methylation status of ORs: CMR if $u=1$, DMR if $-1 < u < 1$, and CUMR if $u = -1$. The measure v represents the ratio of samples in an OR, resulting in reliability of methylation patterns. To merge adjacent ORs with the same methylation pattern, the algorithm chooses an OR with the maximal v and merges with adjacent ORs if their distance is greater than 200 bp and the v values of the adjacent OR are greater than 0.5. This merge is repeated until no ORs with $= 0.5$ remains. The methylation status of a merged region is determined by u_1 and u_2, which are the u values of hotspot and adjacent ORs: CMR if $u_1 = 1$ and $u_2 = 1$, CUMR if $u_1 = -1$ and $u_2 = -1$, and DMR otherwise. DMRs are divided into hyper- or hypomethylated regions based on $u_1, u_2 > 0$ or $u_1, u_2 < 0$.

To quantitatively assess the methylation difference of identified regions, CpG_MPs used modified Shannon entropy proposed in QDMR (94) with the average methylation levels of CpGs in the identified regions. The lower entropy of DMRs represents the larger methylation change among samples, as described in QDMR. However, the entropy of CMRs and CUMRs are differently interpreted, that is, the higher entropy represents the more consistent methylation. CpG_MPs applied the QDMR method to determine the threshold of entropy for significant DMRs, CMRs, or CUMRs.

9.3.2.4 BSmooth BSmooth has proposed a detection method for differential methylated regions using a smoothing technique, which smoothes the methylation of a CpG by incorporating the methylation of adjacent CpGs. Instead of simply averaging the methylation of CpGs in a window, BSmooth estimates smoothed methylation profile for a window with $= 2$ kbp and $= 70$ CpGs using a local likelihood smoother (96), which is suitable for binary data with a non-additive Gaussian error model and is implemented in R package "locfit". The local likelihood smoother

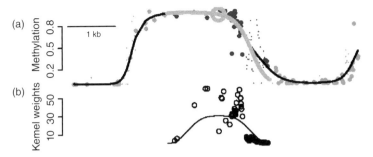

Figure 9.15   Example showing how to smooth methylation scores. Source: Hansen et al. 2011 (36). Open access.

fits the methylation of a CpG in a given window by approximating weighted logistic link function by a second-degree polynomial using tricube kernel and binomial distribution. Figure 9.15a shows the smoothing result (the thick grey line) for a CpG (the thick gry circle) with CpGs within a window (the thick black dots), and Figure 9.15b shows the kernel curve and points for actual weight of CpGs. The smoothing took into account the read coverage: the more read, the bigger weight. The black line in Figure 9.15a represents the final smooth methylation profile estimated for all CpGs.

BSmooth has applied a signal-to-noise statistic (SNS) to identify DMRs between groups of samples and take into account biological variation. For a window, a SNS is defined as $(\mu_T - u_C)/(\sigma_T + \sigma_C)$ where μ_T and σ_T (u_C and σ_C) represent the mean and standard deviation, respectively, of the treatment (control respectively) group. Furthermore, the denominator was approximated by $\sigma \times \text{sqrt}(1/n_1 + 1/n_2)$ where n_1 and n_2 represent the number of samples in treatment and control groups, respectively. The σ is a location-dependent standard deviation to represent biological variation across samples in a window and was estimated by smoothing using a running mean with a window size of 101 the standard deviation, which was computed empirically across two groups and floored at their 75 percentile. DMRs are determined if their SNS is greater than c or less than $-c$ where c is a cutoff examined based on the marginal empirical distribution. It was recommended to use only CpGs with some coverage in most or all samples and to retain DMRs with three or more CpGs, average methylation difference = 0.1, and at least one CpG every 300 bp.

The above-described algorithm was modified for cancer data because of its heterogeneity, that is, cancer cells are highly variable while normal cells have similar methylation. The algorithm was also modified for detecting large hypomethylated regions in cancer as reported in Hansen et al. (36). To these ends, the window size of a DMR was increased to have 500 CpGs of at least 40 kbp wide, and the cutoff for SNS was decreased. These modifications led to including more CpGs into DMRs and detecting large-scale changes by smoothing large, wider windows. The algorithm estimated σ only using normal samples, as described earlier, and then identified small DMRs with < 10 kbp after identifying large-scale DMRs using a hybrid model for an

SNS, that is, $SNS_1 + SNS_2$, where they represent the SNS of small- and large-scale DMRs, respectively.

9.3.2.5 BSpipe BSpipe is a comprehensive package to analyze BS-seq for sequence mapping, methylation quantification based on sliding windows or genomic features, DMR identification and annotation, methylation profiling using PCA and unsupervised clustering, and visualization on the UCSC genome browser and IGV.

Sequence Mapping BSpipe maps bisulfite-treated reads to reference sequences using *in silico* conversion. Briefly, reference sequences are converted and indexed by an in-house script and indexing programs for each of the Watson and Crick strands. Similarly, BS-seq reads are converted once or twice, depending on sequencing protocols, and mapped onto two reference indexes. Therefore, each read can be aligned two or four times. BestSam identifies the best mapping among all alignments for each read by realigning the original reads to the original references and recalculating mismatches and indels, except mismatches due to bisulfite conversion. If there are two or more best alignments in different locations, the read is reported as a multiple mapped read. BestSam finally generates an augmented SAM (MSAM) file including the mapping mode and the positions of methylated and unmethylated cytosines. The mapping mode specifies which base C or G in a read is converted by bisulfite treatment and used for measuring methylation. This pipeline summarizes mapping results such as the numbers of mapped reads, along with the number of errors and the number of mapping locations and those in each reference. BSpipe can use any mapping program by configuring to specify a parameter setting–indexing program, mapping program, seed length, allowing mismatches and indels, SAM output format, and mapping strands if available. Currently as a default, the program supports Bowtie (51), BWA (50), and SOAP2 (52).

Methylation quantification A sorted, binary MSAM (MBAM) file is generated and used to count the number of reads with methylated and unmethylated cytosines for each cytosine position in both strands of the reference genome. This module generates a GFF file in which each line specifies the methylation of a cytosine with the following bases, the number of reads, and the sum of base quality scores with methylated and unmethylated cytosines. Finally, the module generates a BED file for a cytosine pattern, for example, CpG, in which each line specifies the total number of reads (read coverage) and methylation fraction (number of reads with cytosine divided by the read coverage). Users can specify particular patterns of interest, for example, CpH, CpHpG, and CpHpH. BSpipe supports a sliding window approach to combine the methylation of adjacent CpGs because they work together in biological regulation. It outputs the number of CpGs and the mean and standard deviation of methylation for fix-length windows with a step. It can also quantify the methylation of genomic feature such as gene promoters, exons, introns, CpG islands (CGIs), and repeats for the entire region or fix-length windows.

Sample comparison After profiling the methylation of various types of regions, the most important task is to discover outliers or to see the global methylation changes among samples. BSpipe has several modules for this purpose. First, it measures the Pearson or Spearman's correlation coefficient for each pair of samples and generates a heat map to cluster samples based on a matrix of the correlation coefficients. Second, PCA is performed and samples are plotted using the first and second components. Finally, BSpipe conducts hierarchical clustering and k-means to identify differential methylation in an unsupervised manner.

Differential methylation among groups BSpipe identifies DMRs in descriptive way by methylation difference between two samples or groups. It also assesses the significance of DMRs between two groups using statistical tests such as Student t, Wilcoxon, and Kolmogorov–Smirnov tests, and among three or more groups using ANOVA. BSpipe generates adjusted p-values for multiple comparisons, and diagnostic plots such as histogram plots of p-values and adjusted p-values, scatter plots between two samples, and a volcano plot and heat map for DMRs.

External programs and parallel execution BSpipe prepares the input of and runs external programs such as QDMR and BSmooth to enable user's complementary analysis. To take advantage of a multicore computer or a computer cluster, BSpipe coupled with grid engines, for example, sun grid engine, through a configuration file. It splits input files in the FASTQ format into small multiple files and then submits a job for each split file to perform from sequence mapping to methylation quantification. Finally, the methylation of CpGs is merged. BSpipe also supports annotation and visualization of DMRs as depicted in the next section.

9.4 CONCLUSION

Enrichment-based and bisulfite treatment-based approaches described in this chapter have successfully profiled DNA methylation and identified genome-wide differential methylation, for example, References 18, 97–99. As analysis results, sequence mapping and methylation profile and differential methylation of CpGs, tiling windows, or genomic features are stored in various formats such as BAM, wiggle, GFF, and BED, which can be visualized on the UCSC genome browser (68), gbrowser (69), IGV (70), or IGB (71).

Identified DMRs are annotated with genomic features such as genes, CGIs (CpG Islands), noncoding RNAs, and repeats in the NCBI and UCSC databases. Specifically, a gene is regarded as having upstream, exon, intron, and downstream components. Users can set the length of the upstream and downstream part of the genes. Coding exons are further classified into 5′ UTR, 5′ CDS, middle CDS, 3′ CDS, and 3′ UTR. If a window overlaps two or more categories, then users can prioritize them. Furthermore, the annotation distinguishes the strands, that is, categories for each strand, and then merges them into bi-upstream, upstream, bi-downstream, downstream, and the categories for gene body. In the comparison with CpG islands (CGIs), DMRs are stratified into north CpG shelves and shores, inside CGIs, and south shelves and shores. For each of the other features such as repeats and noncoding RNAs, a DMR is classified as an inside or outside feature. The annotation results can be summarized in tables and pie charts.

To precisely uncover the function of genes regulated by DMRs, functional analysis should be performed with GO terms, pathways, and protein–protein interactions using PANTHER (100), DAVID (101), or IPA (102). The DMRs are segregated based on the location to a gene: promoter, downstream, body, exon, or intron. They are also intergrated into epigenomic marks for histone modification, transcription factors, chromatin structure, and nucleosome positioning in ENCODE (103), epigenome ATLS (104), and TCGA (105) by overlapping or distance using BEDTools (106) or FeatureCount (107).

REFERENCES

1. Smith ZD, Meissner A. DNA methylation: roles in mammalian development. Nat Rev Genet 2013;14(3):204–220.
2. Okano M et al. DNA methyltransferases Dnmt3a and Dnmt3b are essential for de novo methylation and mammalian development. Cell 1999;99(3):247–257.
3. Lister R et al. Human DNA methylomes at base resolution show widespread epigenomic differences. Nature 2009;462(7271):315–322.
4. Baylin SB. DNA methylation and gene silencing in cancer. Nat Clin Pract Oncol 2005;2 Suppl 1:S4–s11.
5. Jones PA, Baylin SB. The epigenomics of cancer. Cell 2007;128(4):683–692.
6. Jones PA, Baylin SB. The fundamental role of epigenetic events in cancer. Nat Rev Genet 2002;3(6):415–428.
7. Egger G et al. Epigenetics in human disease and prospects for epigenetic therapy. Nature 2004;429(6990):457–463.
8. Irier HA, Jin P. Dynamics of DNA methylation in aging and Alzheimer's disease. DNA Cell Biol 2012;31(S1):S42–S48.
9. Jayaraman S. Epigenetics of autoimmune diabetes. Epigenomics 2011;3(5):639–648.
10. Movassagh M, Vujic A, Foo R. Genome-wide DNA methylation in human heart failure. Epigenomics 2011;3(1):103–109.
11. Robertson KD. DNA methylation and human disease. Nat Rev Genet 2005;6(8):597–610.
12. Kriaucionis S, Heintz N. The nuclear DNA base 5-hydroxymethylcytosine is present in Purkinje neurons and the brain. Science 2009;324(5929):929–930.
13. Tahiliani M et al. Conversion of 5-methylcytosine to 5-hydroxymethylcytosine in mammalian DNA by MLL partner TET1. Science 2009;324(5929):930–935.
14. He YF et al. Tet-mediated formation of 5-carboxylcytosine and its excision by TDG in mammalian DNA. Science 2011;333(6047):1303–1307.
15. Ito S et al. Tet proteins can convert 5-methylcytosine to 5-formylcytosine and 5-carboxylcytosine. Science 2011;333(6047):1300–1303.
16. Ito S et al. Role of Tet proteins in 5mC to 5hmC conversion, ES-cell self-renewal and inner cell mass specification. Nature 2010;466(7310):1129–1133.
17. Koh KP et al. Tet1 and Tet2 regulate 5-hydroxymethylcytosine production and cell lineage specification in mouse embryonic stem cells. Cell Stem Cell 2011;8(2):200–213.
18. Lee E-J et al. Targeted bisulfite sequencing by solution hybrid selection and massively parallel sequencing. Nucleic Acids Res 2011;39(19):e127.

19. Pastor WA, Aravind L, Rao A. TETonic shift: biological roles of TET proteins in DNA demethylation and transcription. Nat Rev Mol Cell Biol 2013;14(6):341–356.

20. Wu H, Zhang Y. Mechanisms and functions of Tet protein-mediated 5-methylcytosine oxidation. Genes Dev 2011;25(23):2436–2452.

21. Lian CG et al. Loss of 5-hydroxymethylcytosine is an epigenetic hallmark of melanoma. Cell 2012;150(6):1135–1146.

22. Callinan PA, Feinberg AP. The emerging science of epigenomics. Hum Mol Genet 2006;15 Suppl 1:R95–R101.

23. Laird PW. Principles and challenges of genome-wide DNA methylation analysis. Nat Rev Genet 2010;11(3):191–203.

24. Lister R, Ecker JR. Finding the fifth base: genome-wide sequencing of cytosine methylation. Genome Res 2009;19(6):959–966.

25. Zilberman D, Henikoff S. Genome-wide analysis of DNA methylation patterns. Development 2007;134(22):3959–3965.

26. Taiwo O et al. Methylome analysis using MeDIP-seq with low DNA concentrations. Nat Protoc 2012;7(4):617–636.

27. Ficz G et al. Dynamic regulation of 5-hydroxymethylcytosine in mouse ES cells and during differentiation. Nature 2011;473(7347):398–402.

28. Wu H et al. Genome-wide analysis of 5-hydroxymethylcytosine distribution reveals its dual function in transcriptional regulation in mouse embryonic stem cells. Genes Dev 2011;25(7):679–684.

29. Serre D, Lee BH, Ting AH. MBD-isolated Genome Sequencing provides a high-throughput and comprehensive survey of DNA methylation in the human genome. Nucleic Acids Res 2010;38(2):391–399.

30. Rauch TA, Pfeifer GP. The MIRA method for DNA methylation analysis. Methods Mol Biol 2009;507:65–75.

31. Laurent L et al. Dynamic changes in the human methylome during differentiation. Genome Res 2010;20(3):320–331.

32. Li YR et al. The DNA methylome of human peripheral blood mononuclear cells. PLoS Biol 2010;8(11):e1000533.

33. Hodges E et al. Directional DNA methylation changes and complex intermediate states accompany lineage specificity in the adult hematopoietic compartment. Mol Cell 2011;44(1):17–28.

34. Berman BP et al. Regions of focal DNA hypermethylation and long-range hypomethylation in colorectal cancer coincide with nuclear lamina-associated domains. Nat Genet 2012;44(1):40–62.

35. Hon GC et al. Global DNA hypomethylation coupled to repressive chromatin domain formation and gene silencing in breast cancer. Genome Res 2012;22(2):246–258.

36. Hansen KD et al. Increased methylation variation in epigenetic domains across cancer types. Nat Genet 2011;43(8):768–775.

37. Meissner A et al. Reduced representation bisulfite sequencing for comparative high-resolution DNA methylation analysis. Nucleic Acids Res 2005;33(18):5868–5877.

38. Ball MP et al. Targeted and genome-scale strategies reveal gene-body methylation signatures in human cells. Nat Biotechnol 2009;27(4):361–368.

39. Deng J et al. Targeted bisulfite sequencing reveals changes in DNA methylation associated with nuclear reprogramming. Nat Biotechnol 2009;27(4):353–360.

40. Hodges E et al. High definition profiling of mammalian DNA methylation by array capture and single molecule bisulfite sequencing. Genome Res 2009;19(9):1593–1605.

41. Komori HK et al. Application of microdroplet PCR for large-scale targeted bisulfite sequencing. Genome Res 2011;21(10):1738–1745.

42. Lee EJ et al. Analyzing the cancer methylome through targeted bisulfite sequencing. Cancer Lett 2013;340(2):171–178.

43. Huang Y et al. The behaviour of 5-hydroxymethylcytosine in bisulfite sequencing. PLoS ONE 2010;5(1):e8888.

44. Booth MJ et al. Quantitative sequencing of 5-methylcytosine and 5-hydroxymethylcytosine at single-base resolution. Science 2012;336(6083):934–937.

45. Booth MJ et al. Oxidative bisulfite sequencing of 5-methylcytosine and 5-hydroxymethylcytosine. Nat Protoc 2013;8(10):1841–1851.

46. Yu M et al. Base-resolution analysis of 5-hydroxymethylcytosine in the mammalian genome. Cell 2012;149(6):1368–1380.

47. Flusberg BA et al. Direct detection of DNA methylation during single-molecule, real-time sequencing. Nat Methods 2010;7(6):461–465.

48. Laszlo AH et al. Detection and mapping of 5-methylcytosine and 5-hydroxymethylcytosine with nanopore MspA. Proc Natl Acad Sci U S A 2013;110(47):18904–18909.

49. Schreiber J et al. Error rates for nanopore discrimination among cytosine, methylcytosine, and hydroxymethylcytosine along individual DNA strands. Proc Natl Acad Sci U S A 2013;110(47):18910–18915.

50. Li H, Durbin R. Fast and accurate short read alignment with Burrows-Wheeler transform. Bioinformatics 2009;25(14):1754–1760.

51. Langmead B et al. Ultrafast and memory-efficient alignment of short DNA sequences to the human genome. Genome Biol 2009;10(3):R25.

52. Li R et al. SOAP: short oligonucleotide alignment program. Bioinformatics 2008;24(5):713–714.

53. Zhang Y et al. Model-based analysis of ChIP-Seq (MACS). Genome Biol 2008;9(9):R137.

54. Zang C et al. A clustering approach for identification of enriched domains from histone modification ChIP-Seq data. Bioinformatics 2009;25(15):1952–1958.

55. Boyle AP et al. F-Seq: a feature density estimator for high-throughput sequence tags. Bioinformatics 2008;24(21):2537–2538.

56. Nix D, Courdy S, Boucher K. Empirical methods for controlling false positives and estimating confidence in ChIP-Seq peaks. BMC Bioinformatics 2008;9(1):523.

57. Down TA et al. A Bayesian deconvolution strategy for immunoprecipitation-based DNA methylome analysis. Nat Biotechnol 2008;26(7):779–785.

58. Chavez L, Dietrich J. MEDIPS: MeDIP-Seq data analysis. in R package version 1.4.0; 2010.

59. Lan X et al. High resolution detection and analysis of CpG dinucleotides methylation using MBD-Seq technology. PLoS ONE 2011;6(7):e22226.

60. Xu H et al. An HMM approach to genome-wide identification of differential histone modification sites from ChIP-seq data. Bioinformatics 2008;24(20):2344–2349.

61. Ross-Innes CS et al. Differential oestrogen receptor binding is associated with clinical outcome in breast cancer. Nature 2012;481(7381):389–393.

62. Liang K, Kele? S. Detecting differential binding of transcription factors with ChIP-seq. Bioinformatics 2012;28(1):121–122.

63. Robinson MD, McCarthy DJ, Smyth GK. edgeR: a Bioconductor package for differential expression analysis of digital gene expression data. Bioinformatics 2010;26(1):139–140.

64. Anders S, Huber W. Differential expression analysis for sequence count data. Genome Biol 2010;11(10):R106.

65. Lienhard M et al. MEDIPS: genome-wide differential coverage analysis of sequencing data derived from DNA enrichment experiments. Bioinformatics 2014;30(2):284–286.

66. Wilson G et al. Resources for methylome analysis suitable for gene knockout studies of potential epigenome modifiers. GigaScience 2012;1(1):3.

67. Huang J et al. MeQA: a pipeline for MeDIP-seq data quality assessment and analysis. Bioinformatics 2012;28(4):587–588.

68. Kent WJ et al. The human genome browser at UCSC. Genome Res 2002;12(6):996–1006.

69. Stein LD et al. The generic genome browser: a building block for a model organism system database. Genome Res 2002;12(10):1599–1610.

70. Thorvaldsdóttir H, Robinson JT, Mesirov JP. Integrative Genomics Viewer (IGV): high-performance genomics data visualization and exploration. Brief Bioinform 2013;14(2):178–192.

71. Nicol JW et al. The Integrated Genome Browser: free software for distribution and exploration of genome-scale datasets. Bioinformatics 2009;25(20):2730–2731.

72. Chavez L et al. Computational analysis of genome-wide DNA methylation during the differentiation of human embryonic stem cells along the endodermal lineage. Genome Res 2010;20(10):1441–1450.

73. Eckhardt F et al. DNA methylation profiling of human chromosomes 6, 20 and 22. Nat Genet 2006;38(12):1378–1385.

74. Weber M et al. Chromosome-wide and promoter-specific analyses identify sites of differential DNA methylation in normal and transformed human cells. Nat Genet 2005;37(8):853–862.

75. Pelizzola M et al. MEDME: an experimental and analytical methodology for the estimation of DNA methylation levels based on microarray derived MeDIP-enrichment. Genome Res 2008;18(10):1652–1659.

76. Venkatraman ES, Olshen AB. A faster circular binary segmentation algorithm for the analysis of array CGH data. Bioinformatics 2007;23(6):657–663.

77. Lan X et al. W-ChIPeaks: a comprehensive web application tool for processing ChIP-chip and ChIP-seq data. Bioinformatics 2011;27(3):428–430.

78. Xi Y, Li W. BSMAP: whole genome bisulfite sequence MAPping program. BMC Bioinformatics 2009;10(1):232.

79. Chen P-Y, Cokus S, Pellegrini M. BS Seeker: precise mapping for bisulfite sequencing. BMC Bioinformatics 2010;11(1):203.

80. Krueger F, Andrews SR. Bismark: a flexible aligner and methylation caller for Bisulfite-Seq applications. Bioinformatics 2011;27(11):1571–1572.

81. Harris EY et al. BRAT: bisulfite-treated reads analysis tool. Bioinformatics 2010;26(4):572–573.

82. Harris EY et al. BRAT-BW: efficient and accurate mapping of bisulfite-treated reads. Bioinformatics 2012;28(13):1795–1796.

83. Smith AD et al. Updates to the RMAP short-read mapping software. Bioinformatics 2009;25(21):2841–2842.

84. Pedersen B et al. MethylCoder: software pipeline for bisulfite-treated sequences. Bioinformatics 2011;27(17):2435–2436.

85. Ondov BD et al. An alignment algorithm for bisulfite sequencing using the Applied Biosystems SOLiD System. Bioinformatics 2010;26(15):1901–1902.

86. Kreck B et al. B-SOLANA: an approach for the analysis of two-base encoding bisulfite sequencing data. Bioinformatics 2012;28(3):428–429.

87. He J et al. DMEAS: DNA methylation entropy analysis software. Bioinformatics 2013;29(16):2044–2045.

88. Su J et al. CpG_MPs: identification of CpG methylation patterns of genomic regions from high-throughput bisulfite sequencing data. Nucleic Acids Res 2013;41(1):e4.

89. Hansen K, Langmead B, Irizarry R. BSmooth: from whole genome bisulfite sequencing reads to differentially methylated regions. Genome Biol 2012;13(10):R83.

90. Akalin A. AMP for aligning ERRBS and RRBS reads. Available at http://code.google.com/p/amp-errbs/. Accessed 2016 Mar 17.

91. Akalin A et al. MethylKit: a comprehensive R package for the analysis of genome-wide DNA methylation profiles. Genome Biol 2012;13(10):R87.

92. Darby AC et al. Characteristics of the genome of Arsenophonus nasoniae, son-killer bacterium of the wasp Nasonia. Insect Mol Biol 2010;19:75–89.

93. Xie H et al. Genome-wide quantitative assessment of variation in DNA methylation patterns. Nucleic Acids Res 2011;39(10):4099–4108.

94. Zhang Y et al. QDMR: a quantitative method for identification of differentially methylated regions by entropy. Nucleic Acids Res 2011;39(9):e58.

95. Kadota K et al. ROKU: a novel method for identification of tissue-specific genes. BMC Bioinformatics 2006;7(1):294.

96. Loader C. *Local Regression and Likelihood*. New York: Springer-Verlag; 1999.

97. Jin B et al. Linking DNA methyltransferases (DNMTs) to epigenetic marks and nucleosome structure genome-wide in human tumor cells. Cell Rep 2012;2(5):1411–1424.

98. Pei L et al. Genome-wide DNA methylation analysis reveals novel epigenetic changes in chronic lymphocytic leukemia. Epigenetics 2012;7(6):567–578.

99. Wang X et al. Function and evolution of DNA methylation in Nasonia vitripennis. PLoS Genet 2013;9(10):e1003872.

100. Mi H et al. The PANTHER database of protein families, subfamilies, functions and pathways. Nucleic Acids Res 2005;33 Suppl 1:D284–D288.

101. Huang DW, Sherman BT, Lempicki RA. Systematic and integrative analysis of large gene lists using DAVID bioinformatics resources. Nat Protoc 2008;4(1):44–57.

102. Rho M et al. De novo identification of LTR retrotransposons in eukaryotic genomes. BMC Genomics 2007;8:16.

103. ENCODE Project Consortium. An integrated encyclopedia of DNA elements in the human genome. Nature 2012;489(7414):57–74.

104. Bernstein BE et al. The NIH roadmap epigenomics mapping consortium. Nat Biotechnol 2010;28(10):1045–1048.

105. Cancer Genome Atlas Research Network. Comprehensive genomic characterization defines human glioblastoma genes and core pathways. Nature 2008;455(7216):1061–1068.

106. Quinlan AR, Hall IM. BEDTools: a flexible suite of utilities for comparing genomic features. Bioinformatics 2010;26(6):841–842.

107. Liao Y, Smyth GK, Shi W. featureCounts: an efficient general purpose program for assigning sequence reads to genomic features. Bioinformatics 2013;30(7):923–930.

10

BISULFITE-CONVERSION-BASED METHODS FOR DNA METHYLATION SEQUENCING DATA ANALYSIS

ELENA HARRIS

Department of Computer Science, California State University, Chico, CA, USA

STEFANO LONARDI

Department of Computer Science and Eng., University of California, Riverside, CA, USA

10.1 INTRODUCTION

DNA methylation is the addition of a methyl group at the carbon-5 residue of DNA cytosines. In mammals, DNA methylation is observed mainly at CpG dinucleotides (CpG islands). In stem cells and in plants, it is also extensively found in the context of CpHpG where H is either A, T, or C.

In mammals, methylation is first established *de novo* by methyltransferases Dnmt3a and Dnmt3b during early development and gametogenesis (1). It is propagated during DNA replication via maintenance methyltransferase Dnmt1, which recognizes hemi-methylated CpGs (2, 3). Similar *de novo* methyltransferases and maintenance methyltransferases are found in plants.

DNA methylation is also subject to active demethylation, including two rounds of extensive erasure in gametogenesis and before pre-implantation and selective demethylation at specific regions during development of the organism.

DNA methylation is a common epigenetic mechanism, which preserves cellular states through cell divisions without alterations in genomic sequences. The distinct property of the maintenance methyltransferase implies that the epigenetic information is copied across generations accurately.

Computational Methods for Next Generation Sequencing Data Analysis, First Edition.
Edited by Ion I. Măndoiu and Alexander Zelikovsky.
© 2016 John Wiley & Sons, Inc. Published 2016 by John Wiley & Sons, Inc.
Companion website: www.wiley.com/go/Mandoiu/NextGenerationSequencing

DNA methylation affects the phenotype of a cell primarily by regulating the local transcription potentials of certain genes. CpG methylation at promoters generally represses the transcription of affected genes. DNA methylation across a broad region, together with certain histone modifications, such as H3K9me3, plays an important role in the formation of heterochromatin, which inhibits gene expression within these regions.

Cell differentiation involves characteristic changes in DNA methylation patterns. One working model with certain experimental evidence is as follows: during cell differentiation, lineage-specific genes remain void of DNA methylation or undergo active demethylation if there is methylation originally; instead, genes related to the cell pluripotency or that are specific to other lineages gain DNA methylation and therefore become repressed.

DNA methylation has been associated with gene expression, imprinting, transposon silencing, X-chromosome inactivation, embryonic development, and cancer. As a consequence, mechanisms that guide DNA methylation and DNA methylation patterns have been vigorously studied for several decades. Various techniques are available to profile DNA methylation, either genome-wide or targeted at specific regions, such as CpG islands and gene promoters. In general, these techniques are divided into two categories: (i) enrichment based and (ii) bisulfite conversion based. Methods in the first category first enrich genomic regions with high methylation using antibody or methylation status-sensitive endonuclease followed by size selection; then, enriched regions are subject to hybridization via tiling microarrays or by sequencing. Typical methods in this category include MeDIP-Seq, MBD-Seq, and CHARM. Readers are referred to the review (4) for an introduction to these methods and their relative advantages and drawbacks. The second group of bisulfite-conversion-based methods is particularly attractive for methylome profiling because these methods have enabled life scientists to produce methylation maps at single-base resolution.

Bisulfite-conversion-based methods involve the following three steps:

1. chemical conversion of non-methylated cytosines to uracils by treating DNA with sodium bisulfite,
2. PCR amplification (that changes uracils to thymines), and
3. sequencing size-selected DNA fragments.

Initially, the sequencing of bisulfite-treated DNA fragments was carried out via the Sanger method, but the cost of this procedure restricted bisulfite-conversion-based methods to small-scale analysis. Recently, progress in next-generation sequencing technology made it possible to determine the methylation status of a whole genome (BS-seq). In eukaryotes, it was first used for relatively small genomes, such as that of *Arabidopsis thaliana* (5, 6) and reduced representation of large genomes (7, 8). It is now practical to profile whole large genomes such as human and other mammalian genomes (9, 10).

Compared to enrichment-based methods, BS-seq proves to be the most informative approach. It allows the examination of methylation status at single-base resolution, whereas enrichment-based methods only give averaged profile over a region. BS-seq is therefore capable of revealing subtle methylation patterns; for example, application of BS-seq to human stem cells revealed for the first time the extensive methylation of non-CpG sites (9). Another advantage of BS-seq is its capacity to yield quantitative measurement of DNA methylation frequency, which enables investigation of specific contiguous methylation patterns such as partially methylated regions, hypomethylated regions or allele-specific methylated regions (11–13). Finally, whole genome BS-seq achieves the most comprehensive coverage of genomes compared with other methods (14).

10.2   THE PROBLEM OF MAPPING BS-TREATED READS

The first step in methylation analysis with BS-seq is mapping bisulfite-treated sequenced reads (hereafter called *BS-reads*) back to the reference genome. In BS-reads mapping, potential matches of T in a read to C in a reference must be allowed in addition to the usual mapping of C in a read to C in a reference. Moreover, BS-mapping must take into account distinct types of bisulfite libraries: the first type, called *directional*, produces sequenced reads that are bisulfite-converted versions of two original genomic strands (9) and the second type, called *non-directional*, yields reads that correspond to four possible strands resulting from PCR steps (5). Figure 10.1 shows the effect of BS conversion and subsequent PCR amplification. There are four strands of PCR product: PCR_1^+, PCR_1^-, PCR_2^+, and PCR_2^-. In directional bisulfite libraries, the addition of special methylated adapters before bisulfite treatment ensures that only PCR_1^+ and PCR_2^- corresponding to the original genomic strands are sequenced. The nondirectional libraries result in sequencing all four strands of PCR amplification.

Recall that sequencing proceeds from the $5'$-end toward the $3'$-end. For paired-end reads, to produce the $3'$-end of a strand with Illumina sequencing instruments, the $5'$ end of the reverse-complement of the strand is sequenced. Figure 10.1 shows the paired-end sequenced reads resulting from all four PCR strands. In single-end sequencing, only $5'$ mates shown on Figure 10.1 are sequenced. The order of mates in paired-end reads is essential for correct BS-mapping. We illustrate this idea by discussing one of the mapping approaches, in which sequenced reads are mapped directly to the positive strand of the genome , DS^+. With directional bisulfite libraries, when the $5'$ mate maps onto DS^+, only T–C mismatches are legal (i.e., T in a read is allowed to map onto C in the genome). For paired-end reads, the reverse-complement of the corresponding $3'$ mate is mapped onto DS^+ with T–C mapping, indicating the unmethylated status of the cytosine. If the reverse-complement of the $5'$ mate maps onto DS^+, then only A–G mismatches must be allowed indicating the unmethylated status of the cytosine on DS^-, the base pair corresponding to G on DS^+; the corresponding $3'$ mate in paired-end reads maps directly onto DS^+ while allowing A–G mismatches. It is important to emphasize that only one kind of BS-mismatches (either T–C or A–G, but not both) must be allowed per single alignment of a read to a reference; moreover, in paired-end alignment, mates have the same kind of BS-mismatches.

With nondirectional libraries and with the same approach of mapping reads directly to DS^+, in addition to the listed above mappings, one must consider also

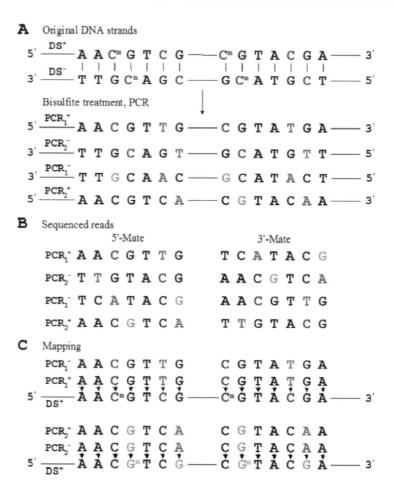

Figure 10.1 (a) DNA is treated with sodium bisulfite followed by PCR amplification, which results in unmethylated cytosines converted into thymines and methylated cytosines remained unchanged. (b) Directional bisulfite libraries generate reads from PCR_1^+ and PCR_2^- strands corresponding to the original positive and negative genomic strands, respectively, and nondirectional bisulfite libraries also generate reads from PCR_1^- and PCR_2^+ strands. (c) Mapping of sequenced BS-reads to the positive genomic strand is shown. The reads from PCR_1^+ and PCR_1^- strands contribute to estimation of methylation states of the cytosines on the positive genomic strand, and the reads from PCR_2^- and PCR_2^+ strands contribute to estimation of methylation states of the cytosines on the negative genomic strand.

the following additional cases. When the 5′ mate maps to DS^+ with allowed A–G mismatches, then a perfect alignment indicates that the read comes from PCR_2^+ strand and the reverse-complement of the corresponding 3′ mate maps with A–G mismatches to DS^+. The reverse-complement of the read maps with T–C allowable mismatches to DS^+, this means that the read comes from PCR_1^- strand and its 3′ mate maps to DS^+ with allowable T–C mismatches.

For the proper estimation of methylation status and level of cytosines, it is essential to determine the PCR strand from which each read comes from. Reads originating from PCR_1^+ and PCR_1^- indicate methylation of cytosines on the positive genomic strand, while reads generated from PCR_2^+ and PCR_2^- account for methylation of the cytosines on the negative genomic strand, as shown in Figure 10.1c.

The accuracy of mapping is essential for inferring properly methylated cytosines. Sequencing errors, adapter contamination, end-repair, and single-nucleotide polymorphisms (SNPs) may affect erroneous methylation calls (15). To avoid these problems, preprocessing procedures are used that check for base quality of reads and trim low-quality ends, and identify and remove the adapter sequences from the reads. Postmapping processing is also necessary for accurate methylation inference; these procedures usually remove extraneous copies of the reads mapped onto the same starting base in the reference (the so-called PCR- or copy-duplicates) and prevent double count of methylated reads from overlapped paired-end read mates.

10.3 ALGORITHMIC APPROACHES TO THE PROBLEM OF MAPPING BS-TREATED READS

The most time-efficient methods for mapping short reads either use hashing or data structures based on the Burrows–Wheeler transform or BWT (16) to reduce the space search. As mentioned earlier, BS-mapping introduces additional challenges because BS-treated reads do not match exactly a reference genome due to conversion of a fraction of cytosines to thymines. To allow for a T in a read to match a C in a genome (hereafter called BS-mismatches), different approaches have been introduced, all of which could be summarized into two categories: (i) all possible combinations of conversion of Ts to Cs in a read are considered; (ii) reference/read conversions are applied. First, we discuss how these approaches are used with hashing and then consider methods involving the BWT.

In both hashing and BWT, letters A, C, G, and T are converted to the corresponding bit representations 00, 01, 10, and 11, respectively. For the purpose of reducing search space, it is practical first to identify potential genomic regions from which each read was originated. A *seed* of length k of a read (also called a k-mer or a *q-gram*) is a substring of the read consisting of k consecutive bases. In hashing, seeds from a reference are hashed into a table: the binary representation of a seed is used as the hashing index, and the genomic position of the seed is stored in the corresponding bin. For each read, a subset of its seeds is hashed into the table and the read is aligned to only those genomic positions that share seeds with the read. To allow for BS-mismatches, some tools use all possible combinations of Ts converted to Cs in each seed and use both genomic strands for hashing, for instance VerJinxer (17), BSMAP (18), Pash (19), and BiSS (20). This method preserves the genome complexity, but introduces a large number of candidates affecting time efficiency. To further reduce search space, some tools (e.g., Pash) use heuristics that verify the distance between the genomic positions, to which different seeds of a read have been hashed, and keep only those candidate positions that are within appropriate distance from each other.

Other tools adopted various reference-to-reads conversions in order to allow for BS-mismatches such as RMAP-bs (21) and BRAT (22). For example, RMAP-bs uses wildcards, which allow certain bases in reads and genome to match to each other. Alternatively, one can convert the reference or the reads to allow BS-mismatches. BRAT introduced different bit representations of a genome and reads; it maintains two separate bit versions, *ta*-representation and *cg*-representation, one bit per letter each. In the *ta*-representation, all Cs are converted to Ts and all Gs are converted to As and bit 1 is assigned to Ts and bit 0 to As. In the *cg*-representation, Cs and Gs are converted to 1 and As and Ts to 0. This genome/reads bit representation allows mapping reads and their reverse-complements directly to the positive genomic strand and distinguishes between BS-mismatches and non-BS-mismatches (the alignment using this representation is discussed later).

Methods that employ the BWT are restricted to using reference/reads conversions and to the best of our knowledge do not use the approach, in which all combinations of Ts are converted to Cs in a read: BS-Seeker (23); Bismark (24); BRAT-bw (25). In addition to reference/reads conversion, tools based on the BWT also use multiseed approach to improve accuracy of mapping due to sequencing errors. These methods are more time efficient and in some cases (with large genomes when hashing is done on a reference) also more space efficient than hashing-based approaches. In hashing, the length of seeds that are used for indexing is fixed and usually is kept short because of space limitations: this results in a greater number of genomic positions, to which each read and its reverse-complement are aligned. With the BWT, the length of seeds is not fixed in advance: the longer is the seed matched using the BWT, the fewer are the genomic positions that need to be checked.

There are different genome/reads conversions used by the tools based on the BWT. Here, we will provide one such example used by BRAT-bw. Two different BWT indices are built on a positive strand of a reference: in the first index, all Cs are converted to Ts, and in the second all Gs are converted to As. The reads and their reverse-complements are also converted resulting in four versions of a single read; Cs are converted to Ts in a read and in the reverse-complement of the read, and Gs are converted to As in a read and in the reverse-complement of the read. Original reads with Cs converted to Ts and their reverse-complements with the same conversion are mapped to the AGT-reduced genome index, and the versions of original reads and their reverse-complements with Gs converted to As are mapped to the ACT-reduced genome index.

Whether one uses hashing or the BWT, once the potential genomic positions are identified, a full-length alignment is performed to estimate the accuracy of each mapping. Different tools choose different alignment strategies. Some tools employ a version of Smith–Waterman algorithm (26) that allows for insertions and deletions of nucleotides in a read (Pash, BiSS); other tools use the Hamming distance to count the number of non-BS-mismatches (RMAP-bs, BRAT-bw, BSMap); the tools that use Bowtie (27) as an underlying mapping tool use Bowtie's output to score the alignments (Bismark, BS-Seeker). If a read is mapped to two or more genomic locations with the same score, it is considered ambiguous and usually is not considered for further methylation estimation analysis.

To illustrate the full-length alignment of a read to the reference, we will describe BRAT's strategy (also used in BRAT-bw). The reads and their reverse-complements are mapped only to the positive genomic strand, DS^+. As mentioned earlier, BRAT uses *ta*-and *cg*-bit representations for genome and reads. Here, we give an example for a read and the genome's *k*-mer of the same length that shows all possible mappings of a letter in a read to a letter in a genome. Let AAAACCCCGGGGTTTT be a sequenced read and ACGTACGTACGTACGT be a 16-mer in the genome. Then *ta*-representation of the read (*ta-read*) is 0000111100001111 and *cg*-representation of the read (*cg-read*) is 0000111111110000. Similarly, the *ta*-representation of the genome's 16-mer (*ta-gen-mer*) is 0101010101010101 and *cg*-representation of the genome's 16-mer (*cg-gen-mer*) is 0110011001100110. BRAT uses fast bit-wise logical operations XOR, OR, AND, and NOT to allow BS-mismatches in an alignment. To allow for T in a read to map to C in a reference, the following statement is used:

((NOT (*ta-read* OR *cg-read*)) AND *cg-gen-mer*) OR

(*cg-read* XOR (*cg-read* AND *cg-gen-mer*)) OR

(*ta-read* XOR *ta-gen-mer*)

For example, if one applies this logic statement to our example of aligning the read AAAACCCCGGGGTTTT to the genome's 16-mer ACGTACGTACGTACGT, the result is 0111101111011010, where 1 at i^{th} position indicates a mismatch between the ith letter in the read and the ith letter in the genome, and 0 corresponds to a match. We can observe that the only matches that are allowed are A–A, C–C, G–G, T–T, and T–C.

To allow for A in a read to map to G in a reference, BRAT performs the following operation:

((*ta-read*AND (NOT *cg-read*)) AND *cg-gen-mer*) OR

(*cg-read* XOR (*cg-read* AND *cg-gen-mer*)) OR

(*ta-read* XOR *ta-gen-mer*)

When applying this procedure to our example, we obtain 0101101111011110, which demonstrates that the only matches that are allowed in this case are A–A, A–G, C–C, G–G, and T–T.

To distinguish between different PCR strands, BRAT uses the strategy described earlier. When the 5′ mate is mapped to DS^+ allowing only T–C mismatches, then the read comes from PCR_1^+ strand; mapping of 5′ mate to DS^+ allowing A–G mismatches indicates that the origin of the read is PCR_2^+ strand; if the reverse-complement of 5′ mate is mapped to DS^+ allowing A–G mismatches, then the read comes from PCR_2^- strand; and finally, if the reverse-complement of a read is mapped to DS^+ allowing T–C mismatches, then the read is generated from PCR_1^- strand. The reads that are resulted in the same-scored best alignments in two or more out of total four possible mappings are disregarded from the further analysis of methylation status estimation. For directional bisulfite libraries, only two full-length alignments are performed for each read: an original read to DS^+ allowing only T–C mismatches, and the reverse-complement of the read to DS^+ allowing only A–G mismatches. As with non-directional libraries, if both of these mapping result in the same best-score, the read is disregarded from further consideration.

10.4   METHYLATION ESTIMATION

After bisulfite-treated reads have been mapped to the reference genomes, copy/PCR duplicate reads removed, and ambiguously mapped reads discarded, the methylation state at each cytosine can be estimated. Each read mapped to a cytosine represents a DNA strand from a single cell. If T in a read is mapped to C in the genome, then the corresponding cytosine was not methylated in the cell from which the read was derived; alternatively, a C in a read aligned to C in the genome determines methylated cytosine in the cell's DNA. The methylation status of a cytosine typically varies across a population of cells. Therefore, the total count of Cs and Ts mapped to a cytosine in the reference is used to infer the methylation status of the cytosine. One of the common ways to estimate methylation status is to calculate the methylation level at a cytosine, that is, the ratio of the number of Cs mapped to the cytosine to the total number of reads mapped to that position. Another approach uses the binomial probabilistic model to reduce the percentage of false positives (6). The probability of observing y methylated cytosines at a genomic position with n mapped reads is calculated according to pre-estimated cumulative probability of observing C in a read mapped to C in a genome due to sequencing errors and cytosine conversion failure rate. The threshold for different values of n is chosen in such a way as to keep the false discovery rate below a given threshold.

10.5   POSSIBLE BIASES IN ESTIMATION OF METHYLATION LEVEL

In the scientific literature, some argue that mapping reads allowing T–C mismatches and not allowing C–T mismatches may introduce a bias in the correct estimation of methylation level because reads not containing Cs have more chance of mapping to different genomic positions (15). To avoid this bias, Bismark considers a read ambiguous (mapped to multiple locations) if the read has been mapped with the same best score to two or more locations using reduced three-letter alignments. BS-Seeker uses a similar approach; it disregards such reads, and for the rest of the reads, it re-calculates the number of non-BS-mismatches and keeps only uniquely mapped reads with the best alignment score.

One might counter-argue that allowing C–T matches also introduces a bias toward overestimating the number of unmethylated cytosines in reads mapped to the cytosine in a reference. A read containing cytosines that could have been unambiguously aligned with methylation aware alignment (allowing C–C and T–C matches only) might become ambiguous and be disregarded from further analysis when both T–C and C–T mappings are allowed; the contribution of C from this read to the count of total number of cytosines mapped to the cytosine in the genome is lost, affecting the estimation of methylation level. BRAT-bw offers an option to users to select between the two approaches: to allow C–T mismatches or not.

Another unavoidable bias in the estimation of methylation level when using nondirectional bisulfite library is as follows. Consider the $5'$ mate AACGTCA from PCR_2^+ strand in Figure 10.1. It maps unambiguously to DS^+ with zero mismatches when A–G mappings are allowed, and it maps to DS^+ with one mismatch when T–C mappings are allowed. Thus, the best alignment is considered the one with zero mismatches, and A–G mapping correctly contributes to methylation status of the cytosine on the negative genomic strand. Now consider the $5'$ mate AACGTC (the last A is missing). This read maps with zero mismatches to DS$^+$ when T–C matches are allowed and when A–G matches are allowed, and thus, the read is considered ambiguous and disregarded from further analysis. By disregarding this read, we affect the methylation level estimation: (i) if the read comes from PCR$_2^+$ strand, the number of Gs mapped to Gs will be affected, and the methylation level of cytosines on the negative genomic strand will be underestimated; (ii) if the read is originated from PCR$_1^+$ strand, then the total count of Cs mapped to Cs will be affected, that is, the methylation level will also be underestimated for cytosines on the positive genomic strand. Of course, if we do not disregard such a read, then the methylation level might be overestimated, which is worse scenario when the DNA is mostly unmethylated.

10.6   BISULFITE CONVERSION RATE

When estimating methylation level of cytosines or inferring methylation status of cytosines, it is important to realize that the bisulfite conversion rate is not 100%. Ideally, during bisulfite treatment all unmethylated cytosines should be converted to uracils; however, in reality some unmethylated cytosines will not convert. There are two common ways for the estimation of the bisulfite conversion rate. If it is known that cytosines are unmethylated in non-CpG context, one can estimate the bisulfite conversion rate by counting the number of Cs mapped to Cs in non-CpG context. Alternatively, one can use controls (*spike-ins*) using the DNA of species with a known methylation state, though the conversion properties of different DNAs might differ and this should be taken into account too.

10.7   REDUCED REPRESENTATION BISULFITE SEQUENCING

Reduced representation bisulfite sequencing (RRBS) approach has been developed and used for comparative methylation analysis between multiple samples (7, 8). First, DNA is digested with the methylation-insensitive MspI restriction enzyme, and then DNA fragments of the specified size (40-220bp) are selected and treated with sodium bisulfite, amplified by PCR and sequenced. The sequenced reads cover the DNA regions enriched for CpG sites. To improve mapping accuracy of the sequenced reads resulting from RRBS, specialized alignment tools have been developed such as RRB-SMAP (28) and BS-Seeker2 (29). Both tools take advantage of the knowledge of the restriction enzymes' sites and DNA fragments size to select the corresponding genomic regions for building index for the alignment of RRBS reads. This saves computational resources and improves the accuracy of the mapping.

10.8   ACCURACY AS A PERFORMANCE MEASUREMENT

Two common major performance measurements of an aligner for BS-reads are mapping accuracy and methylation call accuracy. Mapping accuracy is usually defined as a ratio of correctly mapped unique reads to the total number of uniquely mapped reads by an aligner. Methylation call accuracy is measured in a variety of ways, but the simplest way is to calculate the ratio of the cytosines correctly identified as methylated or correctly identified as unmethylated to the total number of cytosines.

First, we will discuss mapping accuracy in detail and then methylation call accuracy. To measure the mapping accuracy of an aligner, a set of artificial reads is generated and their originating positions are recorded. Then the reads are aligned by an aligner, and only uniquely mapped reads (the reads with the unique best alignment score) are considered for the estimations of mapping accuracy. A read is mapped correctly if it is mapped to the same reference and the same strand and to a position located within some fixed number of base pairs to that of the original position. Thus, some flexibility for a position within the same reference and strand is allowed due to different approaches that aligners use during mapping reads. For example, some aligners allow for local alignments, when the best-score portion of a read is aligned to the genome but the ends of the read are not. In this case, the unaligned ends of the read are usually trimmed and the mapping position is adjusted correspondingly, so there will be a difference between the original position and the mapping position for the reads mapped with local alignment. Another reason why a mapping position might differ from the original one is the presence of insertions and deletions (indels) that could be handled differently by different aligners. Both local alignment and alignment with indels are needed in order to map the reads with sequencing errors, adapter contamination, SNPs, and indels.

Before analyzing BS-reads, it is recommended to select the appropriate aligner with high mapping efficiency (the total number of uniquely mapped reads) and high mapping accuracy. Forcing reads to align at incorrect locations in order to align as many reads as possible is not the best strategy, since this can affect the validity of further analysis. It is important to carefully study the options set by a user for each aligner and run tests with different option settings to identify the best combination of options that will result in high mapping efficiency without compromising mapping accuracy.

To demonstrate the trade-off between different parameters and mapping accuracy, we measured mapping accuracy for three different BS-mapping tools: BRAT-bw, BS-Seeker2, and Bismark, using several sets of parameters and 120,000 artificial, 80-bp long reads generated from human genome GRCh38 (5000 reads from each chromosome). We introduced 2% of total bases of sequencing errors, 1% of total bases of SNPs, 1% of total reads of indels of maximum length of 10 bp, and 10% of total reads adapter sequences of maximum length of 10 bp at the 3′-end of the reads, and in addition the bisulfite-conversion rate was set to 97%. The SNPs were introduced uniformly, randomly along the length of the reads, but sequencing errors were introduced according to the probabilities of errors along the read lengths provided by Illumina instruments: the first 2/3 of the reads' length received the smallest number of sequencing errors, the next 1/6 of length received 2.5 times more errors, and the last 1/6 of length received 28.5 times more than the first 2/3 of the reads. The distribution of sequencing errors reflects the common pattern of observations of more sequencing errors at the ends of reads. The threshold on allowable mapping position proximity compared to the original position of reads was set to 50 bp; in other

TABLE 10.1   The Command-Line Options Used with the Three Tools from the Experiment for Mapping Accuracy Analysis on 120,000 Synthetic 80-bp Long Reads Generated from Human genome GRCh38

Aligner	Options	Label in Figure 10.2	% Uniquely aligned reads	%Mapping accuracy
BRAT-bw	K 2, m 10	1	79.27	97.06
BRAT-bw	K 6, m 10	2	83.26	98.84
BRAT-bw	K 4, m 10	3	87.20	98.10
BRAT-bw	K 6, m 10, F 1	4	89.11	98.49
BRAT-bw	K 6, m 10, $-C$	5	88.87	98.44
Bismark	L, 20, -1; D 30, N 1	1	85.68	93.37
Bismark	Default	2	71.90	98.88
Bismark	L, 0, -1.5; D 30	3	90.23	98.01
Bismark	L, 20, -1; D 30	4	89.79	97.73
BS-Seeker2	sensitive; ma 1; mp 3,3; m 20; G,6,3	1	91.02	80.17
BS-Seeker2	Default	2	85.24	88.36
BS-Seeker2	G,6,10; ma 1; mp 3,3; m 10	3	91.26	92.81
BS-Seeker2	very-fast; ma 1; mp 3,3; m 10; G,6,10	4	90.04	93.10

The numbers correspond to the numbers on the symbols on Figure 10.2.

words, if a read was uniquely mapped within 50 bp of the original position on the same chromosome and same strand, it was considered correctly mapped. Table 10.1 shows the parameters used with each tool and Figure 10.2 shows mapping accuracy (x-axis) together with the total number of uniquely mapped reads (y-axis) calculated for the corresponding parameter sets. BRAT-bw has only four user-defined options that regulate mapping accuracy and the number of mapped reads: K, m, C, and F. The option K sets the number of seeds used with each read, the larger the value of K, the more reads are mapped and the more accurate the mapping. The option m sets the maximum allowable non-BS-mismatches in a read (mismatches other than BS-mismatches). The larger m, the more reads are mapped, but the quality of mapping may be depreciated. With the option F set to 1, BRAT-bw guarantees to find all mapping locations with 1 mismatch in the first 32 bp of the read; the tool is slower with F equal to 1 with slightly more reads found and with a slight improvement in accuracy. Finally, option C allows C in a read to map to T in a genome; with this option, more reads are mapped, but it is questionable whether these matches (C in a read to T in a genome) should be allowed.

BS-Seeker2 is the most powerful tool among the three discussed in terms of flexibility of user's control because it supports all options available with Bowtie2, and Bowtie2 has many different ways to control the quality of the alignment and the number of mapped reads. Using the default parameters is not the best choice for any data set as can be observed on Figure 10.2: BS-Seeker2 had the smallest mapping efficiency and the lowest mapping accuracy with default parameters. We observed that setting consistent values for the minimum alignment score (controlled with the option

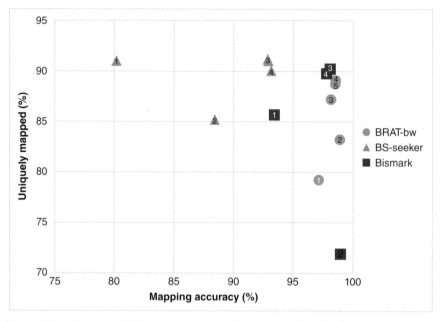

Figure 10.2 Mapping efficiency and accuracy of alignment using simulated reads. The total number of uniquely mapped reads as a function of mapping accuracy for the three tools is shown. 120,000 simulated 80-bp long reads from human genome GRCh38 were generated according to the following conditions: (1) 2% of total bases of sequencing errors were introduced; (2) 1% of total bases of SNPs were randomly generated; (3) indels of random length with the maximum length of 10 bp were introduced to 1% of total reads; and finally (4) adapter sequences of random length with the maximum length of 10 bp were inserted at 3′-end of reads to 10% of total reads. The numbers inside the symbols correspond to the number of option sets used with these three tools and shown in Table 10.1.

score-min) and the value of maximum mismatches allowed (option m) improves mapping accuracy of BS-Seeker2. Moreover, permitting a smaller number (option m) of mismatches also improves mapping accuracy.

Bismark currently does not offer users to control all options of Bowtie2 (might not be the case in future). We observed that setting the option N to allow one mismatch in a seed slows down Bismark, but does not map more reads and does not improve mapping accuracy. To control the quality of alignment, the option score-min and D and R can be used to map more reads with better mapping accuracy (the larger values of D and R slightly slow down the tool, but improve its performance in terms of mapping accuracy and the number of mapped reads). Now we will consider methylation call accuracy. As with measuring mapping accuracy, the starting point is to generate artificial reads, in which methylation status of each cytosine contained is set according to a chosen probabilistic distribution. Then sequencing errors and/or indels are introduced to the reads, and aligners are used to map reads back to the reference. Once reads are aligned, programs that calculate the methylation level at each cytosine from the mapped reads are used to count the number of Cs and the total number of Ts and Cs aligned to each cytosine. The simplest threshold for calling methylation status of a cytosine is using a fixed cutoff threshold. For example, with a cutoff threshold of 0.5, a cytosine is unmethylated if its methylation level is less than 0.5, and otherwise, methylated. Before counting methylation call accuracy, usually a threshold on the minimum number of reads covering a cytosine is used to avoid erroneous methylation calls that are due to small read coverage. The number of correctly identified methylation statuses for cytosines is counted only for the cytosines having sufficient read coverage.

A more sophisticated way of analyzing aligners' performance in terms of methylation call accuracy is to determine Recall and FDR for different values of cutoff threshold for calling a cytosine methylated, where Recall is the ratio of true positive calls to the sum of true positive and false positive and FDR is the ratio of false positive calls to the sum of true positive and false positive. For example, Figure 10.3 shows Recall versus FDR calculated for the three aligners and five different cutoff values: 0.2, 0.4, 0.5, 0.6, and 0.8, which are used to define the methylation status of each cytosine. For this experiment, we used chromosome Y of human genome version GRCh38 and generated artificial reads of length 80 bp long. First, we randomly

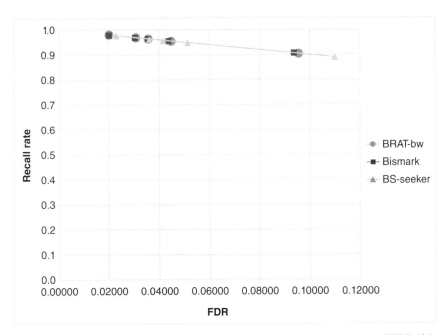

Figure 10.3   Performance in methylation call accuracy. Recall as a function of FDR (false discovery rate) is shown for the three tools and five different cutoff thresholds for determining the methylation status of cytosines for the data set of artificial 80-bp long reads generated from chromosome Y of human genome GRCh38. A total of 10 million reads were generated with 2% sequencing errors and 1% SNPs introduced randomly. Only cytosines covered by at least 10 reads were considered in this analysis. The total number of cytosines in chromosome Y was 9,262,721 on both strands of which BRAT-bw had 4,860,740 cytosines covered with at least 10 reads, Bismark had 4,900,224, and BS-Seeker2 had 4,656,122.

assigned a methylation level to each cytosine using four values, namely, 0.2, 0.4, 0.6, and 0.8. Then during read generation, if a read covered a cytosine at position i, then we used the methylation level of the cytosine at position i to randomly set the methylated status of the cytosine in the read. To clarify, assume that a read covers cytosine at position 100 and methylation level of this cytosine is 0.4, and then the cytosine in the read is set methylated (remains C) with a probability of 0.4 or is converted to T with a probability of 0.6. For each cytosine in the reference, we stored the number of methylated cytosines in the reads covering it and the total number of reads covering it. We used these values to calculate the true methylation level of each cytosine as the ratio of the reads with the corresponding cytosine methylated and the total number of reads covering the cytosine. Once methylated status for each cytosine in the reads was set and the true methylation level for each cytosine in the chromosome Y was calculated, we introduced 2% of sequencing errors and 1% of SNPs to the reads as described earlier. The reads were aligned with each aligner, and then the methylation extractors of the tools were run to count the resulting methylation level (each tool had its own program for methylation extraction). For the analysis described later, we only used cytosines covered by at least 10 reads to estimate methylation level calculated by an aligner. Finally, we calculated Recall and FDR for five different cutoff thresholds (0.2, 0.4, 0.5, 0.6, 0.8) and plotted the results on Figure 10.3. For example, with a cutoff threshold of 0.2, a cytosine was considered methylated if its true methylation level at that position was equal to or greater than 0.2, and unmethylated otherwise. The aligner's methylation call was true positive if it produced methylation level equal to or greater than 0.2 for the corresponding cytosine; and similarly, the aligner's methylation call was false positive if true methylation level was less than 0.2, but aligner's resulting methylation level was equal to or greater than 0.2. In this experiment, we used the following options with each tool: BRAT-bw (K 6, m 10), Bismark (D 30; L, 0, −0.6),

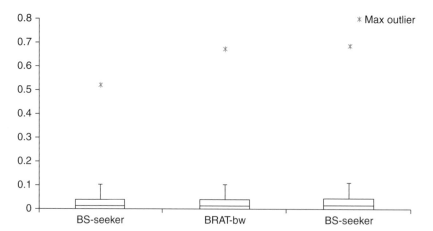

Figure 10.4 The distribution of a random sample of 325,000 absolute values of errors for each tool is shown. Absolute values of error are measured as absolute values of the difference between the true methylation level at each cytosine covered by at least 10 reads and the methylation level calculated by an aligner for the corresponding cytosines.

and BS-Seeker2 (G,6,10; ma 1; mp 3,3). The mapping accuracy of each tool was 97.27% for BRAT-bw, 96.58% for Bismark, and 90.1% for BS-Seeker2. Figure 10.3 shows that the performance of all three tools is comparable with BS-Seeker having negligibly inferior results.

Furthermore, to make analysis of methylation calls accuracy more comprehensive, the distribution of absolute errors between the true methylation level and the calculated methylation level by the aligners can be visualized as in Figure 10.4, which shows such distribution for the three aligners obtained in the same experiment with chromosome Y and 80-bp-long artificial reads. The total number of cytosines (on the top and bottom strands) that were covered by at least 10 reads by each aligner were 4,860,740 for BRAT-bw, 4,900,224 for Bismark, and 4,656,122 for BS-Seeker2. We randomly selected 325,000 errors from each data set and plotted them on Figure 10.4. The performance of the three tools was very similar in this experiment.

REFERENCES

1. Okano M, Bell DW, Haber DA, Li E. DNA methyltransferases Dnmt3a and Dnmt3b are essential for de novo methylation and mammalian development. Cell 1999;99(3):247–257.

2. Bestor T, Laudano A, Mattaliano R, Ingram V. Cloning and sequencing of a cDNA encoding DNA methyltransferase of mouse cells: the carboxyl-terminal domain of the mammalian enzymes is related to bacterial restriction methyltransferases. J Mol Biol 1988;203(4):971–983.

3. Bestor T. Activation of mammalian DNA methyltransferase by cleavage of a Zn binding regulatory domain. EMBO J 1992;11(7):2611–2617.

4. Laird PW. Principles and challenges of genomewide DNA methylation analysis. Nat Rev Genet 2010;11(3):191–203.

5. Cokus SJ, Feng S, Zhang X, Chen Z, Merriman B, Haudenschild CD, Pradhan S, Nelson SF, Pellegrini M, Jacobsen SE. Shotgun bisulphite sequencing of the Arabidopsis genome reveals DNA methylation patterning. Nature 2008;452(7184):215–219.

6. Lister R, O'Malley RC, Tonti-Filippini J, Gregory BD, Berry CC, Millar AH, Ecker JR. Highly integrated single-base resolution maps of the epigenome in Arabidopsis. Cell 2008;133(3):523–536.

7. Meissner A, Gnirke A, Bell GW, Ramsahoye B, Lander ES, Jaenisch R. Reduced representation bisulfite sequencing for comparative high-resolution DNA methylation analysis. Nucleic Acids Res 2005;33(18):5868–5877. DOI: 10.1093/nar/gki901.

8. Meissner A, Mikkelsen TS, Gu H, Wernig M, Hanna J, Sivachenko A, Zhang X, Bernstein BE, Nusbaum C, Jaffe DB, Gnirke A, Jaenisch R, Lander ES. Genome-scale DNA methylation maps of pluripotent and differentiated cells. Nature 2008;454(7205):766–770. DOI: 10.1038/nature07107.

9. Lister R, Pelizzola M, Dowen RH, Hawkins RD, Hon G, Tonti-Filippini J, Nery JR, Lee L, Ye Z, Ngo QM, Edsall L, Antosiewicz-Bourget J, Stewart R, Ruotti V, Millar AH, Thomson JA, Ren B, Ecker JR. Human DNA methylomes at base resolution show widespread epigenomic differences. Nature 2009;462(7271):315–322. DOI: 10.1038/nature08514.

10. Molaro A, Hodges E, Fang F, Song Q, McCombie WR, Hannon GJ, Smith AD. Sperm methylation profiles reveal features of epigenetic inheritance and evolution in primates. Cell 2011;146(6):1029–1041.

11. Hodges E, Molaro A, Dos Santos CO, Thekkat P, Song Q, Uren PJ, Park J, Butler J, Rafii S, McCombie WR, Smith AD, Hannon GJ. Directional DNA methylation changes and complex intermediate states accompany lineage specificity in the adult hematopoietic compartment. Mol Cell 2011;44(1):1–12.

12. Fang F, Hodges E, Molaro A, Dean MD, Hannon GJ, Smith AD. The genomic landscape of human allele-specific DNA methylation. Proc Natl Acad Sci U S A 2012;109(19):7332–7337.

13. Li Y, Zhu J, Tian G, Li N, Li Q, Ye M, Zheng H, Yu J, Wu H, Sun J, Zhang H, Chen Q, Luo R et al. The DNA methylome of human peripheral blood mononuclear cells. PLoS Biol 2010;8(11):e1000533.

14. Harris RA, Wang T, Coarfa C, Nagarajan RP, Hong C, Downey SL, Johnson BE, Fouse SD, Delaney A, Zhao Y, Olshen A, Ballinger T, Zhou X, Forsberg KJ, Gu J, Echipare L, O'Geen H, Lister R, Pelizzola M, Xi Y, Epstein CB, Bernstein BE, Hawkins RD, Ren B, Chung WY, Gu H, Bock C, Gnirke A, Zhang MQ, Haussler D, Ecker JR, Li W, Farnham PJ, Waterland RA, Meissner A, Marra MA, Hirst M, Milosavljevic A, Costello JF. Comparison of sequencing-based methods to profile DNA methylation and identification of monoallelic epigenetic modifications. Nat Biotechnol 2010;28(10):1097–1105.

15. Krueger F, Kreck B, Franke A, Andrews SR. DNA methylome analysis using short bisulfite sequencing data. Nat Methods 2012;9(2):145–151.

16. Burrows M, Wheeler D. A block sorting lossless data compression algorithm. Technical Report #124. Palo Alto (CA): Digital Equipment Corporation; 1994.

17. Zeschnigk M, Martin M, Betzl G, Kalbe A, Sirsch C, Buiting K, Gross S, Fritzilas E, Frey B, Rahmann S, Horsthemke B. Massive parallel bisulfite sequencing of CG-rich DNA fragments reveals that methylation of many X-chromosomal CpG islands in female blood DNA is incomplete. Hum Mol Genet 2009;18(8):1439–1448.

18. Xi Y, Li W. BSMAP: whole genome bisulfite sequence MAPping program. BMC Bioinformatics 2009;10(1):232–240.

19. Coarfa C, Yu F, Miller CA, Chen Z, Harris RA, Milosavljevic A. Pash 3.0: A versatile software package for read mapping and integrative analysis of genomic and epigenomic variation using massively parallel DNA sequencing. BMC Bioinformatics 2010;11(1):572.

20. Dinh HQ, Dubin M, Sedlazeck FJ, Lettner N, Mittelsten Scheid O, von Haeseler A. Advanced methylome analysis after bisulfite deep sequencing: an example in Arabidopsis. PLOS ONE 2012;7(7):e41528.

21. Smith AD, Chung WY, Hodges E, Kendall J, Hannon G, Hicks J, Xuan Z, Zhang MQ. Updates to the RMAP short-read mapping software. Bioinformatics 2009;25(21):2841–2842.

22. Harris EY, Ponts N, Levchuk A, Le Roch K, Lonardi S. BRAT: bisulfite-treated reads analysis tool. Bioinformatics 2010;26(4):572–573.

23. Chen PY, Cokus SJ, Pellegrini M. BS Seeker: precise mapping for bisulfite sequencing. BMC Bioinformatics 2010;11(1):203.

24. Krueger F, Andrews S. Bismark: a flexible aligner and methylation caller for Bisulfite-Seq applications. Bioinformatics 2011;27(11):1571–1572.

25. Harris EY, Ponts N, Le Roch KG, Lonardi S. BRAT-BW: efficient and accurate mapping of bisulfite-treated reads. Bioinformatics 2012;28(13):1795–1796.

26. Smith TF, Waterman MS. Identification of common molecular subsequences. J Mol Biol 1981;147:195–197.

27. Langmead B, Trapnell C, Pop M, Salzberg SL. Ultrafast and memory-efficient alignment of short DNA sequences to the human genome. Genome Biol 2009;10(3):R25.

28. Xi Y, Bock C, Müller F, Sun D, Meissner A, Li W. RRBSMAP: a fast, accurate and user-friendly alignment tool for reduced representation bisulfite sequencing. Bioinformatics 2012;28(3):430–432.

29. Guo W, Fiziev P, Yan W, Cokus S, Sun X, Zhang MQ, Chen PY, Pellegrini M. BS-Seeker2: a versatile aligning pipeline for bisulfite sequencing data. BMC Genomics 2013;14(1):774.

PART III

TRANSCRIPTOMICS

11

COMPUTATIONAL METHODS FOR TRANSCRIPT ASSEMBLY FROM RNA-SEQ READS

STEFAN CANZAR

Center for Computational Biology, McKusick-Nathans Institute of Genetic Medicine, Johns Hopkins University School of Medicine, Baltimore, MD, USA; Toyota Technological Institute at Chicago,Chicago, IL, USA

LILIANA FLOREA

Center for Computational Biology, McKusick-Nathans Institute of Genetic Medicine, Johns Hopkins University School of Medicine, Baltimore, MD, USA

11.1 INTRODUCTION

A major goal in bioinformatics is to identify the genes and their transcript variations, collectively defining the transcriptome of a cell or species. Unlike the genome, which is largely identical for all cells of an organism, the transcriptome of each cell may be different depending on the tissue, developmental stage, and disease or normal condition. Different sets of genes are being expressed, and also each gene may present a different mRNA isoform with its distinct combination of exons, produced by alternative splicing or alternative transcription start or polyadenylation. Recently, next-generation sequencing of cellular RNA (RNA-seq) has started to allow surveying the cellular transcriptome in great detail, under a variety of conditions. A single sequencing instrument can produce hundreds of millions of short (50–250 bp) reads, at a fraction of the time and cost required by traditional (Sanger) sequencing. The wealth of data has allowed better characterization of cancer transcriptomes (1,

Computational Methods for Next Generation Sequencing Data Analysis, First Edition.
Edited by Ion I. Măndoiu and Alexander Zelikovsky.
© 2016 John Wiley & Sons, Inc. Published 2016 by John Wiley & Sons, Inc.
Companion website: www.wiley.com/go/Mandoiu/NextGenerationSequencing

2), finding splicing aberrations in disease, as well as discovering new classes of
RNAs (3) and new splicing variations (4-6). However, the short reads are challenging
to analyze and interpret, posing major challenges to bioinformatics tools. We review
current work in combining short read information to infer gene and transcript models
and their sequences, also known as *transcript assembly*, and discuss limitations and
challenges.

A typical transcriptomic analysis starts with generating the data. In a common
scenario, total RNAs or mRNAs are converted into a cDNA library, which is then
fragmented and each fragment is sequenced to produce short reads from one end
(single-end) or both ends (paired-end). Transcript assembly then involves piecing
together the short reads to form gene and transcript models and sequences. Overlaps
between reads, which suggest that they may have been sampled from the same region,
provide clues about local assembly. In the case of paired-end reads, the fragment
size can be statistically characterized and this information can provide long-range
connectivity in the assembly process.

The general assembly problem, known from genome sequencing, is inherently
difficult (7). Significant complications arise from ambiguous reads that belong to
highly similar regions of the genome, resulted from recent duplications or repeats.
Although transcripts are shorter and therefore easier to assemble, there are specific
challenges. For instance, there is no *a priori* knowledge of the number of transcripts
to be reconstructed. Then, each gene can be sampled by just a few reads, or hundreds
of thousands of reads, as RNA-seq has been shown to capture a range of expression
levels spanning 5 orders of magnitude. Read coverage is not uniform even within the
boundaries of the same transcript or gene due to biases in library preparation and
sequencing (8, 9). Furthermore, each read may be assigned to any of several genes in
a family that share similar sequences or to several isoforms of the same gene. Biological phenomena such as overlapping and nested genes and trans-splicing further
increase the difficulty of assembly, while artifacts due to intronic reads or incorrect
mapping can create spurious exons and introns. Lastly, analyzing the large volume of
data in an RNA-seq experiment demands tools that are scalable and highly efficient.

There are two main classes of transcript assembly methods: *de novo*, which
assemble reads based solely on sequence overlap, and genome-based, which first
align the reads to a reference genome and then assemble the overlapping alignments
(Figure 11.1). Genome-based methods are generally more accurate. However, in
the absence of a genome sequence or if the genome is highly fragmented, *de novo*
assembly can be effectively used to determine a representative set of transcripts,
albeit the assembly process is more complex and error prone. We review the two
approaches, specific methods and design principles, in the following sections.

11.2   *DE NOVO* ASSEMBLY

This class of methods assemble reads based on the overlap of their sequences.
Intuitively, if the end sequence of one read is similar to the start of another, this rela-
tionship establishes an order between the two reads within the originating sequence.
With traditional (Sanger) sequences, these order relationships could be represented

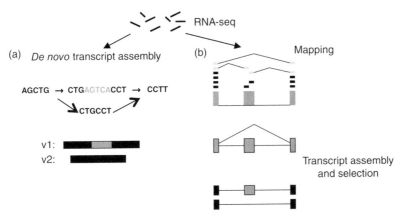

Figure 11.1 Classes of transcript reconstruction methods. (a) *De novo* methods assemble reads based on sequence overlaps, represented as either an overlap graph or a de Bruijn graph. (b) Genome-based methods first map reads to a reference genome allowing for introns, then combine read alignments into a graph (overlap, connectivity, or splicing graph). Transcripts are enumerated from the graph and a final subset is selected using a variety of methods.

TABLE 11.1 *De Novo* **Transcriptome Assembly Methods and Their Properties**

Tool	Approach	Contigging	Full Coverage
Multiple-K (10)	DBG, $k = 19, \ldots, 29$	Merge and reduce	No
Rnnotator (11)	DBG, $k = 19, \ldots, 41$	Minimus assembly	No
Oases (12)	DBG, $k = 19, \ldots, 35$	Velvet assembly	No
Trans-ABySS (13)	DBG, $k = 26, \ldots, 50$	Hierarchical merge and reduce	No
Trinity (14)	DBG, $k = 25$	Na	Yes
EBARdenovo (15)	Extension and bridging	Bridging	Yes

DBG = de Bruijn graph.

as edges in an overlap graph (OG) connecting the vertices, which represent reads (see Section 11.3.1.1 for details). Given the vastly larger number of reads in a typical RNA-seq experiment, this approach is no longer feasible. Instead, another data structure, the de Bruijn graph, has been used to compactly represent all read sequence information. Once the graph is constructed, it is analyzed to extract contigs representing full-length or partial transcript sequences (transfrags). Lastly, the set of contigs is filtered to eliminate redundancy and artifacts, and reads are sometimes assigned to transfrags to estimate their expression levels. A number of programs for *de novo* RNA-seq read assembly have emerged (Table 11.1), which share the three-stage structure but differ significantly in their computational techniques.

11.2.1 Preprocessing of Reads

Because even small errors can significantly impact the quality of the assembled contigs, read sequences are evaluated prior to assembly and filtered to remove low-quality sequences and likely artifacts. Vector and adapter sequences and also polyadenylation stretches at the ends of the reads, which could cluster together sequences from unrelated genes, can be detected and removed using a tool such as cutadapt (16). Sequencing errors most often impact the last few positions in a read, creating sequence differences that prevent assembly; they can be removed by clipping the last few bases of reads. Furthermore, most *de novo* assemblers use the topology of the de Bruijn graph to detect and correct sequencing errors that occur randomly. Lastly, error correction methods have been devised and are highly successful in correcting DNA reads prior to assembly (17), thus improving the quality of the assembled sequence. These methods detect likely errors by identifying sequence words that have a smaller than expected number of occurrences in the data set. However, few of these methods have been adapted to RNA-seq data, an area of active method development.

11.2.2 The De Bruijn Graph for RNA-seq Read Assembly

11.2.2.1 The De Bruijn Graph in Sequence Assembly The de Bruijn graph was introduced to bioinformatics by the work of Idury and Waterman (18) and Pevzner et al. (19), who applied it to the problem of assembling genomes. This data structure is based on short sequence segments of a fixed length k (typically $k \geq 19$), called k-mers. Given a sequence or set of input sequences, the corresponding de Bruijn graph has a node for each distinct k-mer that occurs in the input, and an edge between two nodes if the two k-mers are adjacent within an input sequence (Figure 11.2). Consequently, for each edge, the $(k-1)$ length suffix of the originating node (k-mer) coincides with the $(k-1)$ length prefix of the target node. Usually, the nodes are labeled with the k-mer occurrence count.

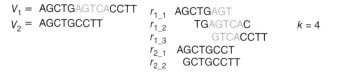

V_1 = AGCTGAGTCACCTT r_{1_1} AGCTGAGT
V_2 = AGCTGCCTT r_{1_2} TGAGTCAC $k = 4$
                     r_{1_3} GTCACCTT
                     r_{2_1} AGCTGCCT
                     r_{2_2} GCTGCCTT

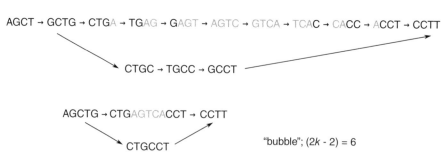

AGCT → GCTG → CTGA → TGAG → GAGT → AGTC → GTCA → TCAC → CACC → ACCT → CCTT

CTGC → TGCC → GCCT

AGCTG → CTGAGTCACCT → CCTT

CTGCCT   "bubble"; $(2k - 2) = 6$

Figure 11.2 De Bruijn graph for two splice variants, v_1, and v_2, sampled by five reads. The expanded graph is shown at the top and the graph with compressed nonambiguous paths at the bottom. The alternative splicing event appears as a "bubble" with one of the two paths of length $(2k - 2)$.

Although similar in concept with the overlap graph, the de Bruijn graph has several properties that differ in a significant way (20). First, the reads are split across several nodes corresponding to adjacent k-mers. Second, a repeat, (i.e., for our purposes, a sequence that occurs a large number of times in the originating sequence, with small variations) will appear as a sequence of adjacent k-mers that occur in a large number of reads. The ends of the repeat region in the graph diverge into the unique portions of the genome or transcript, thus creating a "bubble". Most importantly, the graph can be constructed in time linear in the number of sequences, as compared to the quadratic time required to find all pairwise overlaps among the input reads.

An Eulerian tour of the graph that visits each node precisely once could, in theory, represent a solution to the genome or transcript assembly problem (19). In practice, however, the graph can contain cycles that are introduced whenever the sequence is repeated in the originating genome or transcript sequences, which introduce ambiguity in deciding the correct path. Most frequently, the solution involves enumerating nonambiguous paths (contigs) along with sequences of the repeats, their order and orientation resolved to the extent possible by using contiguity information from paired-end reads.

11.2.2.2 *Adaptation to Transcriptome Assembly*

While transcriptome assembly using de Bruijn graphs emerged as a natural extension of genome assembly, there are significant differences between the two problems. First, instead of the graph encoding just one long genomic sequence, several isoforms can share the same subpath in the graph while differing in the alternatively spliced regions. Second, genes and their isoforms have varying expression levels, and different sets of parameters may be optimal for each assembly problem.

In de Bruijn graphs, the k-mer size significantly affects the balance between sensitivity and accuracy. Lower k values increase sensitivity by allowing some low expression isoforms to be assembled, but can introduce artifacts. Larger k values are more accurate in reconstructing transcripts with high read depths, but may miss low expression isoforms. Starting from the assumption that a single k value is unlikely to render a universally effective solution, several programs were developed that use multiple k values in parallel to produce independent assemblies. The millions of contigs are then merged across assemblies and filtered to remove redundancies and spurious constructs. A simple approach was implemented in the additive Multiple-K method (10). In a more complex merging scheme, Trans-ABySS (13) first builds a de Bruijn graph and assembly problem for each value $k = 26, \ldots, 50$. It then combines the results between pairs of assemblies in a hierarchical manner, starting with consecutive $(k, k + 1)$ pairs and subsequently merging pairs of intermediate assemblies. In a variation, Oases (12) performs an assembly for each $k = 19, \ldots, 35$, then builds a second-stage de Bruijn graph from the merged set of contigs, which is analyzed to produce the final set of transfrags. In yet another variation, Rnnotator (11) employs the genome assembler Velvet (21) with $k = 19, \ldots, 41$ to produce multiple assemblies and then applies a conventional read assembler, Minimus (22), to resolve redundancies. The outcome is a nonredundant set of contigs that need to be filtered for artifacts.

11.2.3   Contig Assembly

Once a de Bruijn graph is constructed, it needs to be analyzed to produce longer trans-frags. The first step is to form contigs, each representing a portion of a splice variant, by combining node labels along unambiguous paths. Where available, paired-end read information (n.b., which is not used in constructing the de Bruijn graph) can be used to further extend contigs. For instance, the mate distance distribution can be calculated from paired-end reads that are included in the same contig and can then be applied to recruit additional compatible contigs for extension, as implemented in Trans-ABySS. A minimum number of supporting read pairs are required to complete a contig merge. Other programs, such as Rnnotator, employ two *de novo* genome assemblers, the de Bruijn graph-based Velvet and the overlap-based Minimus, to first determine a basic set of contigs and then merge these into longer sequences. Similarly, Oases is fashioned on the Velvet genome assembler, which it employs to determine an initial set of contigs. Contigs are classified as either long (length $\geq 50 + k - 1$), which are more reliable, or short (length $< 50 + k - 1$). In subsequent steps, long contigs connected by single reads or read pairs are grouped into loci, each representing a gene or a portion of a gene, and short contigs are later attached to these loci. An overlap graph is constructed at each locus, with contigs as vertices. Lastly, each locus graph is transitively reduced by eliminating long-distance connections between two contigs if there is a distinct path connecting them.

De Bruijn graph-based assemblers designed for genome assembly aim to produce long and largely nonoverlapping contigs, and are not suited to find splice isoforms that share a considerable portion of the original sequence. In the graph, the alternatively spliced portion of transcripts will appear as a bubble with specific length properties. One common approach, pioneered by Trans-ABySS, is to search for characteristic bubbles where one of the contigs has a length of exactly $2k - 2$, corresponding to an insert in one of the sequences (Figure 11.2). These bubbles are first removed and then added to the final assembly to reconstruct splice variants. A more systematic approach is taken by Oases, which analyzes the topology of locus graphs to determine simple patterns such as chains, bubbles, and forks, as well as complex ones. Oases then determines a parsimonious set of putative transcripts that can explain all contigs, detected as heavily weighted paths through the locus graph.

One important observation is that read information is lost when assembling the contigs from k-mers. Of all *de novo* assemblers to date only Trinity (14) and EBAR-denovo (15), which uses a more traditional extend-and-bridge model, guarantee that transcripts are fully supported by the set of original reads (Table 11.1).

11.2.4   Filtering and Error Correction

Assemblers perform error detection and filtering, and sometimes error correction, at all stages in the algorithm. The main classes of artifacts are redundancies resulted from incomplete merging of reads and contigs, fragmented transcripts, chimeric constructs, and collapsing of paralogs. These artifacts occur due to a combination of sequencing errors, repeat or duplicated sequences, chimeric libraries, and vector and linker contamination. Overlapping gene models further contribute to the complexity.

The first checks and corrections take place before assembly, as described in Section 11.2.1. Later, sequencing errors, allelic differences and repeat sequences with minor variations form bubbles with characteristic lengths in the de Bruijn graphs. The bubbles can be "popped" by removing the variant with lower coverage, while recording both variants. Even later, contigs with low coverage or that do not overlap with other contigs in multiple k-assemblies are removed as spurious, whereas those entirely subsumed in other contigs are removed as redundant (13). In another variation, the TourBus algorithm implemented in Oases detects alternative paths between two nodes in the de Bruijn graph and removes the lower coverage path if the two path sequences are highly similar. Additionally, contig edges in the locus graphs are removed if they account for less than 10% of the coverage of all outgoing edges from the current node; the rationale is that on high-coverage regions, spurious errors are likely to reoccur. In Rnnotator, single base errors in the assembled contig are resolved by aligning reads back to the contig to generate a consensus nucleotide sequence. Lastly, if the genome sequence is available, mapping the contigs back to the genome using a spliced alignment algorithm such as BLAT (23) or sim4 (24) can be used to detect gene fusions and spurious contigs that do not align end-to-end.

11.2.5   Variations

There are several notable variations and exceptions to the above algorithms. Trinity (14) is a package for *de novo* transcriptome assembly of RNA-seq reads that combines concepts and techniques from conventional overlap-based and from de Bruijn graph-based assembly. Its three stages are implemented as stand-alone programs. The first stage, Inchworm, constructs a set of linear contigs that minimally represent all k-mers in the read set. At each iteration, it starts from the most frequent k-mer and greedily extends the current contig with compatible k-mers to build a heaviest path. The second stage, Chrysalis, builds clusters of contigs that correspond to portions of splice isoforms or unique portions of paralogous genes and which share at least one $(k-1)$-mer between any two sequences. For each group, it builds a de Bruijn graph and then maps the reads back to the cluster with which it shares the most k-mers. The third stage, Butterfly, traverses the paths in the de Bruijn graph and scores them based on read coverage, and finally selects a subset of paths that are fully supported by reads and read pairs using a dynamic programming algorithm. Trinity is designed to detect highly expressed transcripts with high accuracy using a single k-mer size and stringent criteria for the selected transcripts.

Recently, EBARdenovo (15) introduced an extend-bridge-repeat detection algorithm to accurately assemble transcripts, by incorporating tests for detecting and correcting repeats and chimeric constructs. It starts from a seed read, which it extends conservatively with reads in both directions if a candidate read's mate can be found in the current or neighboring contig. If two or more extensions can be found, they are backtracked and the contig is split, as repeat-containing. Putative chimeric contigs are detected by comparing the reads against the contig consensus sequence.

As a class, *de novo* assembly methods can be flexibly used when the species genome is not available, as well as to complement genome-based annotations. In practice, their main difficulties are in distinguishing among paralogs within the same gene family, correctly reconstructing full-length isoforms of genes, and detecting low-expression splice variants. Formally, the assembly problem they attempt to solve is complicated because repeats and duplicated sequences introduce cycles in the underlying graph representation. As a consequence, an array of heuristic techniques have been developed and applied successfully. In the following, we present genome-based methods and their more tractable mathematical foundations.

11.3 GENOME-BASED ASSEMBLY

Methods developed for the case where a high-quality reference genome is available mostly adhere to the following general scheme. First, a set of candidate isoforms are identified that are believed to include the expressed (true) transcripts. For that purpose, read alignment information is compiled into a compact graph representation such that expressed isoforms correspond to paths in the graph. We describe the different types of graphs and the sets of candidate isoforms they encode in Section 11.3.1.

Once the graph is constructed, the number of possible paths typically exceeds the number of true isoforms, sometimes dramatically. Therefore, transcript reconstruction methods must select a *small* subset of isoforms that are able to explain the observed reads *well*. These principles, which we refer to as *minimality* and *accuracy* in the rest of the review, are incorporated into an optimization problem; we discuss these mutually concurrent objectives in Sections 11.3.2 and 11.3.3. A third objective is *completeness*: algorithms predict sets of isoforms that are able to explain *all* observations, that is, read mappings (see Section 11.3.4). Table 11.2 summarizes the aspects of candidate isoforms, minimality, accuracy, and completeness for the methods included in this review. In the following sections, we describe general principles underlying current methods for genome-based transcriptome assembly.

11.3.1 Candidate Isoforms

Transcription information contained in the aligned reads can be compactly represented by a *graph* structure; the expressed (true) isoforms are among the *paths* in this graph. Several methods for isoform reconstruction, therefore, resort to the well-established framework of graph theory, for example, in the context of minimum path covers or minimum cost flows (25).

Two different types of graphs have been used to represent the available information at different levels of granularity. An *overlap graph* represents relationships between read alignments, whereas a *connectivity graph* (CG), and similarly a *splicing graph* (SG), captures connectivity properties of genomic positions covered by the reads. Unlike with *de novo* assembly, the reference genome allows for a linear ordering of the set of vertices. Consequently, both the overlap and the splicing graphs are directed and acyclic, a class of graphs for which effective solutions have been developed for a large variety of problems and can be applied to assembly algorithms. We introduce the different graphs in the following two sections.

TABLE 11.2   Overview of Concepts Described in Sections 11.3.1–11.3.4

Tool	Candidates	Minimality	Accuracy	Completeness
TRAPH	SG	L^0	Least squares	No
TRAPH 2	SG	L^0	Least squares	No
Scripture	CG	No	Not explicitly	No
CLIIQ	SG	L^0	Min abs. error	Junction cover
MITIE	SG	L^0	Novel loss function	No
iReckon	SG	(11.1)	Likelihood	No
SLIDE	SG	Modified L^1	Least square	No
IsoInfer	SG	L^0	Least squares	Set cover
IsoLasso	CG	L^1	Least squares	Segment cover
Cufflinks	OG	L^0	Likelihood	Path cover
CLASS	SG	L^0	No	Set cover
TRIP	SG	L^0	Fragment length	Junction cover
Montebello	SG	L^0	Likelihood	No
NSMAP	SG	L^p	Likelihood	No

Candidates are either (maximal) paths in splicing graph (SG), connectivity graph (CG), or overlap graph (OG). Paths are not necessarily enumerated in advance but can be, for example, implicitly encoded by an ILP (MITIE) or generated on demand by a Monte Carlo simulation (Montebello). The different norms measuring minimality can be optimized either jointly, while fitting a (linear) model "(Section 11.3.2.2)," or during a separate optimization phase (Section 11.3.2.1), which does not always find the optimal solution, such as the greedy path decomposition in TRAPH. If completeness is not explicitly modeled ("no" in last column), it might still be implicitly contained in the accuracy measure.

11.3.1.1   Overlap Graphs   Historically, the *overlap graph* was the preferred representation model for read compatibility in genomic sequence assembly (19). It represents order relationships between reads that originate from the same sequence, which needs to be reconstructed. The overlap graph has a node for every input read, and two reads are connected by an edge if they could overlap in the original sequence, that is, if the prefix of one read is similar in sequence with the suffix of the other. The edge orientation then defines the order between the two reads in the original sequence.

In the context of isoform construction from RNA-seq data, the overlap graph as implemented in Cufflinks (6) captures relationships between reads that could have originated from the same transcript. Vertices correspond to reads and are connected by an edge if their alignments to a reference genome overlap and are *compatible*. Unlike with DNA sequencing, however, the definition of read alignment compatibility has to be extended to spliced alignments containing introns. Furthermore, when paired-end reads are used, it is further complicated by the unknown insert sequence between the reads on the cDNA fragment.

Specifically, two read alignments are compatible if they overlap as described earlier and, additionally, they share the same set of splice junctions (possibly empty) on the overlap region. This definition extends to paired-end reads. For example, read pairs p_1 and p_2 in Figure 11.3a are compatible, whereas p_1 and p_3 have inconsistent splice junctions and hence are incompatible. However, unlike with single-end reads, in this case, the unknown region of the fragment may overlap with two fragments that have different and incompatible sets of splice junctions. For instance, in Figure 11.3a, p_2 is compatible with both p_1 and p_3, but p_1 and p_3 are incompatible with each other. In such a case, compatibility or incompatibility cannot be assigned to the three pairs of fragments without introducing a conflict.

To avoid conflicts, Cufflinks removes these "uncertain" reads prior to constructing the overlap graph, so that paths in the graph encode pairwise compatible fragments that together could form complete isoforms. Figure 11.4 shows an overlap graph that encodes three different isoforms.

11.3.1.2 *Connectivity Graphs and Splicing Graphs*   Overlap graphs captured connectivity relationships at the level of reads. More recently, two other types of structures, namely connectivity graphs and splicing graphs, have been used in transcript assembly to capture relationships among genomic or transcribed bases.

Connectivity graphs, as used in Scripture (27) and IsoLasso (26), represent each genomic position by a vertex. Two vertices are connected by an edge if they are either consecutive on the genome or if they are linked by a spliced alignment. Thus, the graph represents the connectivity of bases along the transcript, whose projection on the genome is interrupted by introns.

By comparison, a splicing graph (28) retains only those portions of the genome covered by reads. Then, nonambiguous paths in the graph can be collapsed to a single vertex without losing any structural information. Vertices in the splicing graph

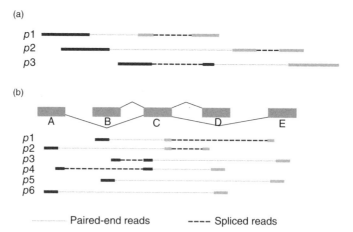

Figure 11.3   Inconsistencies occurring with paired-end reads. Source: Li et al. 2011 (26). Adapted with permission of Mary Ann Liebert, Inc. (a) Read pairs p_1 and p_2, as well as p_2 and p_3, are compatible, whereas p_1 and p_3 cannot originate from the same isoform. Read pair p_2 is thus uncertain. (b) The splicing graph comprises five exons and encodes four paths (isoforms). Exon combinations ACE and BCD are infeasible, since the read pair supporting splice site AC includes exon D and the read pair supporting junction BC additionally includes exon E.

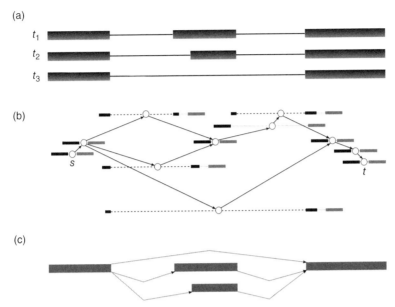

Figure 11.4 Example of splicing graph and overlap graph. (a) Three different isoforms are sampled by the reads. Isoform t_2 uses a different splice site in the middle exon, whereas isoform t_3 skips the exon completely. (b) The overlap graph connects compatible reads and the three possible paths from the leftmost node (source s) to the rightmost node (sink t) resemble the three original isoforms. Pairs of reads are denoted by a gray dotted line, spliced alignments by a black dashed line. (c) The splice graph contains four nodes representing the four different exons and connects them as denoted by the directed edges.

therefore correspond to exon fragments bounded by splice sites, the latter being derived from spliced alignments or extracted from a set of gene annotations. An edge between two vertices is set to indicate consecutive exons or exon fragments in a transcript. Since exons are organized linearly along the genomic axis, in the same orientation as dictated by the gene, there are no cycles. Therefore, the splicing graph is acyclic. Figure 11.4 shows a splicing graph constructed from reads sampled from three different isoforms.

Transcripts are represented as maximal paths in the graph, that is, paths starting at a node with no incoming edge (source) and ending at a node with no outgoing edge (sink). The restriction to *maximal* paths does not allow for transcripts that begin or terminate at an internal exon, resulting from alternative promoters and polyadenylation sites. By adding two additional start and termination nodes to the graph, however, these subpaths can be included in the model.

Methods that identify candidate isoforms as (maximal) paths in a splicing graph include TRAPH (29), CLIIQ (30), MITIE (31), iReckon (32), SLIDE (33), IsoInfer (34), CLASS (35), NSMAP (36), and TRIP (37).

In practice, finding the correct spliced alignment of a read is computationally challenging and error prone, which can significantly affect the sensitivity and accuracy of the transcript prediction process. *Incorrect* spliced alignments (false positives) can be filtered, for instance, by requiring a minimum number of reads supporting the splice junction(s), as implemented in Scripture and CLIIQ. Alternatively, all spliced alignments can be used to build the splicing graph, but the optimization process may select a final set of transcripts that does not necessarily satisfy all input alignments, as in MITIE. In the complementary case of missing alignments (false negatives), transcripts may be only partially sampled and therefore not fully supported by reads, and some methods use external gene annotations to augment the graph (iReckon). The isoform selection model (discussed in Section 11.3.3) then has to take into account the context to distinguish between a splice junction that is simply missed by the read alignments and one that is not present in the current sample.

The number of maximal paths in the graph can be considerably larger than the number of true transcripts present in the sample, even if all edges (i.e., spliced alignments) are correct; see (31) for an example. Longer range connectivity information, such as from paired-end reads, is typically used to restrict the number of transcripts. Strategies to select a small subset of transcripts from among all paths are discussed in the context of the minimality of the solution in Section 11.3.2.

11.3.1.3 *Comparison of Overlap Graphs and Splicing Graphs*   Both types of graphs have their advantages and limitations. Connectivity graphs, and especially splicing graphs, tend to be smaller in size compared to overlap graphs for the same input, as they summarize information from potentially thousands of reads into a small number of genomic segments. This difference can be significant for high-coverage RNA-seq experiments. However, these graphs may also encode combinatorial paths that cannot be fully explained by the input reads, that is, are *infeasible* (26). Figure 11.3b gives an example of such an infeasible path (isoform). If only the subset of feasible maximal paths are concerned, the two data structures are equivalent in expressive power. Indeed, Li et al. (26) showed that the set of isoforms corresponding to feasible maximal paths in the connectivity graph is equal to the set of isoforms corresponding to maximal paths in the overlap graph, provided no uncertain reads exist. Removing uncertain reads, however, can lead to missed splice junctions, and therefore fewer transcripts can be represented in the overlap graph.

11.3.2   Minimality

In reality, only a small subset of the candidate isoforms is expressed in a given cell type. Hence, most methods seek a sparse solution to a large underdetermined isoform selection problem. Disregarding the number of predicted isoforms in the estimation would lead to a large number of isoforms with low but nonzero expression levels, which essentially fit the noise in the data. Strategies to force a small number of isoforms to be selected can be broadly divided into two categories. *Two-stage optimization* methods, summarized in Section 11.3.2.1, separate the isoform selection from the model-fitting stage, which matches the isoforms to the observed read data. In contrast, *regularization*-based approaches simultaneously fit a model and penalize an excessive number of isoforms. We discuss this class of methods in Section

11.3.2.2. One notable exception is Scripture (27), which does not explicitly model minimality as an objective and which has been observed to predict a large number of isoforms in practice.

11.3.2.1 *Two-Stage Optimization* Methods in this class follow a two-stage procedure, first determining a set of transcripts and then assigning reads to estimate their expression levels. More formally, in terms of the optimization criteria introduced in the beginning of Section 11.3, these methods treat the *minimality* of the solution separately from *accuracy*. Programs in this category include Cufflinks (6), CLASS (35), CLIIQ (30), TRAPH (29), and IsoInfer (34). In the following, we explain the strategies and the underlying mathematical foundations for each of these approaches.

We start with the most prominent example in this class, Cufflinks (6), which aims at finding the minimum number of transcripts or paths in the associated overlap graph, which explain *all* input reads. More formally, Cufflinks models the isoform selection problem as a MINIMUM PATH COVER problem in a directed acyclic graph. Given a directed graph $G = (V, E)$, in this case the overlap graph, find a smallest set of directed paths such that every vertex $v \in V$ belongs to at least one path. Although the MINIMUM PATH COVER problem is computationally hard (n.b., NP-hard), it can be solved efficiently for directed acyclic graphs such as the overlap graph. Analog to Fulkerson's proof of Dilworth's theorem (38), a bipartite graph G' can be constructed from an overlap graph $G = (V, E)$ such that, by König's theorem, G' has a matching of size m if and only if there exists a path cover in G of size $n - m$, where $n := |V|$. Additionally, Cufflinks assigns a cost to edges in G' such that coverage inhomogeneities along a path are penalized. The minimum path cover in G is thus obtained in (6) from a *min-cost* maximum cardinality matching in G', which can be computed in time $\mathcal{O}(V(E + V \log V))$.

Similarly, CLASS (35) aims to determine the minimum number of transcripts that collectively satisfy all contiguity constraints from spliced and paired-end reads. Formally, it casts the transcript selection problem as a SET COVER problem, where maximal paths (transcripts) in the splicing graph are represented as sets of constraints that can potentially originate from each of the isoforms. A *constraint* is defined as a class of reads that share the same splicing and subexon pattern. In a variation, weights assigned to the sets (isoforms) can incorporate knowledge on gene structure from external databases. Since SET COVER is itself an NP-hard problem in its general form, CLASS approximates the optimal solution to the set cover instance by a fast greedy algorithm.

In another example, CLIIQ (30) formulates an integer linear program (ILP) that seeks to determine the minimum number of isoforms required to explain the observed reads *well*. More precisely, it minimizes the number of expressed transcripts such that, for all exons and exon junctions, the estimated number of reads lies within a small factor of $(1 + \varepsilon)$ of the observed number of reads, with ε being a user-defined threshold. The solution is subject to the constraints: (i) for every exon junction, the estimated number of reads is at least the minimum observed number of reads over all junctions and (ii) the estimated expression level of an isoform can be at most a factor of $(1 + \varepsilon)$ larger than the minimum expression level of its exons and exon junctions.

The optimal solution to the ILP is then used as an upper bound for the number of isoforms allowed to minimize the total error in a second stage.

Compared to Cufflinks and CLIIQ, which minimize the number of isoforms that are subsequently used to fit the data, TRAPH (29, 39) first attempts to minimize the error of the prediction before it decomposes the estimated number of reads into isoforms. In its original version (29), TRAPH reduces the isoform reconstruction problem to a MINIMUM-COST NETWORK FLOW problem. A flow in a network can be decomposed into paths from the sources to the sinks (25), or equivalently, the predicted number of reads on exons and junctions can be attributed to individual isoforms in a consistent way. Finding a decomposition into a minimum number of paths, however, is NP-hard even for DAGs. In (29), the flow is therefore decomposed by repeatedly removing a path of maximum flow from a source to a sink.

A rather straightforward way of incorporating parsimony into the reconstruction model is by means of enumeration. The general idea is to formulate the isoform reconstruction problem for a fixed number of isoforms k, solve the resulting model for a predefined set of small ks, and choose the best one by hand. In the follow-up work (39), TRAPH merges the minimum-cost network flow problem and the k path decomposition into a problem that asks for a set of k paths from the sources to the sinks along which an amount of flow is sent that minimizes the overall cost. Similarly, IsoInfer (34) fits a quadratic program formulation for all possible combinations of candidate isoforms of a fixed cardinality k, successively increasing k until the prediction is sufficiently *accurate*. Such a strategy is known as "best subset variable selection" in the literature (40).

11.3.2.2 Simultaneous Selection and Fitting

In contrast to methods discussed in the previous section, *regularization*-based approaches solve the isoform discovery and the abundance estimation problem simultaneously.

Unlike the "best subset selection" strategies described in the previous section, which use a preselected subset of isoforms to fit the read data, *regularization*-based methods fit a model that includes all candidate isoform abundances as variables and augment the objective function by a penalty term to force a small number of isoforms in the solution. This *regularization* penalty reflects the complexity of the prediction. The most immediate measure of complexity would simply be the number of nonzero variables, that is, the number of expressed transcripts predicted. The objective function including the L^0-norm of the isoform abundances, however, is nonconvex and the resulting optimization problem is computationally intractable. MITIE (31), for example, resorts to a forward stepwise regression strategy, adding greedily one variable (isoform) at a time. TRIP and Montebello (41) account for the computational complexity of the L^0-norm by NP-complete integer programming and a Monte Carlo simulation, respectively.

In computational biology, the (convex) L^1-norm has been shown to be a good replacement for the L^0-norm. In the context of linear regression (see Section 11.3.3), this so-called *Lasso* regression does not only shrink the expression levels of isoforms toward zero but also *selects* variables, that is, it sets variables (isoforms) to zero one at a time to increase penalty terms. SLIDE (33) and IsoLasso (26) both employ a Lasso strategy to predict a sparse set of expressed isoforms. SLIDE slightly modifies the

261

penalty term to favor isoforms comprising a higher number of exons. To determine the regularization parameter λ, which balances the relative importance of the *accuracy* of the prediction and its sparsity, SLIDE and IsoLasso fit the model for several values of λ and chose the best one according to some (stability) criterion.

On the other hand, NSMAP (36) applies nonconvex L^p regularization and iReckon (32) utilizes the more complex regularization term

$$- \lambda \cdot e^{\sum_i \sqrt[4]{\theta_i}}, \tag{11.1}$$

which penalizes lowly expressed isoforms more than highly expressed ones. Parameter λ is determined by gradually increasing λ until the fraction of reads that cannot be assigned to an expressed isoform exceeds a given threshold.

Notice that several methods, including iReckon and SLIDE, cope with the abundance shrinkage caused by the regularization penalty by re-running the optimization step without the penalty term using only the predicted transcripts.

11.3.3 Accuracy

In this section, we discuss the *accuracy* aspect of the prediction, that is, how well the predicted isoforms and their assigned abundances explain the observed reads. Most approaches fall into one of two categories described in the following sections.

11.3.3.1 Linear Regression
Many methods use a linear model to estimate the number of reads originating from segments of the genome, thus relying on the assumption that reads are uniformly sampled from the expressed transcripts.

In its general form, the estimated number of reads (or read pairs) originating from a certain segment (or a pair of segments) is proportional to the sum of the expression levels of transcripts containing that segment, weighted by the number of genomic positions at which a corresponding fragment can start. Random noise ε is added to fully explain the observations b:

$$b_j = \sum_{p \in \mathcal{P}(G)} f_{p,j}\theta_p + \varepsilon_j, \tag{11.2}$$

where b_j captures the number of reads observed in segment j, $\mathcal{P}(G)$ is the set of all maximal paths in splicing graph G, θ_p is the abundance (to be determined) of the isoform represented by path p, and $f_{p,j}$ is a weight coefficient that distributes reads among different segments along transcript p, with $f_{p,j} = 0$ if p does not contain segment j. Note that nonuniform sampling of reads along a transcript can be captured by coefficients f at most on the level of segments. In what follows, we abstract from precise definitions and units of observations, coefficients, and variables in (11.2) that vary among methods.

In general, methods based on a connectivity or splicing graph summarize individual reads into a read *count* for each segment. Segments typically correspond to the nodes and edges of the splicing graph. While the b_j's in (11.2) are obtained in IsoInfer, IsoLasso, MITIE, and TRAPH from the number of reads falling into segment j,

CLIIQ only counts reads spanning the middle point of segment j. SLIDE goes one step further and introduces paired-end bin counts. That is, a paired-end read with its left mate falling into a segment i and its right mate being mapped to a segment j does not contribute to two independent observations b_i and b_j, but to an observation $b_{i,j}$ counting the number of reads exhibiting this mapping signature.

Coefficients $f_{p,j}$ have to reflect the nature of the observations. TRAPH, MITIE, IsoInfer/IsoLasso essentially count the number of starting positions that imply reads falling into a given segment j. Although this number also depends on the read length and transcript p, $f_{p,j}$ is determined from the relative length of segment j alone, such that $f_{p_1,j} = f_{p_2,j}$ for $p_1 = p_2$. Notice that the minimum-cost network formulation of TRAPH and the ILP formulation of MITIE do not consider transcripts as objects and thus cannot readily account for transcript-dependent coefficients f. In contrast, CLIIQ infers the number of starting positions such that the resulting read spans the middle point of the segment from the read length and the transcript length. The paired-end bin counts allow SLIDE to apply a more involved design matrix, incorporating an estimated cDNA fragment length distribution, modeled as truncated exponential, in the determination of $f_{p,j}$. Simply speaking, $f_{p,j}$ will be smaller if p implies a distance between segment pair j that is unlikely according to the fragment length distribution.

A commonly applied form of estimating the unknown parameters of a linear model (11.2), here the expression levels θ_p of candidate isoforms p, is by means of a *least squares* approach. The optimal solution minimizes a loss function L that is the sum of squared errors ε_j, where ε_j is the difference between the observed number of reads b_j and the number of reads expected by the model (see (11.2)):

$$L(b, \theta) = \sum_j \varepsilon_j^2 = \sum_j \left(b_j - \sum_{p \in P(G)} f_{p,j} \theta_p \right)^2 \qquad (11.3)$$

Isoform reconstruction methods that are based on a loss function of the form (11.3) include TRAPH, SLIDE, IsoInfer, and IsoLasso. CLIIQ, in contrast, captures the abundance estimation problem by a linear model that minimizes the sum of absolute errors $\sum_j |\varepsilon_j|$.

For the loss functions, least squares and least absolute error to yield maximum likelihood (ML) estimates of the θ_p's, read counts must be Gaussian and the errors must be Laplace distributed, respectively. Following a previous observation (42) that a negative binomial distribution with appropriately linked variance and mean provides a more suitable model for read count measurements, MITIE introduced a log-likelihood-based loss function thus employs such a distribution and additionally models noise by a Poisson distribution. To make the resulting loss function accessible to the optimization scheme applied in (31), the authors approximate the loss function by two quadratic functions: one used for negative errors and the other used for non-negative residuals.

11.3.3.2 Likelihood-Based Methods

If read counts are normally distributed, least squares estimates are equivalent to maximum likelihood (ML) estimates. More generally, methods such as Cufflinks and iReckon compute a maximum likelihood assignment of sequenced fragments to isoform candidates. They employ a statistical model of the RNA-seq experiment, parameterized by the transcript abundances θ, which defines the probability of obtaining RNA-seq fragments from transcripts.

Cufflinks extends the model of Jiang and Wong (43) proposed for the isoform expression estimation problem to paired-end reads, incorporating a fragment length distribution. Since it specifies the probability of obtaining a fragment by a linear function of transcript abundances, the corresponding log-likelihood function is concave and its unique maximum can be determined by a hill-climbing method. Transcripts assigned ML estimates close to zero by an *expectation maximization* (EM) algorithm are accounted for in Cufflinks by an additional regularization step. Maximum *a posteriori* estimates are determined by sampling from the posterior distribution using the multivariate Gaussian with mean equal to the MLE as proposal distribution.

iReckon is based on a similar likelihood function as introduced in (44) for the estimation of gene expression under RNA-seq read mapping uncertainty. An EM algorithm tackles the interdependence of the assignment of reads to isoforms and their abundance estimation by an iterative approach. Based on the current estimates of the model parameters, that is, the transcript abundances, (expected) assignments of fragments to isoforms are computed (E-step), which in turn allow to re-evaluate the estimates of transcript abundances (M-step). In contrast to Cufflinks, however, iReckon's model does not only comprise a selected (minimal) set of isoforms but also all candidate transcripts. To prevent overfitting, iReckon thus augments the objective by a regularizing penalty term (see Section 11.3.2.2), which requires an adaption of the M-step to a numerical optimization scheme and which necessitates random restarts through the loss of concavity.

Montebello (41) applies a Monte Carlo simulation to find a set of isoforms with abundances that are scored high according to a L^0-penalized likelihood. NSMAP (36), on the other hand, employs a prior Laplace distribution and derives maximum a posteriori estimates for the transcript abundances. Instead of augmenting the statistical model of (43) by an L^1 regularization term, nonconvex L^p regularization, $0 < p < 1$, is imposed.

11.3.4   Completeness

In the most general sense, completeness refers to the property that all observations, or reads, can be explained by at least one predicted transcript. In particular, if the set or even number of isoforms is determined in the absence of a measure of *accuracy*, two-stage optimization approaches such as Cufflinks, CLASS, and CLIIQ counterbalance the *minimality* objective by a notion of *completeness*: The isoform reconstruction problem is cast into an instance of a *covering* problem, requiring every fragment (Cufflinks) or constraint (CLASS) to be explained by one of the reconstructed transcripts.

Since CLIIQ in a first stage does not attempt to minimize a loss function but minimizes the number of isoforms needed to bound the estimation error, it introduces a covering constraint that requires the estimated number of reads to be at least the minimum number of observed reads on any exon junction. Similarly, TRIP accounts for a relatively coarse measure of accuracy by requiring every splice junction to appear in at least one predicted transcript.

In contrast, the set cover instance solved by IsoInfer merges the solutions obtained for subinstances of the original isoform reconstruction problem. To cope with the combinatorial complexity of the problem, IsoInfer decomposes the instance into subinstances and solves them independently. The solution to the set cover instance combines the solutions to the subinstances in such a way that a small subset of the most frequent transcripts explain (cover) all the reads and all the provided transcription start and polyadenylation sites.

When isoforms and their expression levels are obtained through minimizing a loss function that measures the deviation of the estimated number of reads from the observed read data (see Section 11.3 for an example), *incompleteness* (underestimation) is penalized in a global attempt to balance deviations in both directions, by using positive or negative values of ε in (11.2) simultaneously across all observations b_j. Despite minimizing (11.3), however, Isolasso sets additional lower bounds for *completeness* by introducing constraints that require the estimated number of reads on each observed segment to be above a certain threshold.

11.3.5 Extensions

In the previous sections, we described the general principles underlying current methods for the transcript assembly problem. In practice, many more features play an important role in the applicability of a software tool in a specific context. Clearly, the extent to which a method depends on an existing annotation, from being completely independent of any annotation, such as Cufflinks, to incorporating the known transcription start sites and polyadenylation sites, as is the case for iReckon, determines the suitability of a method for the problem at hand. Methods also differ considerably in the way they employ read pairing information. Some methods may ignore this information completely (TRAPH) or use it in the abundance estimation alone (Cufflinks), while others apply read mapping signatures for filtering or for additional support (MITIE, CLASS, Scripture), and yet others link it to an empirically estimated fragment length distribution (SLIDE).

As an additional design consideration, few methods so far, in particular MITIE and CLIIQ, are able to *jointly* analyze multiple RNA-seq data sets. Lastly, models additionally incorporate solutions to modeling biological phenomena or data artifacts, including unspliced pre-mRNA and intron retention events (32), PCR amplification biases (32) or a GC content bias (33), and reads mapping to multiple locations on the genome (31).

11.4 CONCLUSIONS

When a high-quality reference genome is available, genome-based methods that first map reads to the genome and then assemble alignments into exon–intron transcript structures are generally superior to *de novo* approaches, although the latter can be very accurate (13, 14). Genome-based methods allow for better resolution of repeat and paralogous sequences, as well as overlapping gene models, and offer higher sensitivity, particularly in capturing low-coverage transcripts. As a drawback, they depend critically on the accuracy of read mapping, a challenging task that needs to take into account sequencing errors, polymorphisms, and splicing, and that may be adversely impacted by the quality of the genome. Conversely, *de novo* approaches are favored when there is no genome sequence available or when the genome is fragmented or of low quality. Specific scenarios such as the presence of trans-spliced genes, genes in polymorphic regions that are not included in the reference genome, or gene fusion and rearrangement events characteristic of cancer samples are also best addressed with *de novo* techniques. As a drawback, *de novo* assembly is computationally more challenging and is significantly more sensitive to sequencing errors. It is also more prone to overclustering of paralogs and to producing chimeric constructs. Given the mix of strategies being employed by genome and transcriptome analysis projects, both approaches are likely to be needed and to be widely used in the foreseeable future. Transcript reconstruction methods and their mathematical foundations need to continually adapt to provide more accurate solutions and to adapt to the characteristics and biases of the evolving sequencing technologies.

ACKNOWLEDGMENT

This work was supported in part by the National Institutes of Health grant R01-HG006677 to Steven L. Salzberg and the National Science Foundation award ABI-1159078 to Liliana Florea.

REFERENCES

1. Eswaran J, Cyanam D, Mudvari P, Reddy SD, Pakala SB, Nair SS, Florea L, Fuqua SA, Godbole S, Kumar R. Transcriptomic landscape of breast cancers through mRNA sequencing. Sci Rep 2012;2:264.

2. Seo J-S, Ju YS, Lee W-C, Shin J-Y, Lee JK, Bleazard T, Lee J, Jung YJ, Kim J-O, Shin J-Y, Yu S-B, Kim J, Lee E-R, Kang C-H, Park I-K, Rhee H, Lee S-H, Kim J-I, Kang J-H, Kim YT. The transcriptional landscape and mutational profile of lung adenocarcinoma. Genome Res 2012;22(11):2109–2119.

3. Djebali S, Davis CA, Merkel A, Dobin A, Lassmann T, Mortazavi A, Tanzer A, Lagarde J, Lin W, Schlesinger F, Xue C, Marinov GK, Khatun J, Williams BA, Zaleski C, Rozowsky J, Roder M, Kokocinski F, Abdelhamid RF, Alioto T, Antoshechkin I, Baer MT, Bar NS, Batut P, Bell K, Bell I, Chakrabortty S, Chen X, Chrast J, Curado J, Derrien T, Drenkow J, Dumais E, Dumais J, Duttagupta R, Falconnet E, Fastuca M, Fejes-Toth K, Ferreira P, Foissac S, Fullwood MJ, Gao H, Gonzalez D, Gordon A, Gunawardena H, Howald C, Jha S, Johnson R, Kapranov P, King B, Kingswood C, Luo OJ, Park E, Persaud K, Preall JB, Ribeca P, Risk B, Robyr D, Sammeth M, Schaffer L, See L-H, Shahab A, Skancke J, Suzuki AM, Takahashi H, Tilgner H, Trout D, Walters N, Wang H, Wrobel J, Yu Y, Ruan X, Hayashizaki Y, Harrow J, Gerstein M, Hubbard T, Reymond A, Antonarakis SE, Hannon G, Giddings MC, Ruan Y, Wold B, Carninci P, Guigo R, Gingeras TR. Landscape of transcription in human cells. Nature 2012;489(7414):101–108.

4. Pan Q, Shai O, Lee LJ, Frey BJ, Blencowe BJ. Deep surveying of alternative splicing complexity in the human transcriptome by high-throughput sequencing. Nat Genet 2008;40(12):1413–1415.

5. Wang ET, Sandberg R, Luo S, Khrebtukova I, Zhang L, Mayr C, Kingsmore SF, Schroth GP, Burge CB. Alternative isoform regulation in human tissue transcriptomes. Nature 2008;456(7221):470–476.

6. Trapnell C, Williams BA, Pertea G, Mortazavi A, Kwan G, van Baren MJ, Salzberg SL, Wold BJ, Pachter L. Transcript assembly and quantification by RNA-Seq reveals unannotated transcripts and isoform switching during cell differentiation. Nat Biotechnol 2010;28(5):511–515.

7. Weber JL, Myers EW. Human whole-genome shotgun sequencing. Genome Res 1997;7(5):401–409.

8. Hansen KD, Brenner SE, Dudoit S. Biases in Illumina transcriptome sequencing caused by random hexamer priming. Nucleic Acids Res 2010;38(12):e131.

9. Li J, Jiang H, Wong W. Modeling non-uniformity in short-read rates in RNA-Seq data. Genome Biol 2010;11(5):R50.

10. Surget-Groba Y, Montoya-Burgos JI. Optimization of *de novo* transcriptome assembly from next-generation sequencing data. Genome Res 2010;20:1432–1440.

11. Martin J, Bruno V, Fang Z, Meng X, Blow M, Zhang T, Sherlock G, Snyder M, Wang Z. Rnnotator: an automated de novo transcriptome assembly pipeline from stranded RNA-Seq reads. BMC Genomics 2010;11(1):663.

12. Schulz MH, Zerbino DR, Vingron M, Birney E. Oases: robust *de novo* RNA-seq assembly across the dynamic range of expression levels. Bioinformatics 2012;28(8):1086–1092.

13. Robertson G, Schein J, Chiu R, Corbett R, Field M, Jackman SD, Mungall K, Lee S, Okada HM, Qian JQ, Griffith M, Raymond A, Thiessen N, Cezard T, Butterfield YS, Newsome R, Chan SK, She R, Varhol R, Kamoh B, Prabhu A-L, Tam A, Zhao YJ, Moore RA, Hirst M, Marra MA, Jones SJM, Hoodless PA, Birol I. *De novo* assembly and analysis of RNA-seq data. Nat Methods 2010;7:909–912.

14. Grabherr MG, Haas BJ, Yassour M, Levin JZ, Thompson DA, Amit I, Adiconis X, Fan L, Raychowdhury R, Zeng Q, Chen Z, Mauceli E, Hacohen N, Gnirke A, Rhind N, di Palma F, Birren B, Nusbaum C, Lindblad-Toh K, Friedman N, Regev A. Full-length transcriptome assembly from RNA-Seq data without a reference genome. Nat Biotechnol 2011;29:644–652.

15. Martin JA, Wang Z. Next-generation transcriptome assembly. Bioinformatics 2013;29(8):1004–1010.

16. Martin M. Cutadapt removes adapter sequences from high-throughput sequencing reads. EMBnet.J 2011;17(1):10–12.

17. Yang X, Chockalingam SP, Aluru S. A survey of error-correction methods for next-generation sequencing. Briefings Bioinf 2012;14(1):56–66.

18. Idury RM, Waterman MS. A new algorithm for DNA sequence assembly. J Comput Biol 1995;2:291–306.

19. Pevzner PA, Tang H, Waterman MS. An Eulerian path approach to DNA fragment assembly. Proc Natl Acad Sci U S A 2001;98(17):9748–9753.

20. Flicek P, Birney E. Sense from sequence reads: methods for alignment and assembly. Nat Methods 2009;6(11):S6–S12.

21. Zerbino DR, Birney E. Velvet: algorithms for de novo short read assembly using de Bruijn graphs. Genome Res 2008;18(5):821–829.

22. Sommer DD, Delcher AL, Salzberg SL, Pop M. Minimus: a fast, lightweight genome assembler. BMC Bioinf 2007;8(1):64.

23. Kent JW. BLAT - the BLAST-like alignment tool. Genome Res 2002;12:656–664.

24. Florea L, Hartzell G, Zhang Z, Rubin GM, Miller W. A computer program for aligning a cDNA sequence with a genomic DNA sequence. Genome Res 1998;8(9):967–974.

25. Ahuja RK, Magnanti TL, Orlin JB. *Network flows: theory, algorithms, and applications*, Upper Saddle River (NJ): Prentice-Hall, Inc.; 1993.

26. Li W, Feng J, Jiang T. IsoLasso: a LASSO regression approach to RNA-Seq based transcriptome assembly. J Comput Biol 2011;18(11):1693–1707.

27. Guttman M, Garber M, Levin JZ, Donaghey J, Robinson J, Adiconis X, Fan L, Koziol MJ, Gnirke A, Nusbaum C, Rinn JL, Lander ES, Regev A. *Ab initio* reconstruction of cell type-specific transcriptomes in mouse reveals the conserved multi-exonic structure of lincRNAs. Nat Biotechnol 2010;28(5):503–510.

28. Heber S, Alekseyev M, Sze S-H, Tang H, Pevzner PA. Splicing graphs and the EST assembly problem. Bioinformatics 2002;18 Suppl 1:S181–S188.

29. Tomescu A, Kuosmanen A, Rizzi R, Mäkinen V. A novel min-cost flow method for estimating transcript expression with RNA-seq. BMC Bioinf 2013;14 Suppl 5:S15.

30. Lin Y-Y, Dao P, Hach F, Bakhshi M, Mo F, Lapuk A, Collins C, Sahinalp SC. CLIIQ: accurate comparative detection and quantification of expressed isoforms in a population. In: Raphael B, Tang J, editors. *Algorithms in Bioinformatics*. Volume 7534 of Lecture Notes in Computer Science. Berlin Heidelberg: Springer-Verlag; 2012. p 178–189.

31. Behr J, Kahles A, Zhong Y, Sreedharan VT, Drewe P, Rätsch G. MITIE: simultaneous RNA-Seq-based transcript identification and quantification in multiple samples. Bioinformatics 2013;29(20):2529–2538.

32. Mezlini AM, Smith EJM, Fiume M, Buske O, Savich GL, Shah S, Aparicio S, Chiang DY, Goldenberg A, Brudno M. iReckon: simultaneous isoform discovery and abundance estimation from RNA-seq data. Genome Res 2013;23(3):519–529.

33. Li JJ, Jiang C-R, Brown JB, Huang H, Bickel PJ. Sparse linear modeling of next-generation mRNA sequencing (RNA-Seq) data for isoform discovery and abundance estimation. Proc Natl Acad Sci U S A 2011;108(50):19867–19872.

34. Feng J, Li W, Jiang T. Inference of isoforms from short sequence reads. J Comput Biol 2011;18(3):305–321.

35. Song L, Florea L. CLASS: constrained transcript assembly of RNA-seq reads. BMC Bioinf 2013;14 Suppl 5:S14.

36. Xia Z, Wen J, Chang C-C, Zhou X. NSMAP: a method for spliced isoforms identification and quantification from RNA-seq. BMC Bioinf 2011;12(1):162.

37. Mangul S, Caciula A, Al Seesi S, Brinza D, Banday AR, Kanadia R. An integer programming approach to novel transcript reconstruction from paired-end RNA-seq reads. Proceedings of the ACM Conference on Bioinformatics, Computational Biology and Biomedicine, BCB2012. New York: ACM; 2012. p 369–376.

38. Fulkerson DR. Note on Dilworth's decomposition theorem for partially ordered sets. Proc Am Math Soc 1956;7(4):701.

39. Tomescu AI, Kuosmanen A, Rizzi R, Mäkinen V. A novel combinatorial method for estimating transcript expression with RNA-seq: bounding the number of paths. In: Darling A, Stoye J, editors. *Algorithms in Bioinformatics*. Volume 8126 of Lecture Notes in Computer Science. Berlin Heidelberg: Springer-Verlag; 2013. p 85–98.

40. Hocking RR, Leslie RN. Selection of the best subset in regression analysis. Technometrics 1967;9(4):531–540.

41. Hiller D, Wong WH. Simultaneous isoform discovery and quantification from RNA-Seq. Stat Biosci 2013;5(1):100–118.

42. Anders S, Huber W. Differential expression analysis for sequence count data. Genome Biol 2010;11(10):R106.

43. Jiang H, Wong WH. Statistical inferences for isoform expression in RNA-Seq. Bioinformatics 2009;25(8):1026–1032.

44. Li B, Ruotti V, Stewart RM, Thomson JA, Dewey CN. RNA-Seq gene expression estimation with read mapping uncertainty. Bioinformatics 2010;26(4):493–500.

12

AN OVERVIEW AND COMPARISON OF TOOLS FOR RNA-SEQ ASSEMBLY

RASIAH LOGANANTHARAJ

Bioinformatics Research Lab, The Center for Advanced Computer Studies, University of Louisiana, Lafayette, LA, USA

THOMAS A. RANDALL

Integrative Bioinformatics, National Institute of Environmental Health Sciences, Research Triangle Park, NC, USA

RNA-seq technology enables the genome-wide study of various aspects of the function and regulation of genes at the mRNA level, including expression analysis, identification/quantification of alternative splicing, and the identification of gene fusion events. The amount of data being generated by RNA-seq is allowing the analysis of the transcriptomes of single cells, an unprecedented level of resolution (1). For expression analyses, RNA-seq is also seen as an alternative to traditional microarray-based technologies (2, 3). A limitation of current microarray technology for both expression analysis and alternative splicing analysis is that each is tied to the identification of known genetic elements defined by the specific platform used. RNA-seq does not have this drawback and also allows for differential expression analysis over a much larger dynamic range. *de novo* RNA-seq has immense potential for biological discovery as it allows for the discovery of novel genes and gene isoforms. This potential also extends to both the discovery and analysis of nontraditional genomic elements such as noncoding RNAs (ncRNAs) whose importance in gene regulation is becoming increasingly appreciated (4). The higher cost of RNA-seq, relative to microarray analysis, remains one of the main considerations in deciding which approach is best for a given analysis, although this is also continually dropping.

Computational Methods for Next Generation Sequencing Data Analysis, First Edition.
Edited by Ion I. Măndoiu and Alexander Zelikovsky.
© 2016 John Wiley & Sons, Inc. Published 2016 by John Wiley & Sons, Inc.
Companion website: www.wiley.com/go/Mandoiu/NextGenerationSequencing

It is common to refer to all post-Sanger sequencing technologies as "next generation." It is becoming more accurate to consider the more mature of these technologies such as Illumina, SOLiD, 454, and other platforms generating short reads as second-generation sequencing. This is in part to contrast them from newer technologies such as PacBio, which do single-molecule real-time (SMRT) sequencing and generate smaller amounts of ultralong sequences with very different error characteristics and often complementary applications (5). Second-generation sequencing technologies, such as Illumina, which will be the sole focus of this study, produce large quantities of short sequence reads. Liu et al. (6) provides a good overview of the various technologies available and their capabilities. Many of these platforms, including Illumina, can produce either single or paired end reads (also known as mate pairs; Figure 12.1a) of various lengths. The specific experimental design of a sequencing project should dictate which of these is most appropriate. The value of paired-end reads for RNA-seq is that they contain both positional and strand information that can be used to more accurately assemble transcripts and to identify novel elements. The second-generation sequencing technologies are continuing to develop and increase in capacity. The development of algorithms to analyze these sequence data sets is thus also ongoing and continuous with no truly defined standardized methodology emerging. Thus, it is of interest to compare and contrast various methods that have been used. In the case of RNA-seq, two types of analyses are common: transcriptomics based on a defined reference sequence (assemble to reference) and transcriptomes

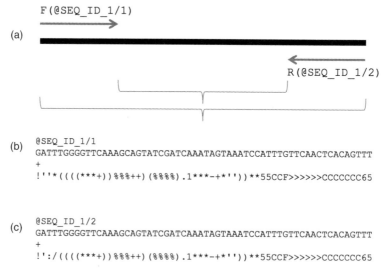

Figure 12.1 Paired-end reads. (a) A paired-end set of reads: the arrows represent the read length, and the position of the arrow indicates strand (top or bottom). The outer bracket represents the length of the fragment being sequenced while the inner bracket represents the mate pair inner distance. (b) Forward read (F, top strand); the header of a read is terminated with the read number followed by 1 or 2 after (/) denoting forward and reverse reads, respectively. (C) Reverse read (R, bottom strand); for example, the quality score at position 3 of the reverse read is encoded by ":"; this represents a phred score of 25 (ASCII_value (":") −33 = 58 − 33 = 25).

RNA > cdna > library prep > sequence >

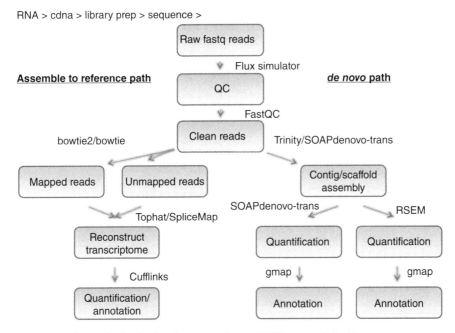

Figure 12.2 Outline for comparison of RNA-seq tools in this study.

that are generated *de novo*. For both types of analyses, multiple analysis algorithms have been developed. The main goal of this study is to compare some of the most widely used algorithms for each type of analysis. A typical workflow of RNA-seq analysis is shown in Figure 12.2 emphasizing the tools we use in this present study.

12.1 QUALITY ASSESSMENT

The sequence reads generated by the new technologies produce, in addition to base-pair calls, quality scores associated with each call. For Illumina, these data are summarized in a sequence format that is unique to this type of sequencing, the fastq format (fasta plus quality). Each read in fastq format has four lines of which the first one is the header beginning with @, followed by a string containing information about the sequencing machine, flowcell, and positional information within a cell; the second line contains base calls. This is followed by a quality score header starting with "+", this header is sometimes empty, but often recapitulates the information in the sequence header, and is followed by a fourth line of ASCII code encoding the quality score corresponding to the reads at each position. The base calling error probability p is represented as phred quality Q as $-10 \log_{10}(p)$. For example, consider the example reads from a paired-end data set as shown in Figure 12.1b and c.

Before any analysis, these reads may need preprocessing of a variety of types depending on how the sequence was generated and the experimental design. At the minimum, the quality of the base calls needs to be considered and calls below a user-defined threshold needs to be removed from either end. A common standard is to trim base call when the median base quality falls below 20. In addition, adapters used in the process of library construction may need to be removed. If samples have been multiplexed, then individual samples will have to be tagged by sequence barcodes. Following the separation of sample reads by these barcodes, they will need to be trimmed from the reads.

Many tools are available for quality control of reads, including FastQC (7), FastX (8), and NGS QC Toolkit (9). The distribution of FastQC is platform independent due to its implementation in Java and can be used both at the command line level and as a GUI on either Windows or Macintosh platforms. We have preferred it for its ease of use and the concise, biologist friendly, graphical presentation of a wide range of quality-related features including base sequence quality, per base sequence content, and k-mer content. FastQC also allows the input of fastq sequence in any of the various quality scorings, fastq sequence in SAM/BAM format, and for the analysis of reads from other sequencing platforms such as 454 and PacBio. For the purpose of illustration, consider the following graphs produced by FastQC as shown in Figure 12.3.

The kmer content variation seen in the reads from position 1 through 16 is often seen in samples, but in conjunction with the results from the Adaptor content graph could show an experimental artifact and would most likely represent adapters and hence the leading edge from 1 through 16 must be clipped before assembly. We have developed a python script to trim or clip reads. The script takes the fastq input file, has three optional parameters, and writes the processed output to a new file. Analysis of the raw sequence is not the only good metric for quality. Assuming one is working with a reference genome, a test alignment to that genome also helps assess quality. If one aligns each RNA-seq data set to Bowtie (Bowtie1) using Tophat2 (10), the output is two BAM files: one file is called accepted_hits.bam (reads that align to reference) and the other file is called unmapped.bam. Several means can be used to compare the contents of these files. Within samtools (11), one can use "view -c" that will count the reads in each of these files. It is desirable to have a high proportion of your sequence actually aligning to the reference genome, if this is in the range of at least 75–80%, then this suggests that your sequence is of reasonable quality. If, however, a high proportion of your data is not mappable to the reference genome, then one may desire to resequence a particular sample or eliminate that data set from downstream analyses as this could be confounding. Simply comparing the file sizes of the above BAM files is also a good proxy for this measurement. One difficulty in *de novo* analyses is that the above type of assessment is not available for truly novel sequence.

12.2 EXPERIMENTAL CONSIDERATIONS

Other than sequence quality, one of the most important issues with RNA-seq, as it always has been with microarray analysis, is the experimental design. Two primary issues are the question of replication (both biological and technical) and sequencing depth. Biological replication (e.g., mRNA from livers of three different mice of the

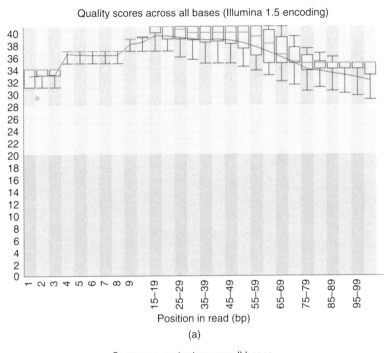

(a)

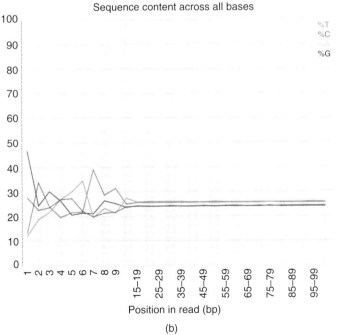

(b)

Figure 12.3 Output from FastQC. (a) The per-base sequence quality scores of the reads clearly indicates high-quality reads with the mean-per-base error rate better than 1 in 1000 (phred score is better than 30). (b) The per-base sequence content. (c) The kmer content in the reads. Please see www.wiley.com/go/Mandoiu/NextGenerationSequencing for a color version of this figure.

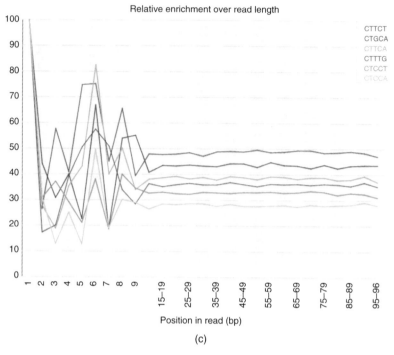

Figure 12.3   *(Continued)*

same genetic background and treatment) and technical replication (for instance, duplicate RNA preps and their subsequent independent library preparation and sequencing reactions from each of the above mice) have long been known to be essential for accuracy in reproducibility in microarray analysis (3). This has not been widespread in RNA-seq analyses due to cost considerations, but the capacity of the sequencing platforms has increased to the point where sufficient depth of coverage is easily reached, so investing in replicate samples is becoming more common. Obviously this is most important when the primary interest is the discovery of differentially expressed (DE) genes. This issue has been discussed in depth (12) and it is being realized that replicates are as important to an analysis as sequencing depth, which reaches a point of diminishing returns, and addition of replicates can increase the power to find differentially expressed genes (36).

Apart from cost, there are a variety of variables to consider in deciding how much sequence is necessary, depending on the overall objective of the analysis. A major determinant is the organism that is being sequenced; the considerations of a transcriptome from a 3 Gb human genome are different from that of the 12 Mb *Saccharomyces cerevisiae* genome in terms of differences in gene sizes, exon–intron structure, isoform abundance, and the extent of gene family expansions. As the output of the current technology is such that any output from current machines will be overkilled for most non-mammalian model organisms, we will only consider this question in terms of mammalian genomes. For mammals, a minimum of 20 million paired-end reads should be done (14) to measure the levels of known genes based on an annotation to a common, relatively conservative collection such as RefSeq. If one is attempting to quantitate low abundance transcripts or is interested in the measurement and/or discovery of new or alternative isoforms, then depth needs to be increased. There does seem to be a point at which returns diminish due to resequencing of common transcripts, and the quality of the experiment will be improved more by including replicate samples rather than increasing depth. It is estimated that 200 million reads would be sufficient to cover the human genome (15), this is easily achievable in a single lane of the HighSeq 2500 (300 million reads per flow cell, accessed on 05/06/2014 (16)). Another way to look at this question is to examine the depth that has been used in publicly available data sets. Sims et al., (17) studying Illumina data sets deposited at the European Nucleotide Archive (ENA), showed that most RNA-seq data sets available from all species through June 2013 contain over one billion base pairs, which would correspond to 5 million paired-end reads of 100 bp.

One of the first widely used measures of transcript/isoform abundance developed for RNA-seq is the concept of RPKM (single-end reads; reads per kilobase per million) and FPKM (paired-end reads; fragments per kilobase per million; (18)). Obviously, 1000 reads mapping to a 10 kb gene body ($5'$ UTR + exons +$3'$ UTR) compared with the same number mapping to a 1 kb gene body does not represent the same abundance level. Normalization methods attempt to take into account both differences in gene length and library sizes between samples being compared. Both RPKM and FPKM are relative measurements of expression, not an absolute measurement. Some have found faults with using FPKM, mainly in introducing biases in genes that are expressed at low levels (19) and there are other ways of measuring abundance (20). This is a very fluid and evolving field, and one reason FPKM is common is due to the popularity of the software packages, many of which we use here, that output this value as the default measurement. This is also a term that biologists are becoming used to understanding and expect to see. Given that, RSEM, which we use below, can output a related value that is also gaining acceptance, TPM (transcripts per million reads). Another important issue in terms of quantitation is determining the minimum accepted abundance value that is believable. Reasons for this include minimizing false discovery rates: the inherent difficulty in reliably comparing abundance measurements when they are based on a very small number of reads or random transcriptional noise. A commonly accepted convention in RNA-seq quantitation is to only accept FPKM values less than or equal to 1.0. A comprehensive analysis of this issue using mammalian transcriptome data sets suggests that a lower cutoff of FPKM $= 0.3$ is often justifiable (21). Alternatively, the use of spike-in samples can be used to give confidence to low expressed genes. The ERCC (External RNA Controls Consortium) has developed a set (22) that is commercially available (23).

12.3 ASSEMBLY

Assembly of sequenced reads into transcripts can be broadly categorized into two types: referenced-based or *de novo*. In a reference-based assembly, the reads are aligned with the reference genome; two classes of reads emerge, those that map well above a stringent threshold (are mapped to a continuous location with only a few mismatches/gaps allowed) and those that are unmappable by those criteria. The latter represent reads that potentially map across splice junctions or simply bad reads. The mapped reads are used to generate a set of all possible exons and thus all possible splice junctions between neighboring exons within a given distance, which would depend on the genome of interest, and the previously unmappable reads are then aligned to this collection. Transcripts, isoforms, and their expression levels can then be derived from these two sets of alignments. On the other hand, in *de novo* assembly, the reads are assembled into contigs without any prior assumptions about annotation or splice junctions, these contigs themselves represent transcripts. Expression values for a given data set can then be derived based on the coverage of these contigs. *de novo* assembly is valuable even for organisms with well-annotated reference genomes as this can allow either for discovery or for a more accurate assembly of a transcriptome from a nonreference isolate of a species, which may be diverged at the sequence level from a sequenced reference genome and annotation. *de novo* assembly is essential for the analysis of many nonmodel organisms for which either no genome exists or the genome is of poor quality.

Based on alignment methodology, the reference-based assemblers are grouped into two major categories, namely hash-table-based or prefix/suffix matching algorithms (24). In the hash-table-based approach, either reads or reference can be indexed. Reads are indexed in assemblers such as MAQ (25), RazerS (26), and RMAP (27) while BFAST (28), Genomemapper (29), Mosik (30), and Novoalign (31) are some examples where the reference is indexed. Linder and Friedel (see Reference 32) have done a comprehensive evaluation of alignment algorithms in the context of RNA-seq and have concluded that FM-index (a compressed full-text substring index) based algorithms outperformed the hash-based approach in run-time or in memory requirement. Therefore, we will consider only the RNA-seq assemblers using FM-based aligners such as Bowtie (33) or Bowtie2 (34). The Tuxedo suite of tools, including Tophat, has been used quite successfully for reference-based RNA-seq assembly. Tophat2 (10) is the most recent release of Tophat that uses Bowtie2 for alignment, whereas Bowtie was implemented in the older releases of Tophat (35). Bowtie2 allows for gap identification and alignment, whereas Bowtie does not. SpliceMap (36) is another popular assembler that uses Bowtie for aligning reads onto reference genome and across splice junctions. In this report, we will compare these two widely used referenced-based assemblers for their ability to reconstruct the correct transcripts from the reads while minimizing the discovery of false ones. To compare the efficacy of *de novo* assembly, we will assemble the same collection of reads with the SOAPdenovo-Trans (37) and Trinity (38), which are among the few often used and well-cited *de novo* assemblers. Importantly, all of the analysis tools we use in our comparisons below are freely available, and a list of such tools for NGS analysis along with those we have used in this work is posted at www.cacs.louisiana.edu/labs/bioinformatics/ngs_os.html.

12.4 EXPERIMENT

The objective of this study is to compare reference-based assemblers with *de novo* based assemblers and assess their relative effectiveness of reconstructing transcripts from RNA-seq data. For the purpose of objective comparison, a controlled experiment of known sets of transcripts has to be carried out. To this end, we simulated human paired-end RNA-seq data based on refSeq transcripts at a coverage level close enough to mimic real mammalian data sets from an Illumina sequencing run. A single lane of the Illumina Hiseq platform can easily generate ∼150 million paired-end reads, which is sufficient for both in-depth quantitation of gene expression in mammalian genomes and is deep enough to allow for the identification of novel events. It is thus convenient to think of a single lane of an Illumina flowcell as a single experimental sample. For simulation, we originally considered some commonly used RNA-seq simulators such as BEER (39) and Flux (40). While BEER provides many desirable parameters to control the simulated reads, it only produces reads with fasta format, and hence it was not considered for this study.

We used Flux simulator (40) version 1.2.1 to simulate paired-end reads from ref-Seq transcripts of human (hg19). The Flux simulator generates reads using the control parameters set in its input file including links to the sorted GTF file for transcripts and the fasta file of the genome of interest using a realistic quality error model reflecting the current sequencing technology of interest. Of the two error models available, we use the quality error model for generating 76 bp reads to scale to the longer reads that were generated. The simulator was set to generate a prespecified number of reads of length 100 bp, and any shorter reads were removed. The paired-end reads are output into a single file and then separated into fastq data sets representing the forward and reverse paired reads that the assemblers expect as input. In addition to producing the reads, Flux also produces a file with the details of the transcriptome profile for each simulated data set, which has information including locus, length of a transcript (say L), transcript id, number of reads (say N) sequenced from individual transcripts, and their genomic coordinates. If the trimmed length of the reads is 85 bp, then the approximate coverage of a transcript will be the number of reads generated by Flux from a transcript times the length of the read divided by the length of the transcript ($N * 85/L$). Five different data sets were generated, each of which were examined with FastQC and we found in all five that the trailing 15 bp that had quality score lower than 20, which we used as our quality cutoff threshold. Therefore, 15 bp of the trailing end of each read was trimmed from the data set. Since these data sets were simulated, the leading edges of the reads were acceptable and we did not observe any impact of adapters. The details of the reads simulated are given in Table 12.1.

TABLE 12.1   Simulated Human Paired-End Reads

Data Set	Number of Pe-Reads	Length of Reads After Trimming (bp)
Set 1	112,322,914	85
Set 2	112,332,180	85
Set 3	149,814,146	85
Set 4	149,844,384	85
Set 5	149,835,654	85

12.5 COMPARISON

These trimmed simulated RNA-seq data sets were assembled by two reference-based assemblers Tophat2 (10) and SpliceMap (36) and two *de novo* assemblers SOAPdenovo-Trans (37) and Trinity (38). Using Bowtie2, the first 500,000 reads from each data set were mapped onto the hg19 refSeq transcripts that were downloaded from Ensembl.org at ftp://ftp.ensembl.org/pub/release-73 to determine the mate pair inner distance for each data set. This calculation is important in terms of properly tuning assemble to reference algorithms and is specific to each analysis as the value generated is dependent both on the size of the cDNA fragments used to prepare the sequencing library and the length of the sequence reads (see Figure 12.1a). The output, an aligned BAM file, was sorted by the SortSam jar module from SamTools (41). Like the fastq sequence format, the SAM/BAM formats have been developed specifically for describing the alignments of fastq reads to reference genomes used in second-generation sequencing (11). SAM (sequence alignment/map——) is a text description of an alignment to reference, while BAM is the compressed, binary version of a SAM file, often more manageable due to the sheer size of the data sets that result from alignments. Most of the widely used tools used to manipulate and analyze these sequence data will recognize one or the other or both of these formats. The CollectInsertSizeMetrices jar tool from the picard package (42) was used to find the mate pair inner distance and standard deviation for each data set. For each simulated data set, Tophat2 was ran with default settings except for the following parameters (inner mate pair distance, standard deviation, GTF file from hg19), which were set as appropriate for each data set. Reads were mapped on to the reference genome for hg19 first, and the consensus regions were then assembled and possible splices between neighboring exons were determined. The initially unmapped reads are then fragmented and then aligned onto the reference and splice junctions. A very good tutorial describing the detailed use of Tophat and Cufflinks for RNA-seq analysis is available (43). For the specific case of detecting novel fusion transcripts, a version of Tophat called Tophat-fusion has been developed (43).

SpliceMap uses a split-reads alignment algorithm. Reads that are less than 50 bp were split into overlapping 25 bp half-reads while longer reads are split into overlapping 50 bp reads. These half reads are aligned to the reference gene with either Bowtie or BWA (44). These alignments are combined together to determine the exons or split junctions. SpliceMap uses an algorithm somewhat similar to Tophat, but generates its analysis without the use of a GTF file describing the transcripts. Each of the five data sets was assembled by Tophat2 and with SpliceMap on a cluster computer with over 18 parallel threads. Tophat creates an aligned BAM file, and the SpliceMap creates an aligned SAM file. Cufflinks, which provides quantification of RNA-seq from the assembled or aligned BAM files, was applied to the aligned BAM/SAM files of our both Tophat and SpliceMap outputs to generate FPKM tracking files of the assembled transcripts. The FPKM (Fragments Per Kilo bases of exons per Million mapped reads) value is one of several means developed to normalize relative transcript abundance in RNA-seq analyses, taking into account that transcript size does vary considerably in most organisms. We have used Cufflinks with the following options (–G hg19_refGenes_canonical.gtf -b hg19canonical.fa -u –compatible-hits-norm) for estimating isoform using the given gtf file (-G option), for improving the estimates of transcript abundance (-b option) and multiple read correction (-u option), and for appropriate compatibility. Identical parameters for Cufflinks were used for each analysis. Cufflinks was originally designed as a tool for the downstream analysis of the output of the Tophat assembler but can be used with the output of other assembly algorithms.

Trinity (38) and SOAPdenovo-Trans (37) were used to generate *de novo* RNA-seq transcriptomes from our simulated data sets. Both are de Bruijn graph-based, *de novo* transcript assembly algorithms (45). SOAPdenovo-Trans allows testing of multiple *k*-mer values, while Trinity uses a fixed value of 25. Each was designed with the goal of the assembly of transcriptomes and quantitation of their transcript abundance. Both attempt to resolve transcript isoforms by resolving de Bruijn subgraphs that potentially contain multiple isoforms. SOAPdenovo-Trans performs transcript quantitation as an option, while the output of Trinity can be used as an input to an algorithm explicitly designed for transcript quantitation, such as RSEM (46), edgeR (47), or DESeq (48).

The five sets of simulated and trimmed paired-end reads were analyzed with both programs using default values except the option to calculate FPKM values was used for the SOAPdenovo-Trans analyses. The Trinity assemblies were then analyzed with RSEM for the determination of FPKM values of the assemblies. Assemblies from both programs were then annotated to hg19 using gmap (49). Trinity is a very memory-intensive program, in this study each run typically took over 6 hours on a server with 256 Mb of RAM while SOAPdenovo-Trans analyzed the same samples in under an hour. For speeding up Trinity with larger data sets, it is very common to increase the memory allocated to the program (or to one or more of the three programs comprising Trinity, inchworm, chrysalis, and butterfly) along with running analyses using multiple processors, all of which can be controlled using command line flags. It is also common with large data sets to do an *in silico* normalization step to reduce the run-time. There is a very active google listserv that discusses current issues in Trinity assembly analyses (50) and a very detailed tutorial available (51).

For the transcriptome output of all *de novo* assemblies, when the aligned length of a contig spans through the length of a known transcript, we consider that a perfect match; otherwise it will be considered to be a fractional match. We have considered 10%, 50%, and 90% fractional matches in annotating contigs to known refSeq transcripts.

12.6 RESULTS

Five different paired-end RNA-seq data were generated with Flux simulator using refSeq transcripts of the human genome (hg19) as a guide. The specific details and the summary are provided in the Experiment section. Each data set, after trimming, was assembled with the reference-based Tophat2 and SpliceMap and with the *de novo* assemblers Trinity and the SOAPdenovo-Trans. For evaluation, true positives are considered to be the transcripts generated with coverage above some specified threshold.

TABLE 12.2 The Average Value of True Positive Rate of Assemblers with Twofold Transcript Coverage

FPKM	Complete Coverage		Soapdenovo Fractional Coverage			Trinity Fractional Coverage		
	Tophat	SpliceMap	90%	50%	10%	90%	50%	10%
0.5	0.8318	0.8166	0.0096	0.0314	0.2194	0.0086	0.0276	0.1905
1	0.7438	0.7254	0.0095	0.0305	0.2078	0.0086	0.0269	0.1820
2	0.6413	0.6231	0.0089	0.0276	0.1854	0.0081	0.0246	0.1645
5	0.4900	0.4746	0.0073	0.0219	0.1462	0.0066	0.0200	0.1349
10	0.3663	0.3551	0.0056	0.0167	0.1119	0.0052	0.0158	0.1073
15	0.2912	0.2826	0.0044	0.0135	0.0912	0.0043	0.0132	0.0892

The identification of transcripts from Trinity and SOAPdenovo-Trans are based on fractional match.

TABLE 12.3 The Average Value of True Positive Rate of Assemblers with Fivefold Transcript Coverage

FPKM	Complete Coverage		Soapdenovo Fractional Coverage			Trinity Fractional Coverage		
	Tophat	SpliceMap	90%	50%	10%	90%	50%	10%
0.5	0.9318	0.9100	0.0109	0.0356	0.2466	0.0098	0.0313	0.2137
1	0.8442	0.8222	0.0109	0.0346	0.2358	0.0097	0.0306	0.2057
2	0.7290	0.7084	0.0101	0.0314	0.2106	0.0092	0.0279	0.1861
5	0.5576	0.5402	0.0083	0.0250	0.1661	0.0076	0.0227	0.1529
10	0.4170	0.4044	0.0064	0.0190	0.1271	0.0060	0.0179	0.1217
15	0.3316	0.3219	0.0050	0.0154	0.1036	0.0049	0.0150	0.1012

The identification of transcripts from Trinity and SOAPdenovo-Trans are based on fractional match.

We have chosen this threshold as being either a two or fivefold level of coverage for comparative purposes throughout. The combined results of the true positive rate are given in Tables 12.2 and 12.3 and are shown in Figures 12.4 and 12.5.

The sensitivity or the true positive rate provides only one aspect of the performance metric. Let us look at the positive transcripts that are common in both the reference-based assemblers and *de novo* assemblers. The results for common transcripts for both two and fivefold coverage are given in Tables 12.4 and 12.5 and graphically in Figures 12.6 and 12.7, respectively.

12.7 SUMMARY AND CONCLUSION

The Flux simulator generated 100-bp, paired-end RNA-seq data sets with a realistic error model based on Illumina sequencing technology as described in the section on Experiment. We have generated five different data sets, each with at least 112 million paired-end reads. Multicore high-performance computers were used to assemble

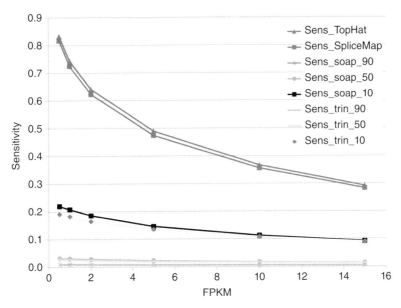

Figure 12.4   The performance relationship in Table 12.2 is illustrated. The sensitivity (Sens) of both reference-based assemblers outperformed those of the *de novo* assemblers. Abbreviations: soap (SOAPdenovo-Trans), trin (Trinity). Please see www.wiley.com/go/Mandoiu/NextGenerationSequencing for a color version of this figure.

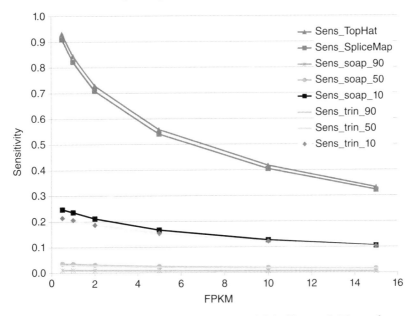

Figure 12.5   The performance relationship in Table 12.3 is illustrated. The performances of all the assemblers are slightly increased with the increased coverage. Abbreviations: soap (SOAPdenovo-Trans), trin (Trinity). Please see www.wiley.com/go/Mandoiu/NextGenerationSequencing for a color version of this figure.

TABLE 12.4 The Percentage of the Transcripts that Are in Common between Assemblers with a Range of Different FPKM Thresholds

FPKM	Per. Common in Tophat and SpliceMap (%)	Per. Common in Trinity and SOAPdenovo-Trans (%)		
		Cover. > 90(%)	Cover. > 50(%)	Cover. > 10(%)
0.5	92.77	67.82	71.00	67.67
1	92.26	67.51	71.18	67.47
2	91.70	68.65	71.41	67.61
5	90.52	72.14	72.03	67.12
10	89.36	70.90	70.42	65.78
15	88.13	70.63	68.99	63.58

This data are for transcripts with twofold coverage.

TABLE 12.5 The Percentage of the Transcripts that Are Common between Assemblers with a Range of Different FPKM Thresholds

FPKM	Per. common in Tophat and SpliceMap (%)	Per. common in Trinity and SOAPdenovo-Trans (%)		
		Coverage > 90(%)	Coverage > 50(%)	Coverage > 10(%)
0.5	94.20	67.76	71.23	68.42
1	92.72	67.45	71.36	67.88
2	91.96	68.58	71.60	67.91
5	90.74	72.08	72.17	67.35
10	89.54	70.82	70.56	65.96
15	88.29	70.78	69.28	63.81

This data is for transcripts with fivefold coverage.

the reads. In this experiment, we have compared widely used and freely available reference-based RNA-seq assemblers with *de novo* assemblers for their efficacy and the accuracy of reproduction of a mammalian transcriptome. The assemble to reference programs consistently had a higher sensitivity by all measures than did the *de novo* assemblers at both levels of transcript coverage tested. We were somewhat surprised by the lack of ability of the *de novo* assemblers to reproduce the transcripts. Even with lowest fractional identification of transcripts (10% or better), the true positive rate is only about 20% while the reference-based assemblers achieve about 90% prediction accuracy. An increased depth of coverage from two to fivefold did not appreciably improve this. This phenomenon had been observed previously in a recent comparison of *de novo* RNA-seq assembly (52). The ability of the two *de novo* assemblers to identify a common set of transcripts from within each of the five data sets was considerably closer to the performance of the assemble to reference algorithms as they dropped off only about 20% in comparison at all FPKM and both coverage levels. Thus, despite identifying significantly fewer transcripts, one should have

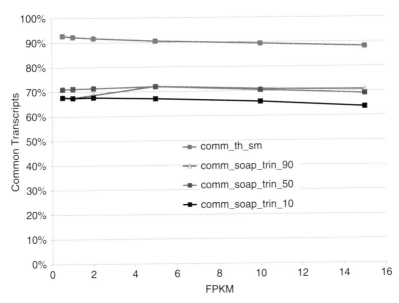

Figure 12.6 Common transcripts in assemblers with twofold coverage. Tophat and SpliceMap share around 90% of the unique true transcripts. SOAPdenovo-Trans and Trinity share around 70% of the unique transcripts in with fractional coverage better than 50% and 90%. Abbreviations: th (Tophat), sm (SpliceMap), soap (SOAPdenovo-Trans), trin (Trinity).

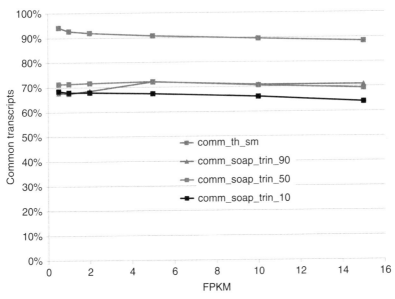

Figure 12.7 Common transcripts in assemblers with fivefold coverage. Tophat and SpliceMap share over 90% of the unique true transcripts. SOAPdenovo-Trans and Trinity share over 70% of the unique transcripts with fractional coverage better than 50% and 90%. Abbreviations: th (Tophat), sm (SpliceMap), soap (SOAPdenovo-Trans), trin (Trinity).

good confidence that those being independently identified by the *de novo* programs are likely to be of value. RNA-seq is a relatively new field, approaches are constantly changing, and while it is maturing in terms of its use in expression analyses, these results suggest that there is a lot of room for improvement in the field of *de novo* RNA-seq assembly. A similar state of the art also exists for *de novo* genome assembly and should not be surprising. This simply emphasizes the advantage of having -a well annotated genome to guide any assembly. The current value of a *de novo* approach to assembly remains in the potential to drive discovery. It will take further improvements, either in the algorithms or the data, before this approach can match that of the assemble to reference methodology.

ACKNOWLEDGMENTS

This article was co-written by Thomas A. Randall in his private capacity. No official support or endorsement by the NIH, National Institute of Environmental Health, is intended or should be inferred.

REFERENCES

1. Saliba AE, Westermann AJ, Gorski SA, Vogel J. Single-cell RNA-seq: advances and future challenges. Nucleic Acids Res 2014;42(14):8845–8860.
2. Guo Y, Sheng Q, Li J, Ye F, Samuels DC, Shyr Y. Large scale comparison of gene expression levels by microarrays and RNAseq using TCGA data. PLoS ONE 2013;8(8):e71462.
3. Wang C, Gong B, Bushel PR, Thierry-Mieg J, Thierry-Mieg D, Xu J, Fang H, Hong H, Shen J, Su Z et al. The concordance between RNA-seq and microarray data depends on chemical treatment and transcript abundance. Nat Biotechnol 2014;32(9):926–932.
4. Fatica A, Bozzoni I. Long non-coding RNAs: new players in cell differentiation and development. Nat Rev Genet 2014;15(1):7–21.
5. Eid J, Fehr A, Gray J, Luong K, Lyle J, Otto G, Peluso P, Rank D, Baybayan P, Bettman B et al. Real-time DNA sequencing from single polymerase molecules. Science 2009;323(5910):133–138.
6. Liu L, Li Y, Li S, Hu N, He Y, Pong R, Lin D, Lu L, Law M. Comparison of next-generation sequencing systems. J Biomed Biotechnol 2012;2012:251364.
7. http://www.bioinformatics.babraham.ac.uk/projects/fastqc/. Available at Accessed 2016 Mar 17.
8. http://hannonlab.cshl.edu/fastx_toolkit/.
9. Patel RK, Jain M. NGS QC Toolkit: a toolkit for quality control of next generation sequencing data. PLoS ONE 2012;7(2):e30619.
10. Kim D, Pertea G, Trapnell C, Pimentel H, Kelley R, Salzberg SL. TopHat2: accurate alignment of transcriptomes in the presence of insertions, deletions and gene fusions. Genome Biol 2013;14(4):R36.
11. Li H, Handsaker B, Wysoker A, Fennell T, Ruan J, Homer N, Marth G, Abecasis G, Durbin R. Genome project data processing S: the sequence alignment/map format and SAMtools. Bioinformatics 2009;25(16):2078–2079.

12. Williams AG et al. RNA-seq data: challenges in and recommendations for experimental design and analysis. Curr Protoc Hum Genet 2014;83:11.13.1-11.13.20.

13. Liu Y, Zhou J, White KP. RNA-seq differential expression studies: more sequence or more replication? Bioinformatics 2014;30(3):301–304.

14. Wang Z, Gerstein M, Snyder M. RNA-Seq: a revolutionary tool for transcriptomics. Nat Rev Genet 2009;10(1):57–63.

15. Tarazona S et al. Differential expression in RNA-seq: a matter of depth. Genome Res 2011;21(12):2213–2223.

16. www.illumina.com/systems.html. Available at Accessed 2016 Mar 17.

17. Sims D et al. Sequencing depth and coverage: key considerations in genomic analyses. Nat Rev Genet 2014;15(2):121–132.

18. Mortazavi A et al. Mapping and quantifying mammalian transcriptomes by RNA-Seq. Nat Methods 2008;5(7):621–628.

19. Oshlack A, Wakefield MJ. Transcript length bias in RNA-seq data confounds systems biology. Biol Direct 2009;4:14.

20. Dillies MA et al. A comprehensive evaluation of normalization methods for Illumina high-throughput RNA sequencing data analysis. Brief Bioinform 2013;14(6):671–683.

21. Ramskold D et al. An abundance of ubiquitously expressed genes revealed by tissue transcriptome sequence data. PLoS Comput Biol 2009;5(12):e1000598.

22. Jiang L et al. Synthetic spike-in standards for RNA-seq experiments. Genome Res 2011;21(9):1543–1551.

23. www.lifetechnologies.com. Available at Accessed 2016 Mar 17.

24. Li H, Homer N. A survey of sequence alignment algorithms for next-generation sequencing. Brief Bioinform 2010;11(5):473–483.

25. Li H, Ruan J, Durbin R. Mapping short DNA sequencing reads and calling variants using mapping quality scores. Genome Res 2008;18(11):1851–1858.

26. Weese D, Emde AK, Rausch T, Doring A, Reinert K. RazerS–fast read mapping with sensitivity control. Genome Res 2009;19(9):1646–1654.

27. Smith AD, Xuan Z, Zhang MQ. Using quality scores and longer reads improves accuracy of Solexa read mapping. BMC Bioinformatics 2008;9:128.

28. Homer N, Merriman B, Nelson SF. BFAST: an alignment tool for large scale genome resequencing. PLoS ONE 2009;4(11):e7767.

29. Schneeberger K, Hagmann J, Ossowski S, Warthmann N, Gesing S, Kohlbacher O, Weigel D. Simultaneous alignment of short reads against multiple genomes. Genome Biol 2009;10(9):R98.

30. Lee WP, Stromberg MP, Ward A, Stewart C, Garrison EP, Marth GT. MOSAIK: a hash-based algorithm for accurate next-generation sequencing short-read mapping. PLoS ONE 2014;9(3):e90581.

31. http://www.novocraft.com/.

32. Lindner R, Friedel CC. A comprehensive evaluation of alignment algorithms in the context of RNA-seq. PLoS ONE 2012;7(12):e52403.

33. Langmead B, Trapnell C, Pop M, Salzberg SL. Ultrafast and memory-efficient alignment of short DNA sequences to the human genome. Genome Biol 2009;10(3):R25.

34. Langmead B, Salzberg SL. Fast gapped-read alignment with Bowtie 2. Nat Methods 2012;9(4):357–359.

35. Trapnell C, Pachter L, Salzberg SL. TopHat: discovering splice junctions with RNA-Seq. Bioinformatics 2009;25(9):1105–1111.

36. Au KF, Jiang H, Lin L, Xing Y, Wong WH. Detection of splice junctions from paired-end RNA-seq data by SpliceMap. Nucleic Acids Res 2010;38(14):4570–4578.

37. Xie Y, Wu G, Tang J, Luo R, Patterson J, Liu S, Huang W, He G, Gu S, Li S et al. SOAPdenovo-Trans: de novo transcriptome assembly with short RNA-Seq reads. Bioinformatics 2014;30(12):1660–1666.

38. Grabherr MG, Haas BJ, Yassour M, Levin JZ, Thompson DA, Amit I, Adiconis X, Fan L, Raychowdhury R, Zeng Q et al. Full-length transcriptome assembly from RNA-Seq data without a reference genome. Nat Biotechnol 2011;29(7):644–652.

39. Grant GR, Farkas MH, Pizarro AD, Lahens NF, Schug J, Brunk BP, Stoeckert CJ, Hogenesch JB, Pierce EA. Comparative analysis of RNA-Seq alignment algorithms and the RNA-Seq unified mapper (RUM). Bioinformatics 2011;27(18):2518–2528.

40. Griebel T, Zacher B, Ribeca P, Raineri E, Lacroix V, Guigo R, Sammeth M. Modelling and simulating generic RNA-Seq experiments with the flux simulator. Nucleic Acids Res 2012;40(20):10073–10083.

41. http://samtools.sourceforge.net/. Available at Accessed 2016 Mar 17.

42. http://picard.sourceforge.net/. Available at Accessed 2016 Mar 17.

43. Trapnell C, Roberts A, Goff L, Pertea G, Kim D, Kelley DR, Pimentel H, Salzberg SL, Rinn JL, Pachter L. Differential gene and transcript expression analysis of RNA-seq experiments with TopHat and Cufflinks. Nat Protoc 2012;7(3):562–578.

44. Li H, Durbin R. Fast and accurate short read alignment with Burrows-Wheeler transform. Bioinformatics 2009;25(14):1754–1760.

45. Compeau PE, Pevzner PA, Tesler G. How to apply de Bruijn graphs to genome assembly. Nat Biotechnol 2011;29(11):987–991.

46. Li B, Dewey CN. RSEM: accurate transcript quantification from RNA-Seq data with or without a reference genome. BMC Bioinformatics 2011;12:323.

47. Robinson MD, McCarthy DJ, Smyth GK. edgeR: a Bioconductor package for differential expression analysis of digital gene expression data. Bioinformatics 2010;26(1):139–140.

48. Anders S, Huber W. Differential expression analysis for sequence count data. Genome Biol 2010;11(10):R106.

49. Wu TD, Watanabe CK. GMAP: a genomic mapping and alignment program for mRNA and EST sequences. Bioinformatics 2005;21(9):1859–1875.

50. https://lists.sourceforge.net/lists/listinfo/trinityrnaseq-users. Available at Accessed 2016 Mar 17.

51. Haas BJ, Papanicolaou A, Yassour M, Grabherr M, Blood PD, Bowden J, Couger MB, Eccles D, Li B, Lieber M et al. De novo transcript sequence reconstruction from RNA-seq using the Trinity platform for reference generation and analysis. Nat Protoc 2013;8(8):1494–1512.

52. Schulz MH, Zerbino DR, Vingron M, Birney E. Oases: robust de novo RNA-seq assembly across the dynamic range of expression levels. Bioinformatics 2012;28(8):1086–1092.

13

COMPUTATIONAL APPROACHES FOR STUDYING ALTERNATIVE SPLICING IN NONMODEL ORGANISMS FROM RNA-SEQ DATA

SING-HOI SZE

Department of Computer Science and Engineering and Department of Biochemistry and Biophysics, Texas A&M University, College Station, TX, USA

13.1 INTRODUCTION

As the amount of data on RNA splicing increases rapidly due to advance in high-throughput sequencing, alternative splicing has become one of the most important mechanisms to study in nonmodel organisms. It is crucial to a variety of biological functions, including sex determination, neural biology, courtship, immunity, chromatin remodeling, growth and development, tissue assembly, and tissue organization (1–5).

13.1.1 Alternative Splicing

In higher organisms, there is usually more than one way to splice an RNA to produce an mRNA (see Figure 13.1). In human, it has been estimated that more than 70% of genes have alternative splicing (6), with over hundreds of variants in some highly alternatively spliced genes (7). Due to alternative splicing, the number of protein variants is much larger than the number of genes. Thus, alternative splicing is a very important regulatory mechanism. Alternative splicing information can be obtained

Computational Methods for Next Generation Sequencing Data Analysis, First Edition.
Edited by Ion I. Măndoiu and Alexander Zelikovsky.
© 2016 John Wiley & Sons, Inc. Published 2016 by John Wiley & Sons, Inc.
Companion website: www.wiley.com/go/Mandoiu/NextGenerationSequencing

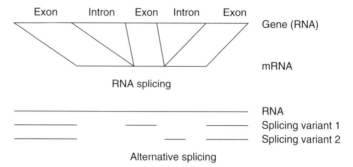

Figure 13.1 Illustration of RNA splicing and alternative splicing. In eukaryotes, genes are made up of exons and introns. Introns are cut out of an RNA, and exons are concatenated together to form an mRNA before translation into a protein. There is usually more than one way to define these exons and introns, leading to alternative splicing.

from experimental annotations, from expressed sequence tag (EST) assemblies (8), or from high-throughput sequencing techniques.

13.1.2 Nonmodel Organisms

In high-throughput sequencing, one popular strategy to obtain alternative splicing information is to apply mapping algorithms such as Bowtie (9), BWA (10), Cufflinks (11), and Scripture (12), in which each read is aligned to the reference genome or transcriptome and alternatively spliced variants are identified by studying these alignments. Mapping algorithms that are designed specifically for splicing junction discovery are also available, such as MapSplice (13) and SpliceMap (14).

In nonmodel organisms, such reference databases are not available, and the alternative is to employ *de novo* sequence assembly algorithms to assemble the reads. While earlier algorithms are designed for genome assembly (15–21), many algorithms are available that are designed specifically for transcriptome assembly. Among the most popular ones are Oases (22), Trans-ABySS (23), and Trinity (24). To obtain further information about the assembled sequences, comparisons are then made against the closest model organism (see Table 13.1).

13.1.3 RNA-Seq Data

The advance in high-throughput sequencing allows the creation of RNA-Seq libraries that contain a large number of reads from all expressed mRNAs. While each library can contain millions of reads with hundreds of nucleotides in each read, the sequencing error rate can be as much as a few percent. The quality of the regions toward the ends of the reads is especially bad, and sequence trimming is often performed before analysis.

Since the reads can contain artifacts and fake fragments due to the way they are constructed, careful rules have to be devised to address them. The reads often contain single-nucleotide polymorphisms (SNPs), in which multiple alleles appear at a

TABLE 13.1   Examples of Publicly Available RNA-Seq Libraries in Nonmodel Organisms, with Organism Denoting the Nonmodel Organism, Closest Model Denoting the Closest Model Organism, Library Denoting the Number of Libraries, and Reference Denoting the Publication that Describes the Libraries

Organism	Closest Model	Library	Reference
Tomicus yunnanensis (beetle)	*Drosophila melanogaster*	1	25
Lucilia sericata (blow fly)	*Drosophila melanogaster*	9	26
Cicer arietinum (chickpea)	*Arabidopsis thaliana*	3	27
Ctenomys sociabilis (rodent)	*Homo sapiens*	10	28
Heterocephalus glaber (naked mole rat)	*Homo sapiens*	13	29

nucleotide position. These small regions of differences are responsible for diversity within a species and are very frequent. Typically, one of the alleles is dominant and the other alleles are much weaker (its nucleotide occurs infrequently). An alternative formulation of the sequence assembly problem calls for the identification of SNPs as well. However, distinguishing SNPs from sequencing errors and other artifacts is a difficult problem.

Recently, the generation of paired-end reads has become a standard, in which each paired-end read associates two reads that are obtained from the same sequence with a known approximate distance distribution between them. The link between each pair of reads puts constraints on how the reads can be assembled, which are employed by various algorithms to improve performance.

13.2   REPRESENTATION OF ALTERNATIVE SPLICING

Alternative splicing information can be represented either as a set of predicted transcripts for each gene or as splicing graphs that retain relationships among different branches. In high-throughput sequencing, a popular strategy of transcriptome assembly in nonmodel organisms is to first construct a de Bruijn graph that contains all branching possibilities from the reads. An additional step is performed to obtain predicted transcripts by identifying probable paths in the de Bruijn graph, which can also be represented as splicing graphs. A similarity search algorithm such as BLAST (30) is then applied against the closest model organism to obtain possible function of these transcripts (see Figure 13.2).

13.2.1   de Bruijn Graph

Given a parameter k, a de Bruijn graph is constructed by taking each k-mer that appears in the reads as a vertex and connecting two vertices by a directed edge if the $(k-1)$-suffix of the first vertex is the same as the $(k-1)$-prefix of the second vertex (31, 32) (see Figure 13.3 for an example). The de Bruijn graph automatically provides an assembly while including information about alternatively spliced variants.

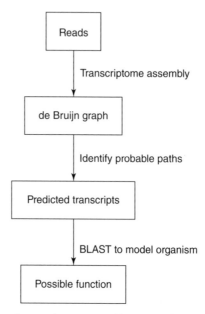

Figure 13.2   Illustration of transcriptome assembly strategy from RNA-Seq data in nonmodel organisms.

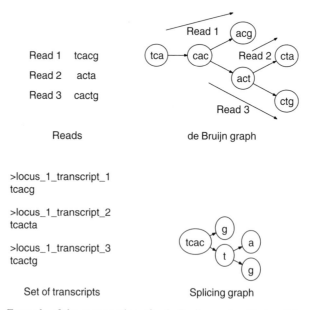

Figure 13.3   Example of the construction of a de Bruijn graph with $k = 3$ from a given set of reads. Each read can be obtained from one of the paths in the de Bruijn graph by sliding a window of size k along the read. The corresponding set of transcripts in FASTA format and splicing graph are also shown.

During the construction of the de Bruijn graph, the most difficult step is to perform k-mer counting to obtain its vertices. Since there are a large amount of distinct k-mers in the reads, this step has high memory requirements. Among the various techniques that are employed to address this problem includes the use of hashing techniques (19, 33), suffix arrays (34), entropy-based compression (35), and disk-based partitioning of the k-mer space (36, 37). After the initial construction of the de Bruijn graph, a compact representation is obtained by collapsing each nonbranching path into a single node. This transformation does not affect the structures of paths in the de Bruijn graph.

The de Bruijn graph can have a very complicated structure due to ambiguities that are caused by repeats and sequencing errors. A repeating sequence can be thousands of nucleotides long, which is longer than the length of a read. There are also repetitive regions throughout a genome, with each instance not exactly the same. Since these regions cannot be easily handled by the de Bruijn graph approach, they can be masked before the assembly. An error correction procedure can also be applied to the reads before construction of the de Bruijn graph to correct some of the sequencing errors. One approach to correct sequencing error in a read is to align it against other reads that overlap with it and use the most frequent (consensus) nucleotide for correction (8).

Various techniques have to be applied to simplify the graph to separate real alternative paths from noise. Low-coverage regions are discarded that are represented very few times, which are likely to correspond to sequencing errors. Short tips and bulges are removed from the graph, which can represent repeating sequences or SNPs. While different paths in the de Bruijn graph may indicate alternative splicing after these steps, a connected component can contain complicated cycles or sequences from more than one gene, and a sequence from one gene can be fragmented into more than one connected component.

13.2.2 A Set of Transcripts

Genome sequence assembly algorithms typically return a large number of unordered contigs, where each contig represents a continuous assembled sequence within the de Bruijn graph, while junction information is ignored. This strategy is employed by algorithms such as Velvet (19) and ABySS (20). For transcriptome assembly, an additional step is performed to obtain predicted transcripts from the de Bruijn graph, leading to the algorithms Oases and Trans-ABySS, respectively. In order to address the highly variable expression levels, these algorithms remove the assumption that sequence coverages are approximately uniform and obtain a set of probable paths in the de Bruijn graph that correspond to transcripts based on utilizing links between paired-end reads and enumerating highly represented paths in decreasing order of coverage.

At the end of the assembly, most transcriptome assembly algorithms return a set of locus, with multiple transcripts for each locus and each transcript within the same locus corresponding to a gene isoform (22, 24) (see Figure 13.3 for an example). While the same locus should contain sequences from the same gene, different predicted locus may contain fragments of the same gene due to the inability to link them together during assembly. Without knowing the location of transcripts on a genome, one drawback of this representation is the lack of explicit locations of the alternative splicing junctions, which can be approximated by comparing sequence overlaps among the transcripts to identify the boundaries of shared sequence blocks.

In difference from reads that are created for genome assembly, RNA-Seq data often contain multiple libraries that correspond to different experimental conditions or developmental stages. It is important to obtain expression estimates for each transcript within each library in order to study differential expression under these conditions. One popular measure is the number of reads per kilobase of transcript per million reads (RPKM) (38, 39), or the number of fragments per kilobase of transcript per million fragments (FPKM) (11) that are used specifically for paired-end reads to account for the possibility that only one end of a paired-end read may be utilized. In these measures, sequence frequencies in the reads are used to obtain expression estimates.

Since transcripts that correspond to different isoforms often have a large amount of sequence overlaps, the RPKM or FPKM measures that are based on transcript lengths may not be very accurate. To obtain better expression estimates for each individual isoform, improved algorithms have been developed based on the expectation-maximization (EM) algorithm. In the presence of a reference genome or a closely related genome, algorithms such as IsoEM (40) can be applied. Alternatively, in the presence of a reference transcriptome or the transcriptome assembly itself, algorithms such as RSEM (41) can be applied.

13.2.3   Splicing Graph

Splicing graph is a data structure to represent alternative splicing, in which each path in the graph represents one alternatively spliced variant and there are no overlapping fragments between adjacent nodes, which is different from the de Bruijn graph (8) (see Figure 13.3 for an example). By locating paths of predicted transcripts within the de Bruijn graph, most transcriptome assembly algorithms also return the simplified de Bruijn graph in addition to the set of predicted transcripts (22, 24), which can be used as an approximate version of a splicing graph.

It is also possible to obtain a splicing graph directly from the de Bruijn graph by performing postprocessing. One such algorithm is described in Reference 26, which resolves tangles in the de Bruijn graph by ignoring edge information within strongly connected components that contain cycles. Junction adjustment is then performed by removing overlapping fragments between adjacent nodes in the de Bruijn graph and moving shared nucleotides across nodes to make the junctions precise.

When only a set of known transcripts is given (such as from experimental annotations), splicing graphs can also be constructed to represent the branching relationships either by representing each exon as a node or by representing each nonredundant subsequence that may only be part of an exon as a node. This later strategy allows a more detailed study of exon extension/shrinking and intron retention/loss. One drawback of the splicing graph representation is that not all paths in a splicing graph correspond to transcripts. There may still be a need to label the paths in the splicing graph that correspond to transcripts in order not to lose any information.

13.3   COMPARISON TO MODEL ORGANISMS

One possible strategy to study differences in alternative splicing and to obtain possible function of predicted transcripts in nonmodel organisms is to compare to the closest model organism by performing translated BLAST search. Since this step only requires the availability of transcriptome sequences in the organism in which comparisons are made but not the full genome, it is also possible to compare to other closely related nonmodel organisms in which an assembled transcriptome is available, which can be obtained either from traditional sequencing, from EST sequencing, or from high-throughput sequencing techniques.

13.3.1   A Set of Transcripts

Since most transcriptome assembly algorithms return a set of predicted transcripts as the main result, translated BLAST search can be performed from each of these transcripts against all known transcripts in a related organism. Alternative splicing among predicted transcripts within the same locus can be studied by identifying the isoform that corresponds to the top BLAST hit from each transcript.

13.3.2   Splicing Graph

Alternatively, since most algorithms also return the simplified de Bruijn graph or a splicing graph as part of the assembly, translated BLAST search can be performed from nodes in these graphs that may correspond to a sequence of exons or part of an exon. Alternatively spliced variants can then be reconstructed along paths in these graphs that contain BLAST hits to a particular isoform.

13.4   ACCURACY OF ALGORITHMS

To compare the performance of the most popular *de novo* transcriptome assembly algorithms Oases, Trans-ABySS, and Trinity, we extract RNA-Seq libraries from a model organism with the known transcripts and compare the assemblies to the annotations in the model organism itself. We obtained reads from three *Drosophila melanogaster* RNA-Seq libraries in Reference 42 at the sequence read archive (43) that correspond to three developmental stages, including 2–16 hours embryos (SRR058885), third instar larvae (SRR059066), and mixed pupae (SRR042298). We trimmed each read by removing all positions including and after the first position with a quality score of less than 15, resulting in a total library size of 1.8 G.

13.4.1   Assembly Results

We fixed the k-mer length of de Bruijn graph to 25 while varying coverage cutoff c. Note that the results are not completely comparable since the algorithms return slightly different structures. Table 13.2 shows that as the coverage cutoff c increases, the assembly conditions become more stringent and all the algorithms returned less

TABLE 13.2   Comparisons of Transcriptome Assemblies of Oases, Trans-ABySS, and Trinity on Three *D. melanogaster* Libraries over Different Values of Coverage Cutoff *c*

							Oases			
c	locus	max length	N50	branched locus	max trans	avg trans	total hits	unique hits	>1-hit locus	max hits
3	32,277	14,818	748	4438	10	4	30,790	10,297	382	8
5	21,246	11,165	880	2674	10	3	20,589	8,979	227	8
10	11,509	6,517	842	1248	10	4	11,276	6,216	127	7

				Trans-ABySS				
c	trans	max length	N50	branched trans	max nodes	avg nodes	total hits	unique hits
3	45,360	7355	543	15,824	85	5	42,617	10,020
5	28,505	7753	617	8,366	66	4	27,235	8,590
10	14,304	8347	589	3,678	34	4	13,869	5,757

							Trinity			
c	locus	max length	N50	branched locus	max trans	avg trans	total hits	unique hits	>1-hit locus	max hits
3	50,656	6165	264	1114	17	2	48,079	9987	139	3
5	35,797	5577	254	603	10	2	34,487	8412	88	3
10	20,281	4660	245	279	19	2	19,759	5946	49	3

For Oases and Trinity, the predicted unit is locus that consists of a set of predicted transcripts. For Trans-ABySS, the predicted unit is transcript (trans) that is a linear concatenation of constituent nodes. The *k*-mer length is fixed to 25, with max length denoting the length of the longest predicted transcript, N50 denoting the N50 value of the longest transcript length within a predicted unit while retaining only predicted transcripts of length at least 100, branched locus/trans denoting the number of predicted units that have branching, max trans/nodes denoting the maximum number of constituent transcripts/nodes within a predicted unit, avg trans/nodes denoting the average number of constituent transcripts/nodes within branched locus/transcripts, total hits denoting the total number of nucleotide BLAST hits from each predicted transcript to annotated *D. melanogaster* transcripts considering only the top hit with *E*-value below 10^{-7} (isoforms are considered to be the same gene and hits from transcripts within the same predicted unit to the same gene are counted only once), unique hits denoting the number of unique hits to different genes, >1-hit locus denoting the number of predicted locus that has hits to more than one gene, and max hits denoting the maximum number of different genes that have hits to a predicted locus.

predicted transcripts and recovered less *D. melanogaster* genes. Oases had the longest assemblies with the highest maximum and median (N50) transcript lengths and recovered the most *D. melanogaster* genes, followed by Trans-ABySS and Trinity. The algorithms returned various percentages of branched structures that could represent alternative splicing. The ratio of the total number of BLAST hits to the number of unique BLAST hits was not very high, indicating some degree of sequence fragmentation. For Oases and Trinity, each locus mostly represents alternatively spliced variants of only one gene.

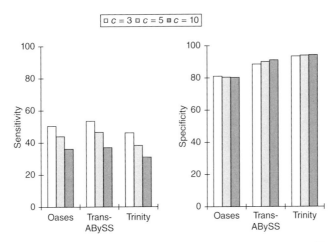

Figure 13.4 Sensitivity and specificity comparisons of Oases, Trans-ABySS, and Trinity with respect to mRNA BLAST results on three *D. melanogaster* libraries over different values of coverage cutoff c with the k-mer length fixed to 25. Sensitivity is defined to be the percentage of nucleotide positions in the *D. melanogaster* transcriptome that are recovered through the top BLAST alignments from each predicted transcript in the assembly considering only *D. melanogaster* gene transcripts that are found in BLAST hits. Specificity is defined to be the percentage of predicted transcript positions in the assembly that are included in the top BLAST alignments considering only positions that have BLAST hits.

13.4.2 mRNA BLAST Results

Figure 13.4 shows that Oases was the least accurate with respect to *D. melanogaster* transcript recovery despite its higher transcript lengths (see Table 13.2). Trans-ABySS had the highest sensitivity, and Trinity had the highest specificity at the expense of lower sensitivity. While the generally low sensitivity is related to the small total library size, all the algorithms had high specificity, which is important to avoid an excessive amount of false positives.

13.4.3 Alternative Splicing Junctions

Figure 13.5 shows that the recovery of alternative splicing junctions was more difficult both in terms of sensitivity and specificity when compared to the sequence recovery results (see Figure 13.4). Oases had higher sensitivity and specificity in recovering alternative splicing junctions than Trans-ABySS. Similar to the sequence recovery results, Trinity had the highest specificity at the expense of lower sensitivity.

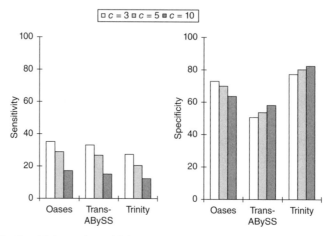

Figure 13.5   Sensitivity and specificity comparisons of Oases, Trans-ABySS, and Trinity with respect to alternative splicing junctions on three *D. melanogaster* libraries over different values of coverage cutoff *c* with the *k*-mer length fixed to 25. Sensitivity is defined to be the percentage of junctions in the *D. melanogaster* gene transcripts that appear somewhere in the assembly. Specificity is defined to be the percentage of junctions in the assembly that appear somewhere in the *D. melanogaster* gene transcripts. Junctions in the *D. melanogaster* gene transcripts are defined by concatenating the two sequences of length *k* that are immediately to the left and to the right of all alternatively spliced locations to obtain a sequence of length 2*k*. Junctions in the assembly are defined by concatenating the two nonoverlapping *k*-mers at the beginning and ending nodes of an edge in the de Bruijn graph to obtain a sequence of length 2*k*. Up to three mismatches are allowed when looking for occurrences of these sequences.

13.5   DISCUSSION

With an increased focus on studying a large number of nonmodel organisms rather than limiting to a small set of model organisms, it has become very important to develop reliable algorithms for transcriptome assembly from RNA-Seq data. Due to the existence of alternative splicing and the need to assemble a large number of libraries together under multiple experimental conditions for differential analysis, transcriptome assembly is often more difficult than genome assembly. Our results show that all the tested algorithms have more difficulties in recovering alternative splicing junctions accurately than in recovering sequences. These algorithms also have trade-offs in terms of running time and memory requirement (44), with Oases generally running faster than Trans-ABySS and Trinity, and Trans-ABySS generally using less memory than Trinity and Oases.

To further improve performance, new techniques have to be developed to address complications regarding repeats and gene families. Simpler repeats without mismatches or indels appear as small knots in the de Bruijn graph, while inexact repeats appear as small cycles with more complicated structures. In order to identify these repeats, additional algorithms have to be developed to extract these special structures within the graph so as to distinguish them from alternative splicing.

Gene families are likely to assemble in a manner that makes them look like highly polymorphic genes. To separate them from alternative splicing, a possible strategy is to first develop algorithms to evaluate patterns of SNP densities and look for overrepresentation of SNPs in known gene families. The resulting information can then be used to identify gene families and to separate the genes within a family to obtain its members.

Recently, algorithms that integrate *de novo* assembly and mapping have been developed. One such algorithm is BRANCH (45), which utilizes exonic information obtained from alignments against the same organism or a related organism to guide the assembly. While these algorithms have improved performance due to the use of additional information, there is a need to continue to improve them in order to facilitate simultaneous comparisons of alternative splicing in a large number of nonmodel organisms.

REFERENCES

1. Hallegger M, Llorian M, Smith CWJ. Alternative splicing: global insights. FEBS J 2010;277:856–866.

2. Keren H, Lev-Maor G, Ast G. Alternative splicing and evolution: diversification, exon definition and function. Nat Rev Genet 2010;11:345–355.

3. Nilsen TW, Graveley BR. Expansion of the eukaryotic proteome by alternative splicing. Nature 2010;463:457–463.

4. Salz HK, Erickson JW. Sex determination in *Drosophila*: the view from the top. Fly (Austin) 2010;4:60–70.

5. Wood MJA, Gait MJ, Yin H. RNA-targeted splice-correction therapy for neuromuscular disease. Brain 2010;133:957–972.

6. Johnson JM, Castle J, Garrett-Engele P, Kan Z, Loerch PM, Armour CD, Santos R, Schadt EE, Stoughton R, Shoemaker DD. Genome-wide survey of human alternative pre-mRNA splicing with exon junction microarrays. Science 2003;302:2141–2144.

7. Graveley BR. Alternative splicing: increasing diversity in the proteomic world. Trends Genet 2001;17:100–107.

8. Heber S, Alekseyev M, Sze S-H, Tang H, Pevzner PA. Splicing graphs and EST assembly problem. Bioinformatics 2002;18:S181–S188.

9. Langmead B, Trapnell C, Pop M, Salzberg SL. Ultrafast and memory-efficient alignment of short DNA sequences to the human genome. Genome Biol 2009;10:R25.

10. Li H, Durbin R. Fast and accurate short read alignment with Burrows-Wheeler transform. Bioinformatics 2009;25:1754–1760.

11. Trapnell C, Williams BA, Pertea G, Mortazavi A, Kwan G, van Baren MJ, Salzberg SL, Wold BJ, Pachter L. Transcript assembly and quantification by RNA-Seq reveals unannotated transcripts and isoform switching during cell differentiation. Nat Biotechnol 2010;28:511–515.

12. Guttman M, Garber M, Levin JZ, Donaghey J, Robinson J, Adiconis X, Fan L, Koziol MJ, Gnirke A, Nusbaum C, Rinn JL, Lander ES, Regev A. *Ab initio* reconstruction of cell type-specific transcriptomes in mouse reveals the conserved multi-exonic structure of lincRNAs. Nat Biotechnol 2010;28:503–510.

13. Wang K, Singh D, Zeng Z, Coleman SJ, Huang Y, Savich GL, He X, Mieczkowski P, Grimm SA, Perou CM, MacLeod JN, Chiang DY, Prins JF, Liu J. MapSplice: accurate mapping of RNA-seq reads for splice junction discovery. Nucleic Acids Res 2010;38:e178.

14. Au KF, Jiang H, Lin L, Xing Y, Wong WH. Detection of splice junctions from paired-end RNA-seq data by SpliceMap. Nucleic Acids Res 2010;38:4570–4578.

15. Dohm JC, Lottaz C, Borodina T, Himmelbauer H. SHARCGS, a fast and highly accurate short-read assembly algorithm for *de novo* genomic sequencing. Genome Res 2007;17:1697–1706.

16. Butler J, MacCallum I, Kleber M, Shlyakhter IA, Belmonte MK, Lander ES, Nusbaum C, Jaffe DB. ALLPATHS: *de novo* assembly of whole-genome shotgun microreads. Genome Res 2008;18:810–820.

17. Chaisson MJ, Pevzner PA. Short read fragment assembly of bacterial genomes. Genome Res 2008;18:324–330.

18. Hernandez D, François P, Farinelli L, Østerås M, Schrenzel J. *De novo* bacterial genome sequencing: millions of very short reads assembled on a desktop computer. Genome Res 2008;18:802–809.

19. Zerbino DR, Birney E. Velvet: algorithms for *de novo* short read assembly using de Bruijn graphs. Genome Res 2008;18:821–829.

20. Birol I, Jackman SD, Nielsen CB, Qian JQ, Varhol R, Stazyk G, Morin RD, Zhao Y, Hirst M, Schein JE, Horsman DE, Connors JM, Gascoyne RD, Marra MA, Jones SJM. *De novo* transcriptome assembly with ABySS. Bioinformatics 2009;25:2872–2877.

21. Li R, Zhu H, Ruan J, Qian W, Fang X, Shi Z, Li Y, Li S, Shan G, Kristiansen K, Li S, Yang H, Wang J, Wang J. *De novo* assembly of human genomes with massively parallel short read sequencing. Genome Res 2010;20:265–272.

22. Schulz MH, Zerbino DR, Vingron M, Birney E. Oases: robust *de novo* RNA-seq assembly across the dynamic range of expression levels. Bioinformatics 2012;28:1086–1092.

23. Robertson G, Schein J, Chiu R, Corbett R, Field M, Jackman SD, Mungall K, Lee S, Okada HM, Qian JQ, Griffith M, Raymond A, Thiessen N, Cezard T, Butterfield YS, Newsome R, Chan SK, She R, Varhol R, Kamoh B, Prabhu A-L, Tam A, Zhao Y, Moore RA, Hirst M, Marra MA, Jones SJM, Hoodless PA, Birol I. *De novo* assembly and analysis of RNA-seq data. Nat Methods 2010;7:909–912.

24. Grabherr MG, Haas BJ, Yassour M, Levin JZ, Thompson DA, Amit I, Adiconis X, Fan L, Raychowdhury R, Zeng Q, Chen Z, Mauceli E, Hacohen N, Gnirke A, Rhind N, di Palma F, Birren BW, Nusbaum C, Lindblad-Toh K, Friedman N, Regev A. Full-length transcriptome assembly from RNA-Seq data without a reference genome. Nat Biotechnol 2011;29:644–652.

25. Zhu J-Y, Zhao N, Yang B. Global transcriptome profiling of the pine shoot beetle, *Tomicus yunnanensis* (Coleoptera: Scolytinae). PLoS ONE 2012;7:e32291.

26. Sze S-H, Dunham JP, Carey B, Chang PL, Li F, Edman RM, Fjeldsted C, Scott MJ, Nuzhdin SV, Tarone AM. A *de novo* transcriptome assembly of *Lucilia sericata* (Diptera: Calliphoridae) with predicted alternative splices, single nucleotide polymorphisms, and transcript expression estimates. Insect Mol Biol 2012;21:205–221.

27. Garg R, Patel RK, Tyagi AK, Jain M. *De novo* assembly of chickpea transcriptome using short reads for gene discovery and marker identification. DNA Res 2011;18:53–63.

28. MacManes MD, Lacey EA. The social brain: transcriptome assembly and characterization of the hippocampus from a social subterranean rodent, the colonial tuco-tuco (*Ctenomys sociabilis*). PLoS ONE 2012;7:e45524.

29. Kim EB, Fang X, Fushan AA, Huang Z, Lobanov AV, Han L, Marino SM, Sun X, Turanov AA, Yang P, Yim SH, Zhao X, Kasaikina MV, Stoletzki N, Peng C, Polak P, Xiong Z, Kiezun A, Zhu Y, Chen Y, Kryukov GV, Zhang Q, Peshkin L, Yang L, Bronson RT, Buffenstein R, Wang B, Han C, Li Q, Chen L, Zhao W, Sunyaev SR, Park TJ, Zhang G, Wang J, Gladyshev VN. Genome sequencing reveals insights into physiology and longevity of the naked mole rat. Nature 2011;479:223–227.

30. Altschul SF, Gish W, Miller W, Myers EW, Lipman DJ. Basic local alignment search tool. J Mol Biol 1990;215:403–410.

31. Pevzner PA. *l*-tuple DNA sequencing: computer analysis. J Biomol Struct Dyn 1989;7:63–73.

32. Idury RM, Waterman MS. A new algorithm for DNA sequence assembly. J Comput Biol 1995;2:291–306.

33. Marçais G, Kingsford C. A fast, lock-free approach for efficient parallel counting of occurrences of *k*-mers. Bioinformatics 2011;27:764–770.

34. Kurtz S, Narechania A, Stein JC, Ware D. A new method to compute K-mer frequencies and its application to annotate large repetitive plant genomes. BMC Genomics 2008;9:517.

35. Conway TC, Bromage AJ. Succinct data structures for assembling large genomes. Bioinformatics 2011;27:479–486.

36. Rizk G, Lavenier D, Chikhi R. DSK: *k*-mer counting with very low memory usage. Bioinformatics 2013;29:652–653.

37. Deorowicz S, Debudaj-Grabysz A, Grabowski S. Disk-based *k*-mer counting on a PC. BMC Bioinformatics 2013;14:160.

38. Mortazavi A, Williams BA, McCue K, Schaeffer L, Wold B. Mapping and quantifying mammalian transcriptomes by RNA-Seq. Nat Methods 2008;5:621–628.

39. Trapnell C, Pachter L, Salzberg SL. TopHat: discovering splice junctions with RNA-Seq. Bioinformatics 2009;25:1105–1111.

40. Nicolae M, Mangul S, Măndoiu II, Zelikovsky A. Estimation of alternative splicing isoform frequencies from RNA-Seq data. Algorithms Mol Biol 2011;6:9.

41. Li B, Dewey CN. RSEM: accurate transcript quantification from RNA-Seq data with or without a reference genome. BMC Bioinformatics 2011;12:323.

42. Daines B, Wang H, Wang L, Li Y, Han Y, Emmert D, Gelbart W, Wang X, Li W, Gibbs R, Chen R. The *Drosophila melanogaster* transcriptome by paired-end RNA sequencing. Genome Res 2011;21:315–324.

43. Sayers EW, Barrett T, Benson DA, Bolton E, Bryant SH, Canese K, Chetvernin V, Church DM, DiCuccio M, Federhen S, Feolo M, Geer LY, Helmberg W, Kapustin Y, Landsman D, Lipman DJ, Lu Z, Madden TL, Madej T, Maglott DR, Marchler-Bauer A, Miller V, Mizrachi I, Ostell J, Panchenko A, Pruitt KD, Schuler GD, Sequeira E, Sherry ST, Shumway M, Sirotkin K, Slotta D, Souvorov A, Starchenko G, Tatusova TA, Wagner L, Wang Y, Wilbur WJ, Yaschenko E, Ye J. Database resources of the National Center for Biotechnology Information. Nucleic Acids Res 2010;38:D5–D16.

44. Zhao Q-Y, Wang Y, Kong Y-M, Luo D, Li X, Hao P. Optimizing *de novo* transcriptome assembly from short-read RNA-Seq data: a comparative study. BMC Bioinformatics 2011;12 Suppl 14:S2.

45. Bao E, Jiang T, Girke T. BRANCH: boosting RNA-Seq assemblies with partial or related genomic sequences. Bioinformatics 2013;29:1250–1259.

14

TRANSCRIPTOME QUANTIFICATION AND DIFFERENTIAL EXPRESSION FROM NGS DATA

OLGA GLEBOVA, YVETTE TEMATE-TIAGUEU, AND ADRIAN CACIULA

Department of Computer Science, Georgia State University, Atlanta, GA, USA

SAHAR AL SEESI

Department of Computer Science and Engineering, University of Connecticut, Storrs, CT, USA

ALEXANDER ARTYOMENKO

Department of Computer Science, Georgia State University, Atlanta, GA, USA

SERGHEI MANGUL

Department of Computer Science, University of California, Los Angeles, CA, USA

JAMES LINDSAY AND ION I. MĂNDOIU

Department of Computer Science and Engineering, University of Connecticut, Storrs, CT, USA

ALEXANDER ZELIKOVSKY

Department of Computer Science, Georgia State University, Atlanta, GA, USA

14.1 INTRODUCTION

RNA sequencing (RNA-Seq) is a widely used cost-efficient technology with several medical and biological applications. This technology, however, presents scholars with a number of computational challenges. RNA-Seq protocol provides full

Computational Methods for Next Generation Sequencing Data Analysis, First Edition.
Edited by Ion I. Măndoiu and Alexander Zelikovsky.
© 2016 John Wiley & Sons, Inc. Published 2016 by John Wiley & Sons, Inc.
Companion website: www.wiley.com/go/Mandoiu/NextGenerationSequencing

transcriptome data at a single transcript level. In this chapter, we focus on the transcriptome quantification problem, which is to estimate the expression level of each transcript. Transcriptome quantification analysis is crucial to determine similar transcripts or unraveling gene functions and transcription regulation mechanisms. Here, we propose a novel simulated regression-based method for isoform frequency estimation from RNA-Seq reads. We present SimReg (1)—a novel regression-based algorithm for transcriptome quantification. Simulated data experiments demonstrate superior frequency estimation accuracy of SimReg comparatively to that of the existing tools, which tend to skew the estimated frequency toward supertranscripts.

Gene expression is the process by which the genetic code (the nucleotide sequence) of a gene becomes a useful product. The motivation behind analyzing gene expression is to identify genes whose patterns of expression differ according to phenotype, disease, experimental condition (e.g., disease and control), or even from different organisms. Important factors to consider while analyzing differentially expressed genes are normalization, accuracy of differential expression detection, and differential expression analysis when one condition has no detectable expression. A very popular domain of application of gene expression analysis is time-series gene expression data (2), which stems from the fact that biological processes are often dynamic.

14.1.1   Motivation and Problems Description

RNA-Seq has become the new standard for the analysis of differential gene expression (3–5) due to its wider dynamic range and smaller technical variance (6) compared to traditional microarray technologies. However, simply using the raw fold change of the expression levels of a gene across two samples as a measure of differential expression can still be unreliable, because it does not account for read mapping uncertainty or capture, fragmentation, and amplification variability in library preparation and sequencing. Therefore, the need for using statistical methods arises. Traditionally, statistical methods rely on the use of replicates to estimate biological and technical variability in the data. Popular methods for analyzing RNA-Seq data with replicates include edgeR (7), DESeq (8), CuffDiff (9), and the recent NPEBSeq (10).

Unfortunately, due to the still high cost of sequencing, many RNA-Seq studies have no or very few replicates (11). Methods for performing differential gene expression analysis of RNA-Seq data sets without replicates include variants of Fisher's exact test (6). Recently, Feng et al. introduced GFOLD (12), a nonparametric empirical Bayesian-based approach, and showed that it outperforms methods designed to work with replicates when used for single replicate data sets.

A simple approach is to select genes using a fold-change criterion. This may be the only possibility in cases where no, or very few, replicates are available. An analysis solely based on fold change, however, does not allow the assessment of significance of expression differences in the presence of biological and experimental variation, which may differ from gene to gene. This is the main reason for using statistical tests to assess differential expression (13).

14.1.2   RNA-Seq Protocol

RNA-Seq is an increasingly popular approach to transcriptome profiling that uses the capabilities of next-generation sequencing (NGS) technologies and provides better measurement of levels of transcripts and their isoforms. One issue plaguing RNA-Seq experiments is reproducibility. This is a central problem in bioinformatics in general. It is not easy to benchmark the entire RNA-seq process (9), and the fact that there are fundamentally different ways of analyzing the data (assembly, feature counting, etc.) make it more difficult. Nevertheless, RNA-Seq offers huge advantages over microarrays since there is no limit on the numbers of genes surveyed, no need to select what genes to target, and no requirements for probes or primers, and it is the tool of choice for metagenomics studies. Also, RNA-seq has the ability to quantify a large dynamic range of expression levels, and this lead to transcriptomics and metatranscriptomics.

Rapid advances in NGS have enabled shotgun sequencing of total DNA and RNA extracted from complex microbial communities, ushering the new fields of metagenomics and metatranscriptomics. Depending on surrounding conditions, for example, food availability, stress, or physical parameters, the gene expression of organisms can vary widely. The aim of transcriptomics is to capture the gene activity. Transcriptomics helps perform gene expression profiling to unravel gene functions. It can tell us which metabolic pathways are in use under the respective conditions and how the organisms interact with the environment. Hence, it can be applied for environmental monitoring and for the identification of key genes. Transcriptomics also play a role in clinical diagnosis and in screening for drug targets or for genes, enzymes, and metabolites relevant for biotechnology (14–16).

Although transcriptomics deals with the gene expression of single species, metatranscriptomics covers the gene activity profile of the whole microbial community. Metatranscriptomics studies change in the function and structure of complex microbial communities as it adapts to environments such as soil and seawater. Unfortunately, as in all "meta" approaches, only a small percentage of the vast number of ecologically important genes has been correctly annotated (17).

Here, we apply RNA-Seq protocol and transcriptome quantification to estimate gene expression and differential gene expression analysis.

RNA-Seq, or deep sequencing of RNAs, is a cost-efficient high-coverage powerful technology for transcriptome analysis. There are various tools and algorithms for RNA-Seq data analysis devoted to different computational challenges, among them are transcriptome quantification and reconstruction. We focus on the problem of transcriptome quantification, that is, on the estimation the expression level of each transcript.

Recent review of computational methods for transcriptome quantification from RNA-Seq data reports several problems with the current state of transcriptome quantification, among them is a significant variation in distributions of expression levels throughout transcriptome reconstruction and quantification tools (18). Transcriptome quantification from RNA-Seq data highly depends on read depth. Due to the sparse read support at some loci, many tools fail to report all/some of the exons or exon–intron junctions.

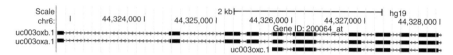

Figure 14.1   Screenshot from Genome browser. Source: Kent 2002 19. Reproduced with permission of Cold Spring Harbor Laboratory Press.

Improving isoform frequency estimation error rate is critical for detecting similar transcripts or unraveling gene functions and transcription regulation mechanisms, especially in those cases when one isoform is a subset of another. Figure 14.1 shows a gene with subtranscripts from human genome (hg19).

14.2   OVERVIEW OF THE STATE-OF-THE-ART METHODS

14.2.1   Quantification Methods

From optimization point of view, the variety of approaches to transcriptome quantification is very wide. The most popular approach is maximizing likelihood using different variants of expectation maximization (EM) (20–22), integer linear program (LP)-based methods (23), (24), min-cost flow (25), and regression (26).

RNA-Seq by Expectation Maximization (RSEM) is an expectation maximization (EM) algorithm that works on the isoform level. The initial version of RSEM only handled single-end reads; however, the latest version (21) has been extended to support paired-end reads, variable-length reads, and incorporates fragment length distribution and quality scores in its modeling. In addition to the maximum likelihood estimates of isoform expressions, RSEM also calculates 95% confidence intervals and posterior mean estimates. RSEM is the best algorithm presented so far, so we compare our tool SimReg to RSEM in Evaluation section.

The main limitation of statistically sound EM approach is that it does not include uniformity of transcript coverage, that is, it is not clear how to make sure that a solution with more uniform coverage of transcripts will be preferred to the one where coverage is volatile. LP and integer LP-based methods overcome this limitation but cannot handle many isoforms simultaneously.

More recently, the authors of (22) proposed a quasi-multinomial model with a single parameter to capture positional, sequence, and mapping biases. Tomescu et al. (27) proposed a method based on network flows for a multiassembly problem arising from transcript identification and quantification with RNA-Seq. This approach is good at keeping overall uniformity coverage but is not suitable for likelihood maximization.

Regression-based approaches are the most related to the proposed method. The most representative of these is IsoLasso approach (26). IsoLasso mathematically models a gene partitions into segments (a segment is a consecutive exon region while a subexon is a nonspliced region).

IsoLasso approach also assumes reads being uniformly sampled from transcripts. The Poisson distribution (28) then used to approximate the binomial distribution for the number of reads falling into each segment or subexon. The following quadratic program (26) is well known as a LASSO approach (29):

$$\text{minimize:} \quad \sum_{i=1}^{M} \left(\frac{r_i}{l_i} - \sum_{j=1}^{N} a_{ji} x_j \right)^2$$

$$\text{subject to:} \quad x_j \geq 0, \, 1 \leq j \leq N, \, \sum_{j=1}^{N} x_j \leq \lambda, \forall t = 1 \, \ldots \, |T| \qquad (14.1)$$

and two more "completeness" constraints (namely that each segment or junction with mapped reads is covered by at least one isoform; and the sum of expression levels of all isoforms that contain this segment or junction should be strictly positive (26)) were added to this program in IsoLasso. The main oversimplification is an assumption that each segment receives from containing transcripts the number of reads proportional to its length. For example, it is not clear how to handle very short subexons and take in account position of a subexon in a transcript. Fragment length distribution also can discriminate one subexon from another. Especially difficult to accurately estimate portions of pair-end reads emitted from each subexon since in fact such reads are frequently emitted by multiple subexons collectively. Furthermore, mapping of the reads into transcripts is frequently ambiguous, which is consciously ignored in Reference 26.

In this chapter, we propose to apply a more accurate simulation of read emission. Our novel algorithm falls into the category of regression-based methods: namely, SimReg is a Monte Carlo based regression method.

In general, one of the main goals of differential expression (DE) analysis is to identify the differentially expressed genes between two or more conditions. Such genes are selected based on a combination of expression level threshold and expression score cutoff, which is usually based on p-values generated by statistical modeling. The expression level of each RNA unit is measured by the number of sequenced fragments that map to the transcript, which is expected to correlate directly with its abundance level (30).

The outcome of DE analysis is influenced by the way primary analysis (mapping, mapping parameters, counting) is conducted (30). In addition, the overall library preparation protocol and quality is also an important factor of bias (31–33). As described in the following chapters, DE analysis methods differ in how to deal with these pre-analysis phases. Furthermore, RNA Seq experiments tend to be underpowered (too few replicates) and we need methods to perform DE under these circumstances.

14.2.2 Differential Expression Methods

In this section, we briefly describe competing differential expression methods, namely, GFOLD (12), Cuffdiff (9), edgeR (7), and DESeq (8).

GFOLD is a generalized fold change algorithm that produces biologically meaningful rankings of differentially expressed genes from RNA-Seq data. GFOLD assigns reliable statistics for expression changes based on the posterior distribution of log-fold change. The authors show that GFOLD outperforms other commonly used methods when used for single replicate data sets.

Cuffdiff uses a beta negative binomial distribution model to test the significance of change between samples. The model accounts for both uncertainty resulting from read mapping ambiguity and cross-replicate variability. Cuffdiff reports fold change in gene expression level along with statistical significance.

Cufflinks includes a separate program, Cuffdiff, which calculates expression in two or more samples and tests the statistical significance of each observed change in expression between them. The statistical model used to evaluate changes assumes that the number of reads produced by each transcript is proportional to its abundance but fluctuates because of technical variability during library preparation and sequencing and because of biological variability between replicates of the same experiment. Cufflinks is a transcript-level fragment count estimates. Cuffdiff uses an algorithm to model the expression of a gene G. Following this algorithm, it is able to get a distribution for the expression of a G. In the presence of replicates, to estimate the distribution of the log-fold change in expression for G under the null hypothesis, Cuffdiff computes the average of these distributions and takes their log ratio. The process is repeated thousand times across the two conditions. To calculate a p-value of observing the real log-fold change, they sort all the samples and count how many of them are more extreme than the log-fold change they actually saw in the real data. This number divided by the total number of draws is the estimate for the p-value (9).

edgeR is a statistical method for differential gene expression analysis, which is based on the negative binomial distribution. Although edgeR is primarily designed to work with replicates, it can also be run on data sets with no replicates. We used edgeR on counts of uniquely mapped reads, as suggested in Reference 34.

edgeR as well as DESeq are downstream count-based analysis tools, like both available as R/Bioconductor packages. The edgeR can be used to analyze replicate data sets (highly recommended) and nonreplicate.

A particular feature of edgeR functionality is empirical Bayes method that permits the estimation of gene-specific biological variation, even for experiments with minimal levels of biological replication. edgeR can be applied to differential expression at the gene, exon, transcript, or tag level. In fact, read counts can be summarized by any genomic feature. edgeR analyses at the exon level are easily extended to detect differential splicing or isoform-specific differential expression.

DESeq and edgeR are two methods and R packages for analyzing quantitative readouts (in the form of counts) from high-throughput experiments such as RNA-seq. After alignment, reads are assigned to a feature, where each feature represents a target transcript, in the case of RNA-Seq. An important summary statistic is the count of the number of reads in a feature (for RNA-Seq, this read count is a good approximation of transcript abundance).

Methods used to analyze array-based data assume a normally distributed, continuous response variable. However, response variables for digital methods such as RNA-seq and ChIP-seq are discrete counts. Thus, both DESeq and edgeR methods are based on the negative binomial distribution.

edgeR and DESeq use a model where they separate out the shot noise (aka counting noise, sampling noise, Poisson noise) that arises from the count nature of the data from the variance introduced by other types of noise (technical variance and biological variance). Then the extra variance is modeled either as uniform (as in the case of edgeR—e.g., all biological/technical variances for all transcripts are set to the same over dispersion from Poisson) or as quasi-correlated with read depth for DESeq. DESeq assumes that there is a correlation between read depth and extra-Poisson noise while edgeR assumes no correlation.

14.3 RECENT ALGORITHMS

14.3.1 SimReg: Simulated Regression Method for Transcriptome Quantification

The proposed method for estimating frequencies of transcripts is based on the novel approach for estimating expected read frequencies. First, we describe the essence of our approach and contrast it with IsoLasso.

As discussed earlier, it is very difficult (if at all possible) to accurately estimate portions of pair-end reads emitted from each subexon. Instead, rather than distinguishing reads by their gene position, we partition reads into *classes* each consisting of reads consistent with each element of a particular subset of transcripts. In other words, two reads are assigned to the same class if they are consistent with exactly the same transcripts. Our second innovation is to use Monte Carlo simulations instead of attempting to formally estimate contributions of each transcript to each read class. For any particular read class R, the expected frequency is estimated based on the frequencies of contributing transcripts as well as portions of reads that fall into the class R. Finally, using the standard regression method, we estimate transcript frequencies by minimizing deviation between expected and observed read class frequencies.

The following is the general description of the proposed simulated regression algorithm (SimReg) consisting of four steps and is described in detail.

14.3.1.1 Splitting the Transcripts and Reads into Independent Connected Components
We assume that alignment of a read to transcript is valid if the fragment length deviates from the mean by less than 4 standard deviations. Our simulations show that the Monte Carlo estimates become accurate enough only when simulated coverage is sufficiently high, that is, approaching 1000×. Such high coverage is time consuming since each simulated read needs to be aligned with each possible transcript. In order to reduce run-time, we split transcripts into small related subsets roughly corresponding to sets of overlapping genes. First, we build the matching graph $M = (\mathcal{T} \cup \mathcal{R}, E)$, where $\mathcal{T}$ and $\mathcal{R}$ are the sets of all transcripts and reads, respectively, and each edge $e = (r, T) \in E$ corresponds to a valid alignment of a read r to a transcript $T \in \mathcal{T}$. Transcript frequencies within each connected component of M do not depend on transcript frequencies within other connected components and can be estimated separately. A significant portion of

---

Algorithm SimReglgorithm

Split transcripts and reads into independent connected components:
   Estimate mean μ and standard deviation σ of read fragment distribution
   Find valid alignment of all observed reads to all transcripts
   Construct matching graph $M = (\mathcal{T} \cup \mathcal{R}, E)$ with edges corresponding to valid alignments
   Find connected components of M
   Find observed read classes R's in $\mathcal{R}$
Estimate transcript frequencies inside each connected component:
for each component C of M **do**
   for each transcript T in C **do**
      Simulate reads with 1000× coverage from T
      Map simulated reads to all other transcripts in C
      Find simulated read classes from reads mapped to the samesubset of transcripts in C
      Find $D_{R,T} = \{d_{R,T}\}$, distributionof reads simulated from T between read classes in $\mathcal{R}$
   end for
   Combine observed read classes and simulated read classes
   Find crude transcript frequencies F'_T in C minimizing deviation between observed read class frequencies $O_R = \{o_R\}$ and expected read class frequencies
   $F'_T \leftarrow arg\min(D_{R,T} \times F'_T - O_R)^2$
end for
Update initial estimates of transcript frequencies:
for each component C of M **do**
   Initialize aimed read class frequencies $A = \{a_R\}$with observed frequencies:
$a_R \leftarrow o_R$
   repeat
      For $i = 0, ...$
      Simulate reads with 100x coverage based on crude transcript frequency F'_T
      $s_R \leftarrow$ simulated frequency of read class R
      Compute deviations between observed and simulated read class frequencies
      $\Delta \leftarrow S - O$
      Update aimed read class frequenciesa_R –
      $A \leftarrow A_R - \Delta/2$
      Compute crude transcript frequencies F'_T based on corrected read class frequencies $\{c_R\}$, that is,
      $F'_T \leftarrow arg\min(D_{R,T} \times F'_T - A_R)^2$
   until $\Delta^2 < \epsilon$
   Obtain transcript frequencies from crude transcript frequencies
end for
Combine transcript frequency estimates from all connected components

---

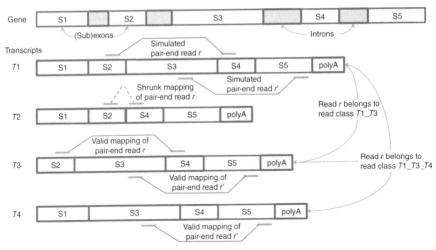

Figure 14.2   Paired reads r and r' are simulated from the transcript $T1$. Each read is mapped to all other transcripts ($T2, T3, T4$). Mapping of the read r into the transcript $T2$ is not valid since the fragment length is 4 standard deviations away from the mean. Then each read is assigned to the corresponding read class—the read r is placed in the read class $T1\_T3$ and the read r' is placed in the read class $T1\_T3\_T4$.

connected components contains just a single transcript for which the next step is trivial. Finally, the observed reads are partitioned into read classes each consisting of reads mapped to the same transcripts (see Figure 14.2).

14.3.1.2   Estimating Transcript Frequencies within Each Connected Component
As discussed earlier, in each connected component C, we simulate reads with 1000× coverage for each transcript (see Figure 14.2). Thus, for a transcript T with the length $|T|$, we generate $N_T = 1000 l_T$ reads, where $l_T = |T| - \mu + 1$ is the adjusted length of T. Similar to observed reads, we allow only alignments with fragment length less than 4σ away from μ. The reads that belong to exactly the same transcripts are collapsed into a single read class. Let $\mathcal{R} = \{R\}$ be all read classes found in the connected component C, and let R_T be the number of reads simulated from the transcript T that fall in the read class R. The first inner loop outputs the set of coefficients $D_{R,T} = \{d_{R,T}\}$, where $d_{R,T}$ is the portion of reads generated from T belonging to R

$$D_{R,T} = \left\{ \frac{|R_T|}{N_T} \right\}$$

Let $F'_T = \{f'_T\}$ be the *crude* transcript frequency, that is, the portions of reads emitted by transcripts in the connected component C. Then the expected read class frequency E_R can be estimated as

$$E_R = D_{R,T} \times F'_T \tag{14.2}$$

Regression-based estimation of f_t''s minimizes squared deviation

$$(D_{R,T} \times F_T' - O_R)^2 = \sum_{R \in R} (e_R - o_R)^2 \tag{14.3}$$

between expected read class frequencies e_R's and observed read class frequencies o_R's. Minimizing (14.3) is equivalent to the following quadratic program that can be solved with any constrained quadratic programming solver.

minimize: $\displaystyle\sum_{R \in R} \left(\sum_{T \in C} d_{R,T} f_T' - o_R \right)^2$

subject to: $\displaystyle\sum_{T \in C} f_T' = 1$ and $f_T' \geq 0, \quad \forall T \in C \tag{14.4}$

14.3.1.3   Update Initial Estimates of Transcript Frequencies   The obtained crude transcript frequency estimation F_T' can deviate from the true crude frequency since the minimization of deviation is done uniformly. Indeed, the deviation in frequency is minimized on the same scale for each read class while different read classes have different size, as well as contribute to different subsets of transcripts. Instead of estimating unknown coefficients, we propose to directly obtain F_T' for which simulated read class frequencies $S_R = \{s_R\}$ match the observed frequencies O_R accurately enough as follows.

Until the deviation between simulated and observed read class frequencies is small enough, we repeatedly

- simulate reads according to F_R',
- find deviation between simulated and observed reads, $\Delta_R = S_R - O_R$,
- obtain read frequencies $C_R = O_R - \Delta_R/2$ corrected half-way in the direction opposite to the deviation,
- update estimated crude transcript frequencies F_T' based on corrected read class frequencies $\{C_R\}$,

Finally, the transcript frequencies f_T's can be obtained from crude frequencies f_T''s as follows:

$$f_T = \frac{f_T'/l_T}{\sum_{T' \in C} f_{T'}'/l_{T'}} \tag{14.5}$$

14.3.1.4   Combining Transcript Frequency Estimates from all Connected Components   Finally, we combine together individual solutions for each connected component. Let f_T^{glob} and f_T^{loc} be the global frequency of the transcript T and local frequency of the transcript T in its connected component C. Then the global frequency can be computed as follows:

$$f_T^{glob} = f_T^{loc} \times \frac{|R_C|/\sum_{T' \in C} f_{T'}^{loc} l_{T'}}{\sum_{C' \in C} \frac{|R_{C'}|}{\sum_{T' \in C'} f_{T'}^{loc} l_{T'}}} \tag{14.6}$$

where C is the set of all connected components in the graph M and $|R_C|$ is the number of reads emitted by the transcripts contained in the connected component C.

14.3.2 Differential Gene Expression Analysis: IsoDE

14.3.2.1 Bootstrap Sample Generation As most differential expression analysis packages, IsoDE starts with a set A of RNA-Seq read alignments for each condition. Bootstrapping can be used in conjunction with any method for estimating individual gene expression levels from aligned RNA-Seq reads, estimation typically expressed in *fragment per kilobase of gene length per million reads* (FPKM). In IsoDE, we use the IsoEM algorithm (20), an expectation maximization (EM) algorithm that takes into account gene isoforms in the inference process to ensure accurate length normalization. Unlike some of the existing estimation methods, IsoEM uses nonuniquely mapped reads, relying on the distribution of insert sizes and base quality scores (as well as strand and read-pairing information if available) to probabilistically infer their origin. Previous experiments have shown that IsoEM yields highly accurate FPKM estimates with lower run-time compared to other commonly used inference algorithms (35).

The first step of IsoDE is to generate M bootstrap samples by randomly resampling with replacement from the reads represented in A. When a read is selected during resampling, all its alignments from A are included in the bootstrap sample. The number of resampled reads in each bootstrap sample equals the total number of reads in the original sample. However, the total number of alignments may differ between bootstrap samples, depending on the number of alignments of selected reads and the number of times each read is selected. The IsoEM algorithm is then run on each bootstrap sample, resulting in M FPKM estimates for each gene. The bootstrap sample generation algorithm is summarized as follows:

1. Sort the alignment file A by read ID
2. Compute the number N of reads and generate a list $\mathcal{L}$ containing read IDs in the alignment file A
3. For $i = 1, \dots, M$ do:
   (a) Randomly select with replacement N read IDs from $\mathcal{L}$, sort selected read IDs, and extract in A_i all their alignments with one linear pass over A (if a read is selected m times, its alignments are repeated m times in A_i)
   (b) Run IsoEM on A_i to get the ith FPKM estimate for each gene

14.3.2.2 Bootstrap-Based Testing of Differential Expression To test for differential expression, IsoDE takes two folders as input, which contain FPKM estimates from bootstrap samples generated for the two conditions to be compared. In case of replicates, a list of bootstrap folders can be provided for each condition (one folder per replicate, normally with an equal number of bootstrap samples)—IsoDE will automatically merge the folders for the replicates to get a combined folder per condition and then perform the analysis as in the case without replicates.

In the following, we assume that a total of M bootstrap samples is generated for each of the compared conditions. We experimented with two approaches for pairing the FPKMs estimated from the two sets of bootstrap samples. In the "matching" approach, a random one-to-one mapping is created between the M estimates of first condition and the M estimates of the second condition. This results in M pairs of FPKM estimates. In the "all" approach, M^2 pairs of FPKM estimates are generated by pairing each FPKM estimate for first condition with each FPKM estimate for second condition. When pairing FPKM estimate a_i for the first condition with FPKM estimate b_j for the second condition, we use a_i/b_j as an estimate for the fold change in the gene expression level between the two conditions. The "matching" approach thus results in $N = M$ fold change estimates, while the "all" approach results in $N = M^2$ fold change estimates.

The IsoDE test for differential expression requires two user-specified parameters, namely the minimum fold change f and the minimum bootstrap support b. For a given threshold f (typically selected based on biological considerations), we calculate the percentage of fold change estimates that are equal to or higher than f when testing for overexpression, respectively, equal to or lower than $1/f$ when testing for underexpression. If this percentage is higher than the minimum bootstrap support b specified by the user, then the gene is classified as differentially expressed (DE), otherwise the gene is classified as non-differentially expressed (non-DE). The actual bootstrap support for fold change threshold f and the minimum fold change with bootstrap support of at least b, are also included in the IsoDE output to allow the user to easily increase the stringency of the DE test.

As discussed in the results section, varying the bootstrap support threshold b allows users to achieve a smooth trade-off between sensitivity and specificity for a fixed fold change f (see, e.g., Figure 14.3). Since different trade-offs may be desirable in different biological contexts, no threshold b is universally applicable. In our experiments, we computed b using a simple binomial model for the null distribution of fold change estimates and a fixed significance level $\alpha = 0.05$. Specifically, we assume that under the null hypothesis the fold changes obtained from bootstrap estimates are equally likely to be greater or smaller than f. We then compute b as x_{min}/N, where $x_{min} = \min\{x : P(X \geq x) \leq \alpha\}$ and X is a binomial random variable denoting the number of successes in N independent Bernoulli trials with success probability of 0.5. For convenience, a calculator for computing the bootstrap support needed to achieve a desired significance level given the (possibly different) numbers of bootstrap samples for each condition has been made available online (see Availability).

The number M of bootstrap samples is another parameter that the users of IsoDE must specify. As discussed in the results section, computing the bootstrap support for all genes takes negligible time, and the overall running time of IsoDE is dominated by the time to complete the $2M$ IsoEM runs on bootstrap samples. Hence, the overall run-times grows linearly with M. Experimental results suggest that the "all" pairing approach produces highly accurate results with relatively small values of M (e.g., $M = 20$), and thus results in practical run-times, independent of the number of replicates. We also note that for studies involving pairwise DE analysis

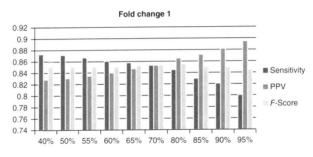

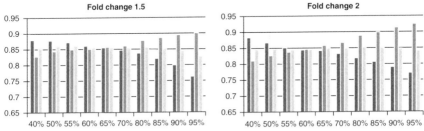

Figure 14.3 Sensitivity, PPV, and F-score of IsoDE-Match ($M = 200$ bootstrap samples per condition) on the Illumina MAQC data, with varying bootstrap support threshold.

of more than two conditions, IsoDE only requires M independently generated bootstrap samples per condition. Since the time for computing pairwise bootstrap support values is negligible, the overall running time will grow linearly with the number of conditions.

14.4 EXPERIMENTAL SETUP

14.4.1 Quantification Methods

14.4.2 Differential Expression Methods

We conducted experiments on publicly available RNA-Seq data sets generated from two commercially available reference RNA samples and a breast cancer cell line.

To compare the accuracy of different methods, we used RNA-Seq data RNA samples that were well characterized by quantitative real-time PCR (qRT-PCR) as part of the MicroArray Quality Control Consortium (MAQC) (36); namely an Ambion Human Brain Reference RNA, Catalog # 6050), henceforth referred to as HBRR and a Stratagene Universal Human Reference RNA (Catalog # 740000) henceforth referred to as UHRR. To assess accuracy, DE calls obtained from RNA-Seq data were compared against those obtained as described in the Methods section from TaqMan qRT-PCR measurements collected as part of the MAQC project (GEO accession GPL4097).

We used RNA-Seq data generated for HBRR and UHRR using three different technologies: Illumina, ION-Torrent, and 454.

The MCF-7 RNA-Seq data was generated (from the MCF-7 ATCC human breast cancer cell line) by Liu et al. (34) using Illumina single-end sequencing with read length of 50 bp. A total of 14 biological replicates were sequenced from two conditions: 7 replicates for the control group and 7 replicates for E2-treated MCF-7 cells. Sequencing of each replicate produced 25–65 millions of mapped reads.

14.4.2.1 Mapping RNA-Seq Reads   MAQC Illumina reads were mapped onto hg19 Ensembl 63 transcript library; all other data sets were mapped onto hg19 Ensembl 64 transcript library. Illumina data sets (MAQC and MCF-7) were mapped using Bowtie v0.12.8 (37). ION Torrent reads were mapped using TMAP v2.3.2, and 454 reads were mapped using MOSAIK v2.1.33 (38). For edgeR, nonunique alignments were filtered out and read counts per gene were generated using coverageBed (v2.12.0). The number of mapped reads per kilobase of gene length used in Fisher's exact test calculation is based on IsoEM FPKMs.

14.4.2.2 IsoDE: Evaluation Metrics   For each evaluated method, genes were classified according to the differential expression confusion matrix detailed in Table 14.1. Methods were assessed using sensitivity, positive predictive value (PPV), F-score, and accuracy, defined as follows:

$$Sensitivity = \frac{(TPOE + TPUE)}{(TOE + TUE)}$$

$$PPV = \frac{(TPOE + TPUE)}{(POE + PUE)}$$

$$Accuracy = \frac{(TPOE + TPND + TPUE)}{(TOE + TND + TUE)}$$

$$F-score = 2 \times \frac{Sensitivity \times PPV}{Sensitivity + PPV}$$

14.4.2.3 Compared Methods   The following four methods that were compared to IsoDE are briefly described.

TABLE 14.1   Confusion Matrix for Differential Gene Expression

	Ground truth		
Predicted	Overexpressed (TOE)	Nondifferential (TND)	Underexpressed (TUE)
Overexpressed (POE)	TPOE		
Nondifferential (PND)		TPND	
Underexpressed (PUE)			TPUE

Fisher's exact test: Fisher's exact test is a statistical significance test for categorical data that measures the association between two variables. The data is organized in a 2×2 contingency table according to the two variables of interest. We use Fisher's exact test to measure the statistical significance of change in gene expressions between two conditions A and B by setting the two values in the first row of the table to the estimated number of reads mapped per kilobase of gene length (calculated from IsoEM estimated FPKM values) in conditions A and B, respectively. The values in the second row of the contingency table depend on the normalization method used. We compared three normalization methods. The first one is total read normalization, where the total number of mapped reads in conditions A and B are used in the second row. The second is normalization by a housekeeping gene. In this case, the estimated number of reads mapped per kilobase of housekeeping gene length in each condition is used. We also test normalization by ERCCs RNA spike-in controls (39). FPKMs of ERCCs are aggregated together (similar to aggregating the FPKMs of different transcripts of a gene), and the estimated number of reads mapped per kilobase of ERCC are calculated from the resulting FPKM value and used for normalization. In our experiments, we used POLR2A as a housekeeping gene.

The calculated p-value, which measures the significance of deviation from the null hypothesis that the gene is not differentially expressed, is computed exactly by using the hypergeometric probability of observed or more extreme differences while keeping the marginal sums in the contingency table unchanged. We adjust the resulting p-values for the set of genes being tested using the Benjamini and Hochberg method (40) with 5% false discovery rate (FDR).

GFOLD: We used GFOLD v1.0.7 with default parameters and fold change significance cutoff of 0.05.

Cuffdiff: In our comparison, we used Cuffdiff v2.0.1 with default parameters.

edgeR: We followed the steps provided in the edgeR manual for RNA-Seq data. calcNormFactors(), estimateCommonDisp(), estimateTagwiseDisp(), and exactTest() were used with default parameter when processing the MCF-7 replicates. When processing MAQC data and a single replicate of MCF-7 data, estimateTagwiseDisp() was not used and the dispersion was set to 0 when calling exactTest(). The results were adjusted for multiple testing using the Benjamini and Hochberg method with 5%.

14.4.2.4 Data Sets and Ground Truth Definition   On MAQC data set, the ground truth was defined based on the available qPCR data from Reference 36. Each TaqMan assay was run in four replicates for each measured gene. POLR2A (ENSEMBL gene ID ENSG00000181222) was chosen as the reference gene, and each replicate CT was subtracted from the average POLR2A CT to give the log2 difference (delta CT). For delta CT calculations, a CT value of 35 was used for any replicate that had CT >35. The normalized expression value of a gene g would be $2^{(CT \text{ of POLR2A})-(CT \text{ of } g)}$. We filtered out genes that (i) were not detected in one or more replicates in each samples or (ii) had a standard deviation higher than 25% for the four TaqMan values in each of the two samples. Of the resulting subset, we used in the comparison genes whose

TaqMan probe IDs unambiguously mapped to Ensemble gene IDs (686 genes). A gene was considered differentially expressed if the fold change between the average normalized TaqMan expression levels bin the two conditions was greater than a set threshold with the *p*-value for an unpaired two-tailed *t*-test (adjusted for 5% FDR) of less than 0.05. We ran the experiment for fold change thresholds of 1, 1.5, and 2.

For experiments with replicates, we used the RNA-Seq data generated from E2-treated and control MCF-7 cells in Reference 34. In this experiment, we compared IsoDE with GFOLD and edgeR. The predictions made by each method when using all 7 replicates for each condition were used as its own ground truth to evaluate predictions made using fewer replicates. The ground truth for IsoDE was generated using a total of 70 bootstrap samples per condition.

14.5 EVALUATION

14.5.1 Transcriptome Quantification Methods Evaluation

14.5.1.1 Results on Simulated Data We tested (1) *SimReg* on several test cases using simulated human RNA-Seq data. The RNA-Seq data was simulated from UCSC annotation (hg18 Build 36.1), using Grinder read simulator (version 0.5.0) (41), with a uniform 0.1% error rate. Experiments on synthetic RNA-seq data sets show that the proposed method improves transcriptome quantification accuracy compared to previous methods.

The following three test cases have been used to validate *SimReg*:

Case 1: Consists of a single gene with 21 transcripts extracted from chromosome 1 (see Figure 14.4). From this gene, we have simulated around 3000 (coverage 100×) paired-end reads of length 100 bp and mean fragment length $\mu = 300$.

Case 2: We have randomly chosen 100 genes from which we have simulated reads using same parameters as in case 1.

Case 3: We have run our tool on the entire chromosome 1, which contains a total of 5509 transcripts (from 1990 genes) from where we have simulated 10*M* paired-end reads of length 100 bp.

We have compared our results with *RSEM*, one of the best tool for transcriptome quantification. Frequency estimation accuracy was assessed using r^2 and the comparison results are presented in Table 14.2. The results show better correlation compared with *RSEM* especially because of those cases of subtranscripts where *RSEM* skewed the estimated frequency toward supertranscripts.

TABLE 14.2 Comparison Results between SimReg and RSEM

Isoform Expression—r^2 Values					
Case 1: 1 Gene		Case 2: 100 Genes		Case 3: chr. 1	
SimReg	RSEM	SimReg	RSEM	SimReg	RSEM
0.958	0.923	0.999	0.93	0.995	0.924

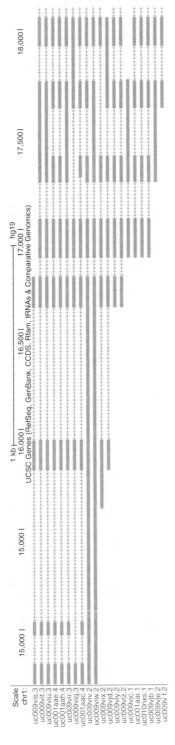

Figure 14.4 Screenshot from Genome browser of a gene with 21 subtranscripts. Source: Kent 2002 19. Reproduced with permission of Cold Spring Harbor Laboratory Press.

317

14.5.1.2   Results on Real Data   For the real data set, we assayed sets of human genes using MAQC, Human Brain Reference (HBR) sample (36), and NanoString nCounter amplification-free detection system (18).

For MAQC, we have correlated (1) our results using the Taqman qRT-PCR values while for NanoString we have used the probe counts provided in Reference 18. Since Taqman qRT-PCR and NanoString counts only measure the expression levels of genes and probes, respectively, we only compare gene (probe) abundance estimations. The expression level of a gene (probe) is obtained by summing up the frequencies of all transcripts in the gene (probe). For both data sets, we have used the Ensembl Homo sapiens genome sequence indexes (GRCH37) provided by Illumina.

There are three 2×50 bp paired-end data sets for Human Brain in SRA in MAQC data set. The average insert size is about 200 bp and the standard deviation about 30 bp. In NanoString data set, we have paired-end reads of length 75 bp and similar characteristics as in MAQC (more details can be found in References 18, 42).

In order to compute the 95% confidence interval (CI), we performed bootstrapping procedure by randomly choosing reads from the given set and returning chosen samples back to the pool. As a result, our chosen subsample may contain several copies of the same reads, whereas some reads are never chosen. We repeat subsampling procedure 200 times. For each sample we compute *MPE* and r^2 for *Cufflinks* (v2.2.0), *RSEM* (v1.2.19), and *SimReg*, and we count how many times our estimates are better than RSEM (since RSEM shows best performance compared to the other tools).

The results presented in Table 14.3 (1) show that *SimReg* has accuracy comparable to that of RSEM on the MAQC data but outperforms *RSEM* in both MPE and r^2 on the NanoString data set. Mean percentage error of SimReg is less than that of RSEM in 90.5% of cases.

All experiments were conducted on a Dell PowerEdge R815 server with quad 2.5 GHz, 16-core AMD Opteron 6380 processors and 256 Gb, RAM running under Ubuntu 12.04 LTS.

SimReg is freely available at http://alan.cs.gsu.edu/NGS/?q=adrian/simreg

TABLE 14.3   Median Percent Error (MPE) and r^2 Together with 95% CI for Transcriptome Quantification on MAQC and NanoString Data Sets (1)

Algorithm	MPE (%)	[95% CI] (%)	r^2 (%)	[95% CI] (%)
		Data Set: MAQC (36)		
SimReg	**77.2**	76.0–79.7	85.7	80.2–89.0
RSEM	78.0	77.4–80.1	86.4	81.1–89.3
Cufflinks	81.3	79.5–85.2	82.5	78.9–85.1
		Data Set: NanoString (42)		
SimReg	**57.0**	55.2–59.7	82.0	80.2–89.0
RSEM	65.8	61.3–68.2	82.6	78.7–85.4
Cufflinks	67.9	62.5–70.1	79.9	75.3–82.4

14.5.2 Differential Expression Methods Evaluation

We compared IsoDE against GFOLD, Cuffdiff, edgeR, and different normalization methods for Fisher's exact test; namely total normalization, housekeeping gene (POLR2A) normalization, and normalization using External RNA Controls Consortium (ERCC) RNA spike-in controls (39). Cuffdiff results were considerably worse on the Illumina MAQC data set, compared to other methods. Consequently, Cuffdiff was not included in other comparisons. edgeR was also not included in further comparisons due to lack of clear definition of uniquely mapped reads for ION-Torrent and 454 data sets, which were mapped using local read alignment tools. ERCC spike-ins were available only for ION Torrent samples; therefore, ERCC normalization for Fisher's exact test was conducted only for ION Torrent data sets.

Table 14.4 shows the results obtained for the MAQC Illumina data set using minimum fold change threshold f of 1, 1.5, and 2, respectively.

TABLE 14.4 Accuracy, Sensitivity, PPV, and F-Score in % for MAQC Illumina Data Set and Fold Change Threshold f of 1, 1.5, and 2

Fold Change	Method	Accuracy (%)	Sensitivity (%)	PPV (%)	F-Score (%)
1	FishersTotal	70.41	70.79	91.24	79.72
	FishersHousekeeping	65.60	65.22	95.05	77.36
	GFOLD	78.13	80.06	92.67	**85.90**
	Cuffdiff	11.37	6.96	**100.00**	13.01
	edgeR	73.03	73.26	95.56	82.94
	IsoDE-Match	**82.63**	**87.46**	83.70	85.54
	IsoDE-All	82.22	87.17	82.82	84.94
1.5	FishersTotal	74.05	78.20	84.85	81.39
	FishersHousekeeping	76.68	73.61	93.67	82.44
	GFOLD	79.15	79.35	90.41	84.52
	Cuffdiff	28.43	8.60	**100.00**	15.85
	edgeR	**80.01**	79.92	92.07	**85.57**
	IsoDE-Match	78.98	86.23	84.62	85.42
	IsoDE-All	79.01	**86.42**	84.49	85.44
2	FishersTotal	78.43	81.86	82.44	82.15
	FishersHousekeeping	81.20	80.00	88.21	83.90
	GFOLD	82.94	78.84	92.37	85.07
	Cuffdiff	40.96	10.47	**100.00**	18.95
	edgeR	**83.67**	81.63	91.17	**86.13**
	IsoDE-Match	82.04	85.58	85.19	85.38
	IsoDE-All	81.20	**86.74**	83.07	84.87

The number of bootstrap samples is $M = 200$ for IsoDE-Match and $M = 20$ for IsoDE-All, and bootstrap support was determined using the binomial model with significance level $\alpha = 0.05$. The best values in the corresponding categories are in bold.

Table 14.5 shows the results obtained from combining the ION Torrent runs for each of the MAQC data sets for the same values of f. Table 14.6 shows the results for the First 454 MAQC data set. For each fold change threshold, the best performing method for each statistic is highlighted in bold.

TABLE 14.5 Accuracy, Sensitivity, PPV, and F-Score in % for Ion Torrent Data Set and Fold Change Threshold f of 1, 1.5, and 2

Fold Change	Method	Accuracy (%)	Sensitivity (%)	PPV (%)	F-Score (%)
1	FisherTotal	71.68	72.76	90.56	80.69
	FisherHousekeeping	67.15	66.87	**94.74**	78.40
	FisherERCC	71.39	72.45	88.97	79.86
	GFOLD	75.77	77.55	90.43	83.50
	IsoDE-Match	**81.75**	**86.38**	82.18	**84.05**
	IsoDE-All	81.46	86.07	82.13	**84.05**
1.5	FisherTotal	74.16	78.39	85.06	81.59
	FisherHousekeeping	76.06	73.23	**92.96**	81.93
	FisherERCC	74.31	78.59	85.45	81.87
	GFOLD	75.47	77.63	87.88	82.44
	IsoDE-Match	77.66	83.94	84.75	**84.34**
	IsoDE-All	**77.81**	**84.13**	84.45	84.29
2	FisherTotal	79.71	83.02	84.00	83.51
	FisherHousekeeping	**81.75**	80.70	88.75	84.53
	FisherERCC	79.42	82.56	84.12	83.33
	GFOLD	80.58	76.74	**90.66**	83.12
	IsoDE-Match	**81.75**	85.81	84.63	**85.22**
	IsoDE-All	81.61	86.28	84.13	85.19

The number of bootstrap samples is $M = 200$ for IsoDE-Match and $M = 20$ for IsoDE-All, and bootstrap support was determined using the binomial model with significance level $\alpha = 0.05$. The best values in the corresponding categories are in bold.

TABLE 14.6 Accuracy, Sensitivity, PPV, and F-Score in % for the First 454 Dataset and Fold Change Threshold f of 1, 1.5, and 2

Fold Change	Method	Accuracy (%)	Sensitivity (%)	PPV (%)	F-Score (%)
1	FisherTotal	34.01	30.50	95.63	46.24
	FisherHousekeeping	24.52	20.12	**94.74**	33.38
	GFOLD	55.62	54.18	92.11	68.23
	IsoDE-Match	75.33	79.57	77.41	78.47
	IsoDE-All	**78.85**	**84.67**	81.04	**82.82**
1.5	FisherTotal	48.18	35.37	89.81	50.75
	FisherHousekeeping	42.48	24.86	**97.74**	39.63
	GFOLD	62.19	58.13	85.39	69.17
	IsoDE-Match	64.09	74.19	72.52	73.35
	IsoDE-All	**72.85**	**79.54**	80.62	**80.08**

TABLE 14.6 (*Continued*)

Fold Change	Method	Accuracy (%)	Sensitivity (%)	PPV (%)	*F*-Score (%)
2	FisherTotal	57.96	39.53	85.43	54.05
	FisherHousekeeping	55.33	29.30	**97.67**	45.08
	GFOLD	69.05	61.16	83.49	70.60
	IsoDE-Match	67.15	76.51	70.30	73.27
	IsoDE-All	**75.18**	**80.93**	78.03	**79.45**

The number of bootstrap samples is $M = 200$ for IsoDE-Match and $M = 20$ for IsoDE-All, and bootstrap support was determined using the binomial model with significance level $\alpha = 0.05$. The best values in the corresponding categories are in bold.

IsoDE has very robust performance, comparable or better than that of the other methods for differential gene expression. Indeed, IsoDE outperforms them in a large number of cases, across data sets and fold change thresholds. Very importantly, unlike GFOLD and Fisher's exact test, IsoDE maintains high accuracy (sensitivity and PPV around 80) on data sets with a small number of mapped reads such as the two 454 data sets. This observation is confirmed on results obtained for pairs of individual ION-Torrent runs, presented in Tables 14.7 and 14.8. This makes IsoDE particularly attractive for such low-coverage RNA-Seq data sets.

TABLE 14.7 Accuracy, Sensitivity, PPV, and *F*-Score in % for ION Torrent Pair HBRR: LUC-140_265 and UHRR: POZ-126_269, with Fold Change Threshold *f* of 1, 1.5, and 2

Fold Change	Method	Accuracy (%)	Sensitivity (%)	PPV (%)	*F*-Score (%)
1	FishersTotal	49.05	46.44	97.09	62.83
	FishersHousekeeping	40.88	37.62	**98.38**	54.42
	FisherERCC	52.55	51.70	88.83	65.36
	GFOLD	59.27	59.29	91.41	71.92
	IsoDE-All	**79.12**	**83.75**	79.91	**81.78**
1.5	FisherTotal	60.29	53.35	90.29	67.07
	FisherHousekeeping	57.08	45.31	**96.34**	61.64
	FisherERCC	57.96	56.02	82.07	66.59
	GFOLD	67.01	65.58	87.72	75.05
	IsoDE-All	**72.55**	**80.50**	80.34	**80.42**
2	FisherTotal	69.05	59.53	86.78	70.62
	FisherHousekeeping	68.90	53.49	**94.26**	68.25
	FisherERCC	66.42	60.23	80.93	69.07
	GFOLD	72.85	66.51	86.93	75.36
	IsoDE-All	**76.64**	**81.40**	81.02	**81.21**

The number of bootstrap samples for IsoDE-All is $M = 20$, and bootstrap support was determined using the binomial model with significance level $\alpha = 0.05$. Best result for each of the metrics is presented in bold.

TABLE 14.8   Accuracy, Sensitivity, PPV, and *F*-Score in % for ION Torrent Pair HBRR: GOG-139_281 and UHRR: POZ-127_270, with Fold Change Threshold *f* of 1, 1.5, and 2.

Fold change	Method	Accuracy (%)	Sensitivity (%)	PPV (%)	F-Score (%)
1	FishersTotal	51.09	48.76	96.04	64.68
	FishersHousekeeping	46.42	43.65	**97.58**	60.32
	FisherERCC	55.04	54.33	88.86	67.43
	GFOLD	60.44	60.84	83.09	70.24
	IsoDE-All	**79.56**	**84.37**	80.50	**82.39**
1.5	FisherTotal	62.34	56.02	90.43	69.19
	FisherHousekeeping	61.61	52.20	**94.46**	67.24
	FisherERCC	60.44	57.17	85.92	68.66
	GFOLD	64.09	62.91	82.87	71.52
	IsoDE-All	**74.60**	**80.69**	82.10	**81.39**
2	FisherTotal	69.05	60.47	85.81	70.94
	FisherHousekeeping	70.36	58.84	**90.03**	71.17
	FisherERCC	67.30	60.00	82.96	69.64
	GFOLD	72.55	65.12	86.96	74.45
	IsoDE-All	**76.50**	**79.77**	81.28	**80.52**

The number of bootstrap samples for IsoDE-All is $M = 20$, and bootstrap support was determined using the binomial model with significance level $\alpha = 0.05$. Best result for each of the metrics is presented in bold.

14.5.2.1   DE Prediction with Replicates

We also studied the effect of the number of biological replicates on prediction accuracy using the MCF-7 data set. We performed DE predictions using an increasing number of replicates. IsoDE was run with a total of 20 bootstrap samples per condition, distributed equally (or as close to equally as possible) among the replicates, as detailed in Table 14.9. GFOLD and edgeR were evaluated for 1 through 6 replicates using as ground truth the results obtained by running each method on all 7 replicates (see the Differential Expression Methods Evaluation section). For IsoDE, we also include the results using $M = 20$ bootstrap samples from all 7 replicates as its ground truth is generated using a much larger number of bootstrap samples ($M = 70$). Figure 14.5 shows the results of the three compared methods for a fold change threshold of 1, results for fold change thresholds 1.5 and 2 are shown in Figures 14.6 and 14.7.

Since for this experiment the ground truth was defined independently for each method, it is not meaningful to directly compare accuracy metrics of different methods. Instead, we focus on the rate of change in the accuracy of each method as additional replicates are added. Generally, all methods perform better when increasing the number of replicates. However, the accuracy of IsoDE varies smoothly and is much less sensitive to small changes in the number of replicates. Surprisingly, this is not the case for GFOLD and edgeR sensitivity, which drops considerably when transitioning from 1 to 2 replicates, most likely due to the different statistical models employed with and without replicates. Although we varied the number

TABLE 14.9 IsoDE setup for experiments with replicates.

# Replicates	Rep1	Rep2	Rep3	Rep4	Rep5	Rep6	Rep7	Bootstraps Per Condition
1	20							20
2	10	10						20
3	7	7	6					20
4	5	5	5	5				20
5	4	4	4	4	4			20
6	4	4	3	3	3	3		20
7	3	3	3	3	3	3	2	20

IsoDE experiments on the MCF-7 dataset was performed as follows. First we generated, for each of the 7 replicates of each condition 20, 10, 6, 5, 4, 3, respectively 2 bootstrap samples. We then used subsets of these bootstrap samples as input for IsoDE to perform DE analysis with varying number of replicates and a fixed total number $M = 20$ of bootstrap samples per condition. In experiment 1 we used 20 bootstrap samples from first replicate of each condition, in experiment 2 we used 10 bootstrap samples for each of the first 2 replicates of each condition, and so on.

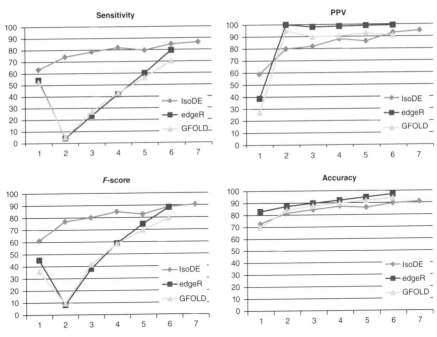

Figure 14.5 Sensitivity, PPV, F-score, and accuracy of IsoDE-All (with 20 bootstrap runs per condition), edgeR, and GFOLD on the Illumina MCF-7 data with minimum fold change of 1 and varying number of replicates.

of replicates without controlling the total number of reads as Liu et al. (34), our results strongly suggest that cost-effectiveness metrics such as those proposed in Reference 34 are likely to depend on to the specific method used for performing DE analysis. Therefore, the analysis method should be taken into account when using such a metric to guide the design of RNA-Seq experiments.

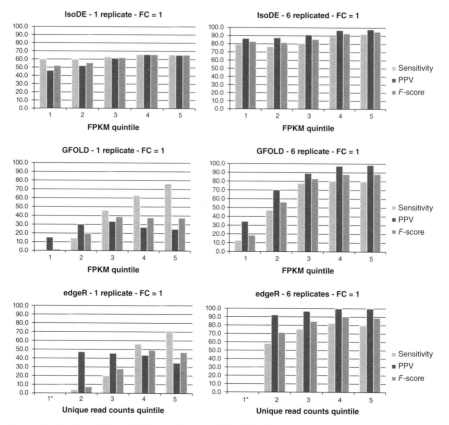

Figure 14.6   Sensitivity, PPV, and *F*-score of IsoDE-All (with 20 bootstrap runs per condition), edgeR, and GFOLD on the Illumina MCF-7 data, computed for quantiles of expressed genes after sorting in nondecreasing order of average FPKM for IsoDE and GFOLD and average count of uniquely aligned reads for edgeR. First quantile of edgeR had 0 differentially expressed genes according to the ground truth (obtained by using all 7 replicates). *there are no gene present in this quantile. edgeR is not able to detect DE genes when their expression level is very low.

14.5.2.2   Effect of Gene Abundance   We also studied the effect of gene abundance on the IsoDE, GFOLD, and edgeR prediction accuracy. We selected the subset of genes that are expressed in at least one of the two RNA samples. We sorted these genes by the average of the gene's expression. We used the FPKM values predicted by IsoEM, the FPKM values predicted by GFOLD, and the number of uniquely mapped reads, for IsoDE, GFOLD, and edgeR, respectively. The genes were then divided into quantiles, for each method independently, where quantile 1 had the genes with the lowest expression levels, and quantile 5 had the genes with the highest expression levels. Sensitivity, PPV, and *F*-score were calculated for each quantile separately.

Figure 14.8 shows that, for results with both 1 and 6 replicates, sensitivity, PPV, and *F*-score of IsoDE are only slightly lower on genes with low expression levels

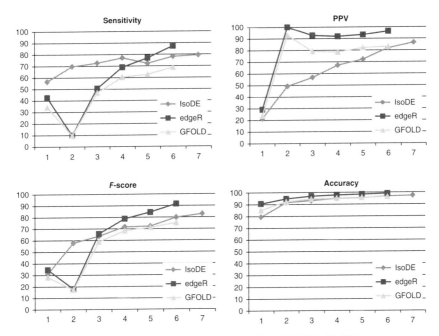

Figure 14.7   Sensitivity, PPV, *F*-score, and accuracy of IsoDE-All (with 20 bootstrap runs per condition), edgeR, and GFOLD on the Illumina MCF-7 data with varying number of replicates and minimum fold change 1.5.

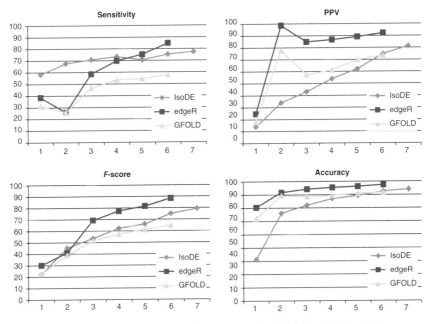

Figure 14.8   Sensitivity, PPV, *F*-score, and accuracy of IsoDE-All (with 20 bootstrap runs per condition), edgeR, and GFOLD on the Illumina MCF-7 data with varying number of replicates and minimum fold change 2.

compared to highly expressed genes (similar results are achieved for intermediate numbers of replicates and higher fold change thresholds). In contrast, GFOLD shows a marked difference in all accuracy measures for genes in the lower quantiles compared to those in the higher quantiles. The sensitivity of edgeR is also lower for genes expressed at low levels; however, its PPV is relatively constant across expression levels.

ACKNOWLEDGMENTS

This work has been partially supported by two Collaborative Research Grant from Life Technologies, awards IIS-0916401 and IIS-0916948 from NSF, and Agriculture and Food Research Initiative Competitive Grant no. 201167016-30331 from the USDA National Institute of Food and Agriculture.

REFERENCES

1. (a) Caciula A, Glebova O, Artyomenko A, Mangul S, Lindsay J, Măndoiu II, Zelikovsky A. Simulated regression algorithm for transcriptome quantification. Bioinformatics Research and Applications: 10th International Symposium, ISBRA 2014, Zhangjiajie, China, June 28–30, 2014, Proceedings, Volume 8492. Springer; 2014. p. 406; (b) Caciula A, Glebova O, Artyomenko A, Mangul S, Lindsay J, Măndoiu II, Zelikovsky A. Deterministic regression algorithm for transcriptome frequency estimation. 2014 IEEE 4th International Conference on Computational Advances in Bio and Medical Sciences (ICCABS), IEEE; 2014. p 1–1.

2. Fujita A, Severino P, Kojima K, Sato JR, Patriota AG, Miyano S. Functional clustering of time series gene expression data by granger causality. BMC Syst Biol 2012;6(1):137.

3. Mortazavi A, Williams BA, McCue K, Schaeffer L, Wold B. Mapping and quantifying mammalian transcriptomes by RNA-Seq. Nat Methods 2008;5(7):621–628.

4. Morozova O, Hirst M, Marra MA. Applications of new sequencing technologies for transcriptome analysis. Annu Rev Genomics Hum Genet 2009;10:135–151.

5. Wang Z, Gerstein M, Snyder M. RNA-Seq: a revolutionary tool for transcriptomics. Nat Rev Genet 2009;10(1):57–63.

6. Bullard J, Purdom E, Hansen K, Dudoit S. Evaluation of statistical methods for normalization and differential expression in mRNA-Seq experiments. BMC Bioinformatics 2010;11(1):94.

7. Robinson MD, McCarthy DJ, Smyth GK. edgeR: a bioconductor package for differential expression analysis of digital gene expression data. Bioinformatics 2010;26(1):139–140.

8. Anders S, Huber W. Differential expression analysis for sequence count data. Genome Biol 2010;11(10):106.

9. Trapnell C, Hendrickson DG, Sauvageau M, Goff L, Rinn JL, Pachter L. Differential analysis of gene regulation at transcript resolution with RNA-Seq. Nat Biotechnol 2012;31(1):46–53.

10. Bi Y, Davuluri RV. NPEBseq: nonparametric empirical bayesian-based procedure for differential expression analysis of RNA-Seq data. BMC Bioinformatics 2013;14(1):262.

11. Barrett T, Troup DB, Wilhite SE, Ledoux P, Evangelista C, Kim IF, Tomashevsky M, Marshall KA, Phillippy KH, Sherman PM et al. NCBI GEO: archive for functional genomics data sets - 10 years on. Nucleic Acids Res 2011;39 Suppl 1:1005–1010.

12. Feng J, Meyer CA, Wang Q, Liu JS, Liu XS, Zhang Y. GFOLD: a generalized fold change for ranking differentially expressed genes from RNA-Seq data. Bioinformatics 2012;28(21):2782–2788.

13. Scholtens D, von Heydebreck A. Analysis of differential gene expression studies. In: Gentleman R, Carey V, Huber W, Irizarry R, Dudoit S, editors. *Bioinformatics and Computational Biology Solutions Using R and Bioconductor*. New York: Springer-Verlag; 2005. p 229–248.

14. Alwine JC, Kemp DJ, Stark GR. Method for detection of specific RNAs in agarose gels by transfer to diazobenzyloxymethyl-paper and hybridization with DNA probes. Proc Natl Acad Sci U S A 1977;74(12):5350–5354.

15. Schena M, Shalon D, Davis RW, Brown PO. Quantitative monitoring of gene expression patterns with a complementary DNA microarray. Science 1995;270(5235):467–470.

16. Wang AM, Doyle MV, Mark DF. Quantitation of mRNA by the polymerase chain reaction. Proc Natl Acad Sci U S A 1989;86(24):9717–9721.

17. Moran MA. Metatranscriptomics: eavesdropping on complex microbial communities. Microbe 2009;4(7):329–335.

18. Steijger T, Abril JF, Engström PG, Kokocinski F, The RGASP Consortium T.J.H., Guigó R, Harrow J, Bertone P. Assessment of transcript reconstruction methods for RNA-Seq. Nat Methods 2013;10:1177–1184.

19. Kent WJ, Sugnet CW, Furey TS, Roskin KM, Pringle TH, Zahler AM, Haussler AD. The human genome browser at UCSC. Genome Res 2002;12(6):996–1006.

20. Nicolae M, Mangul S, Mandoiu II, Zelikovsky A. Estimation of alternative splicing isoform frequencies from RNA-Seq data. Algorithms Mol Biol 2011;6(1):9.

21. Li B, Dewey C. RSEM: accurate transcript quantification from RNA-Seq data with or without a reference genome. BMC Bioinformatics 2011;12(1):323.

22. Li W, Jiang T. Transcriptome assembly and isoform expression level estimation from biased RNA-Seq reads. Bioinformatics 2012;28(22):2914–2921.

23. Lin YY, Dao P, Hach F, Bakhshi M, Mo F, Lapuk A, Collins C, Sahinalp SC. CLIIQ: accurate comparative detection and quantification of expressed isoforms in a population. Proceedings of the 12th Workshop on Algorithms in Bioinformatics; 2012.

24. Mangul S, Caciula A, Al Seesi S, Brinza D, Banday AR, Kanadia R. An integer programming approach to novel transcript reconstruction from paired-end RNA-Seq reads. Proceedings of the ACM Conference on Bioinformatics, Computational Biology and Biomedicine. BCB '12; 2012. New York: ACM. p 369–376.

25. Tomescu AI, Kuosmanen A, Rizzi R, Mäkinen V. A novel min-cost flow method for estimating transcript expression with RNA-Seq. Proceedings of RECOMB-seq 2013; 2013.

26. Li W, Feng J, Jiang T. IsoLasso: a LASSO regression approach to RNA-Seq based transcriptome assembly. J Comput Biol 2011;18:1693–1707.

27. Tomescu AI, Kuosmanen A, Rizzi R, Mäkinen V. A novel min-cost flow method for estimating transcript expression with RNA-Seq. BMC Bioinformatics 2013;14 Suppl 5:15.

28. Jiang H, Wong WH. Statistical inferences for isoform expression in RNA-Seq. Bioinformatics 2009;25(8):1026–1032.

29. Tibshirani R. Regression shrinkage and selection via the lasso. J R Stat Soc, Ser B 1996;58:267–288.

30. Rapaport F, Khanin R, Liang Y, Pirun M, Krek A, Zumbo P, Mason CE, Socci ND, Betel D. Comprehensive evaluation of differential gene expression analysis methods for RNA-Seq data. Genome Biol 2013;14(9):95.

31. Li J, Jiang H, Wong W. Method modeling non-uniformity in short-read rates in RNA-Seq data. Genome Biol 2010;11(5):25.

32. Hansen KD, Brenner SE, Dudoit S. Biases in illumina transcriptome sequencing caused by random hexamer priming. Nucleic Acids Res 2010;38(12):e131.

33. Roberts A, Trapnell C, Donaghey J, Rinn JL, Pachter L. Improving RNA-Seq expression estimates by correcting for fragment bias. Genome Biol 2011;12(3):22.

34. Liu Y, Zhou J, White KP. RNA-Seq differential expression studies: more sequence or more replication? Bioinformatics 2014;30(3):301–304.

35. Li B, Dewey CN. RSEM: accurate transcript quantification from RNA-Seq data with or without a reference genome. BMC Bioinformatics 2011;12(1):323.

36. MAQC Consortium. The Microarray Quality Control (MAQC) project shows inter- and intraplatform reproducibility of gene expression measurements. Nat Biotechnol 2006;24(9):1151–1161. DOI: 10.1038/nbt1239.

37. Langmead B, Trapnell C, Pop M, Salzberg S. Ultrafast and memory-efficient alignment of short DNA sequences to the human genome. Genome Biol 2009;10(3):25. DOI: 10.1186/gb-2009-10-3-r25.

38. Lee W-P, Stromberg MP, Ward A, Stewart C, Garrison EP, Marth GT. MOSAIK: a hash-based algorithm for accurate next-generation sequencing short-read mapping. PLoS ONE 2014;9(3):90581.

39. Reid LH. Proposed methods for testing and selecting the ercc external RNA controls. BMC Genomics 2005;6(1):1–18.

40. Benjamini Y, Hochberg Y. Controlling the false discovery rate: a practical and powerful approach to multiple testing. J R Stat Soc, Ser B 1995;57(1):289–300.

41. Angly FE, Willner D, Rohwer F, Hugenholtz P, Tyson GW. Grinder: a versatile amplicon and shotgun sequence simulator. Nucleic Acids Res 2012;40(12):94. DOI: 10.1093/nar/gks251.

42. Kulkarni MM, Willner D, Rohwer F, Hugenholtz P, Tyson GW. Digital multiplexed gene expression analysis using the NanoString nCounter system. Curr Protoc Mol Biol 2011;94:25B.10.1-25B.10.17.

PART IV

MICROBIOMICS

15

ERROR CORRECTION OF NGS READS FROM VIRAL POPULATIONS

PAVEL SKUMS

Division of Viral Hepatitis, Centers of Disease Control and Prevention, Atlanta, GA, USA

ALEXANDER ARTYOMENKO AND OLGA GLEBOVA

Department of Computer Science, Georgia State University, Atlanta, GA, USA

DAVID S. CAMPO AND ZOYA DIMITROVA

Division of Viral Hepatitis, Centers of Disease Control and Prevention, Atlanta, GA, USA

ALEXANDER ZELIKOVSKY

Department of Computer Science, Georgia State University, Atlanta, GA, USA

YURY KHUDYAKOV

Division of Viral Hepatitis, Centers of Disease Control and Prevention, Atlanta, GA, USA

15.1 NEXT-GENERATION SEQUENCING OF HETEROGENEOUS VIRAL POPULATIONS AND SEQUENCING ERRORS

Highly mutable RNA viruses, such as human immunodeficiency virus (HIV) and hepatitis C virus (HCV), are the major causes of morbidity and mortality in the world. HCV infects approximately 130–170 million people of the world and is a major cause of liver diseases worldwide (1), while approximately 33 million people are currently infected with HIV (2). RNA-dependent polymerases of RNA viruses are error prone, which results in a high mutation rate (3). Owing to the accumulation of mutations, RNA viruses exist in infected hosts as highly heterogeneous populations of genetically close variants commonly termed as quasispecies (4, 5).

Computational Methods for Next Generation Sequencing Data Analysis, First Edition.
Edited by Ion I. Măndoiu and Alexander Zelikovsky.
© 2016 John Wiley & Sons, Inc. Published 2016 by John Wiley & Sons, Inc.
Companion website: www.wiley.com/go/Mandoiu/NextGenerationSequencing

TABLE 15.1   Error Rates for Different NGS Platforms

Platform	Estimated Error Rates	Percentage of Correct or Mapped Reads
454	0.1% mismatches, 0.3% insertions, 0.2% deletions (6); 0.02% mismatches, 0.27% insertions, 0.23% deletions (7); 0.09% mismatches, 0.54% insertions, 0.36% deletions (8); 1.09% (9)	91.7% mapped (6); 82% correct (7); 10.09% correct (8); 85.4% mapped (10); 95% mapped (11)
Ion Torrent	1.78% (12)	15.92% correct (12); 93% mapped (13)
Illumina	0.12% mismatches, 0.004% insertions, 0.006% deletions (6); 0.4% (12); 0.38% (9)	74% mapped (6); 94.7–97.5% mapped (10); 76.45% correct (12); 43% mapped (11)
PacBio	13% (12)	0% correct (12)

Next-generation sequencing (NGS) technologies allow the analysis of an unprecedented number of viral sequences carried in samples of infected individuals, presenting a novel opportunity for studying structure of viral populations and understanding the pathogen epidemiology, evolution, drug resistance, and immune escape. Currently, a number of NGS platforms are available, including 454 GS Junior and 454 GS-FLX Titanium (Roche), Ion PGM and Ion Proton (Life Technologies), MiSeq and HiSeq (Illumina), PacBio RS (Pacific Biosystems), and SOLiD (Life Technologies).

However, the increase in quantity of data had a detrimental effect on its quality. NGS is error prone, with error rates for different technologies being estimated in a number of studies. The estimated values are often different, and some of them are summarized in Table 15.1.

The types and nature of sequencing errors vary for different NGS platforms. For example, pyrosequencing used in the 454 platform introduces more than one nucleotide at a time into the nascent DNA strand on homopolymer regions. The number of the incorporated nucleotides is estimated using a signal calibration (6, 14). However, the signal strength may be misinterpreted, causing underestimation or overestimation of the number of incorporated nucleotides. Thus, sequence reads from the 454 platform frequently have insertions and deletions in homopolymer regions, although nucleotide replacements are also observed. A similar strategy is also used in the Ion Torrent platform, which is based on the detection of an electric rather than light signal generated by a hydrogen ion released during nucleotide incorporation. Thus, the error profile of the Ion Torrent platform is similar to the 454 platform. Unlike the 454 and Ion Torrent platforms, the Illumina technology extends DNA molecules by adding only a single nucleotide at each iteration (6, 14), which results in a higher prevalence of nucleotide replacements than deletions or insertions.

Nevertheless, none of the NGS platforms produces only one type of errors. Error profiles of NGS platforms are fairly heterogeneous. For example, Figure 15.1 illustrates error profiles of reads obtained from 14 HCV Hypervariable Region 1 (HVR1) samples, each containing a single plasmid clone, which were sequenced

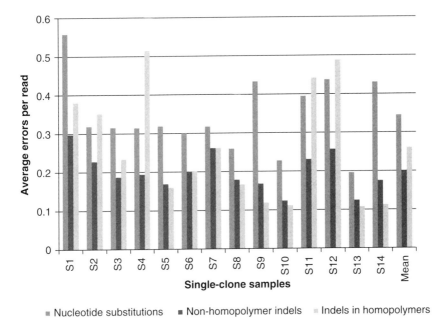

Figure 15.1 Error profile of single-clone samples. Three types of errors are shown: nucleotide replacements, non-homopolymer indels, and indels in homopolymer.

using 454/Roche GS FLX platform (15). Most sequencing errors observed in these samples are insertions and deletions, 54.99% of which are located in homopolymers; however, substitutions are also very abundant.

Most of the early NGS error correction algorithms assumed that sequencing errors are distributed randomly, which led to the conclusion that erroneous reads have low frequencies and, therefore, can be easily filtered using an appropriate frequency cutoff. The cutoff-based filtering is still one of the most widely used approaches to the NGS error correction. However, subsequent studies have shown that the assumption of the random error distribution is not always correct. In particular, error rate is strongly affected by position in the sequence (with probability of error increasing at the read's end), the length of a homopolymer or sequence read, and the spatial localization in PT plates (8, 15, 16). As a result, erroneous reads can have frequencies that are often comparable to such correct haplotypes (Figure 15.2), which significantly impedes the use of frequency cutoffs.

Despite differences in error rates and types among NGS technologies, all studies agree that sequencing errors pose a serious problem, which becomes especially challenging for heterogeneous viral samples. Originally, the emphasis was on obtaining a consensus sequence of an intrahost viral population, and in many studies this is still the main goal of using NGS. However, recent studies show that individual intrahost variants of highly heterogeneous viruses substantially differ in their properties and roles, with minor variants being often responsible for immune escape, drug resistance, and viral transmission (17–22). These observations emphasize the need for in-depth characterization of viral populations and suggest examination of individual intrahost

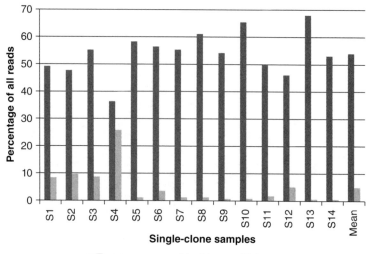

Figure 15.2 Frequency of the true haplotype in single-clone samples 15. In each pair of bars, left bar show the percentage of all reads from true haplotypes and right bar show the frequency of the most common false haplotype. The average percentage of error-free reads (true sequence) in single-clone samples is 54.02%. The most common false haplotype was found with an average frequency of 4.96% but can be as frequent as 25.85% (sample S4). Source: Skums 2012 15. Reproduced with permission of Springer Science+Business Media.

variants rather than consensus sequences. The need for identification and preservation of low-frequency natural variants in NGS reads makes the problem of detection and correction of NGS errors for viral samples especially challenging, because error correction methods must distinguish between real and artificial genetic heterogeneity. As an example, consider a distance graph of reads from the NGS viral data set obtained by sequencing a single-clone amplicon using 454 GS-FLX titanium pyrosequencing (15) (Figure 15.3). Here, only the most frequent haplotype (shown in red) corresponds to the real viral variant, while all other vertices represent sequencing artifacts. It is important to note that a similar pattern of heterogeneity is also expected in a real viral population originating from a single source after several rounds of replication.

Currently, a number of error correction algorithms for NGS data are available. These algorithms are based on different methodologies, including clustering of reads (23–29), multiple sequence alignments (15, 30), and k-mers (15, 31–34). Here, we will review some of the algorithms applicable to heterogeneous viral samples.

15.2 METHODS AND ALGORITHMS FOR THE NGS ERROR CORRECTION IN VIRAL DATA

15.2.1 Clustering-Based Algorithms

The largest family of NGS error correction methods is based on clustering techniques (23–29). The algorithms from this family correct sequencing reads and, if necessary, estimate plausibility of haplotypes by clustering reads or flowgrams. It is assumed

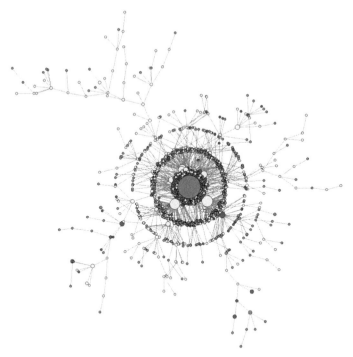

Figure 15.3 Minimum spanning tree of a distance graph G_{dist} of the NGS data set obtained from a single-clone sample. Each node represents a unique haplotype. The diameter of the node is proportional to its frequency. The true haplotype is shown in red, haplotypes with indel errors are shown in yellow, haplotypes with nucleotide substitutions are shown in blue, and haplotypes with both types of errors are shown in green. Here, G_{dist} is a complete weighted graph with the vertices corresponding to unique reads, and the weight of each edge is equal to the distance between the corresponding reads. Please see www.wiley.com/go/Mandoiu/NextGenerationSequencing for a color version of this figure.

that the within-cluster variability is produced by erroneous sequencing of a single haplotype represented by the cluster consensus or centroid. The main challenges associated with this approach are the requirement for efficient clustering of a large number of NGS reads and the lack of information on the number of actual clusters (i.e., the number of real haplotypes). Different algorithms handle these challenges differently.

The first clustering-based approach for the NGS error correction in viral samples was proposed in Reference 28, and its basic scheme is summarized in the algorithm for NGS error correction in viral data using k-means clustering. This algorithm estimates the number of haplotypes as follows: given a multiple sequence alignment of reads, the significance of mutations in each alignment column is tested using the null hypothesis of a single haplotype, which implies that the probability to observe at least l mutations in that column follows the binomial distribution and is equal to

$$P(X \geq l) = \sum_{i=l}^{d} \binom{d}{i} \epsilon^i (1 - \epsilon)^{d-i}, \tag{15.1}$$

where d is the number of reads that overlap the position corresponding to the column under consideration and ϵ is the sequencing error rate. Under the same null hypothesis, the test statistics for pairs of mutations is the probability of having l co-occurrences in two different alignment columns i and j, which is given by the formula:

$$P(Y = l) = \frac{\binom{n_j}{l}\binom{d-n_j}{n_i-l}}{\binom{d}{n_i}}, \tag{15.2}$$

where n_i and n_j are the number of times the mutations under consideration have been observed in columns i and j. Initially, the expected number of haplotypes is defined as the sum of numbers of significant mutations and pairs of mutations; however, this number can be an overestimation of the real haplotypes number, and several approaches to reduce this overestimation are proposed (28).

Algorithm NGS Error Correction in Viral Data Using k-Means Clustering (28)

Require: An alignment of reads, which can be constructed, for instance, using a pairwise alignment with a reference genome

1: Partition aligned reads into windows over the alignment.
2: **for** each window **do**
3:      Estimate k – the number of haplotypes in the window using a statistical testing to identify significant mutations and pairs of mutations, which can be associated with separate haplotypes. Significant mutations and pairs of mutations are found using the binomial and Fisher's exact tests, respectively.
4:      Partition reads into clusters using k-means clustering with the Hamming distance defining clusters' membership and correct reads into their clusters' centers.
5: **end for**

The clustering approach to error correction was further developed in SHORAH (23–25). This approach resolves the problem associated with the unknown number of clusters by assigning reads into existing clusters or introducing new clusters using Dirichlet process mixture (DPM).

SHORAH is based on the following probabilistic model. It assumes that all N reads are produced from haplotypes via random sequencing errors regulated by the error rate parameter θ, while all haplotypes originate from a reference genome via an evolutionary process based on random mutations regulated by the mutation probability parameter γ. Then, the probability of the read r_i being erroneously emitted from the haplotype h_k can be estimated as

$$Pr(r_i|h_k, \theta) = (1 - \theta)^{L-d(r_i,h_k)}\left(\frac{\theta}{3}\right)^{d(r_i,h_k)} \tag{15.3}$$

and the probability to observe the set of reads $\mathcal{R} = \{r_1, \ldots, r_N\}$ given the set of haplotypes $\mathcal{H} = \{h_1, \ldots, h_K\}$ and the set of clusters $\mathcal{C} = \{C_1, \ldots, C_K\}$ (where the haplotype

h_k represents a center of the cluster C_k) is

$$Pr(\mathcal{R}|C, \mathcal{H}, \theta) = \prod_{k=1}^{K} \prod_{r_i \in C_k} Pr(r_i|h_k), \qquad (15.4)$$

where L is the length of the studied genomic region and $d(r_i, h_k)$ is Hamming distance between sequences r_i and h_k (in Reference 23 the different notation is used). Analogously, probability of a haplotype $h_k = (h_{k,1}, \ldots, h_{k,L})$, given the cluster associated with it, can be estimated as

$$Pr(h_k|C_k, \theta) = \prod_{j=1}^{L} Pr(h_{k,j}|C_k, \theta), \qquad (15.5)$$

where probability $Pr(h_{k,j}|C_k, \theta)$ of the jth base of h_k is calculated as

$$Pr(h_{k,j}|C_k, \theta) = \frac{(1-\theta)^{m_{k,j}(h_{k,j})} \left(\frac{\theta}{3}\right)^{n-m_{k,j}(h_{k,j})}}{\sum\limits_{u=1}^{4} (1-\theta)^{m_{k,j}(b_u)} \left(\frac{\theta}{3}\right)^{n-m_{k,j}(b_u)}}, \qquad (15.6)$$

where $(b_1, b_2, b_3, b_4) = (a, c, t, g)$ are the bases and $m_{k,j}(b_u)$ is the number of reads from the cluster C_k having the base b_u in the position j.

Finally, SHORAH requires the calculation of the likelihood $Pr(r_i|\text{ref}, \theta, \gamma)$ of emission of the read r_i by one of the haplotypes originated from the reference genome ref, which is estimated as

$$Pr(r_i|\text{ref}, \theta, \gamma) = \sum_{k=1}^{K} Pr(r_i|h_k, \theta) Pr(h_k|\text{ref}, \gamma) \qquad (15.7)$$

SHORAH iteratively re-assigns reads to clusters or introduce new clusters using formulas (15.3) and (15.7), sample haplotypes base-by-base from the constructed clusters using formula (15.6), and re-estimated parameters θ and γ. Reads are assigned to existing or new clusters according to the probabilities following DPM, which are estimated as follows:

$$Pr(r_i|C_k \setminus \{r_i\}, \mathcal{H}, \theta, \gamma) = \begin{cases} b\dfrac{|C_k \setminus \{r_i\}|}{|C_k| - 1 + \alpha} Pr(r_i|h_k, \theta) & \text{if } |C_k \setminus \{r_i\}| > 0; \\ b\dfrac{\alpha}{|C_k| - 1 + \alpha} Pr(r_i|\text{ref}, \theta, \gamma), & \text{otherwise.} \end{cases} \qquad (15.8)$$

Here, α is the algorithm hyper-parameter, which controls the utilization of existing clusters. SHORAH steps are summarized in SHORAH error correction algorithm.

A different clustering-based approach to error correction was implemented in kGEM (29) (a similar principle was used for clustering of viral populations in Reference 35). Instead of minimizing intracluster distances or distance to cluster centers (like in the k-means algorithm), kGEM maximizes the specially defined

---

Algorithm SHORAH Error Correctionlgorithm

1: Initialize the algorithm by randomly assigning reads to $N/10$ clusters
   $C = \{C_1,\ldots,C_{N/10}\}$;
2: **while** not all reads are permanently assigned to haplotypes **do**
3:     $K \leftarrow |C|$
4:     Estimate haplotypes $\mathcal{H} = \{h_1,\ldots,h_K\}$ base-by-base according to (15.6)
5:     **for** $r_i \in \mathcal{R}$ **do**
6:         **if** r_i has been assigned to the same haplotype for more than 80% of
       iterations **then**
7:             assign r_i to h_k permanently;
8:         **end if**
9:     **end for**
10:    Estimate parameters θ and γ as follows:
       $$\theta \leftarrow \frac{\sum_{k=1}^{K}\sum_{r_i \in C_k} d(r_i, h_k)}{NL}$$
       $$\gamma \leftarrow \frac{\sum_{k=1}^{K} d(r_i, \text{ref})}{KL}$$
11:    recalculate clusters using (15.8)
12: **end while**
Ensure: the set of haplotypes $\mathcal{H}$.

---

likelihood of emission of the observed set of reads from one of k unknown, actual haplotypes. Moreover, during calculations, instead of actual sequences kGEM uses *generalized haplotypes*—the families formed according to the distribution of frequencies of bases at genomic positions.

Formally, let alignment of a given set of unique reads $\mathcal{R} = \{r_1,\ldots,r_N\}$ with multiplicities $\{o_1,\ldots,o_N\}$ be represented by $N \times L$ matrix $\mathcal{A}$, with rows corresponding to reads and columns corresponding to alignment positions. For a given set of reads $C^j \subseteq \mathcal{R}$, a *generalized haplotype* $G^j = g(C^j)$ with a multiplicity $m_j = \sum_{r_i \in C_j} o_i$ is an $5 \times L$ matrix, with columns representing alignment positions and rows representing the bases $\{a, c, t, g, d\}$; the value $G^j_{e,m}$ is the frequency of a base e in mth position among all sequences in C^j. Then, the probability of the read $r_i = (r_i^1,\ldots,r_i^l)$ being emitted by a generalized haplotype G^j can be estimated as

$$\Pr(r_i|G^j) = \prod_{l=1}^{L} G^j_{r_i^l,l} \tag{15.9}$$

and the probability to observe r_i, given the set of k generalized haplotypes $\mathcal{G} = \{G^1,\ldots,G^k\}$ with frequencies $\{f_1,\ldots,f_k\}$, is $\Pr(r_i|\mathcal{G}) = \sum_{j=1}^{k} f_j \Pr(r_i|G^j)$.

The objective of kGEM is to find an optimal set $\mathcal{G}^* = \{G^1,\ldots,G^k\}$ of k generalized haplotypes that most likely emitted $\mathcal{R}$, that is,

$$G^* = \arg\max_{|\mathcal{G}|=k} \prod_{i=1}^{N} Pr(r_i|\mathcal{G})^{o_i}, \tag{15.10}$$

which is equivalent to maximizing the log-likelihood function

$$\ell(G) = \sum_{i=1}^{N} o_i \log \Pr(r_i|G). \tag{15.11}$$

kGEM replaces the problem of maximizing (15.11) with the easier problem of maximizing the log-likelihood of the hidden model

$$\ell_{hid}(G) = \sum_{i=1}^{N} \sum_{j=1}^{k} p_{j,i} \log(f_j \Pr(r_i|G^j)), \tag{15.12}$$

where $p_{j,i}$ is the portion of reads matching the read r_i, which are emitted by the generalized haplotype G^j. It iteratively estimates generalized haplotypes G and the missing data f_j and $p_{j,i}$ using expectation maximization (EM) approach. At each iteration of the algorithm, after EM step is finished, kGEM estimates the probability p_{err}^j of having errors for each estimated haplotype G^j:

$$p_{err}^j = 1 - (1 - \sum_{u=m_j/2}^{m_j} \binom{m_j}{u} \varepsilon^u (1 - \varepsilon)^{m_j - u})^L. \tag{15.13}$$

Haplotypes with $p_{err} > 0.05$ are discarded. The following is the work flow of kGEM summarized in kGEM error correction algorithm.

15.2.2   *k*-Mer-Based Algorithms

k-mer-based algorithms currently are available either as separate applications (Reptile (36), EDAR (34), KEC (15)) or as parts of the general genome assembly frameworks (EULER-DB, EULER-SR (31–33, 37), ALLPATH (38)). The main idea behind this family of algorithms is to switch from consideration of whole-length reads to substrings of reads of a given length k (*k*-mers). For each *k*-mer, its frequency (*k-count*) is calculated. Each read of length l contains $l - k + 1$ consecutive *k*-mers, distribution of *k*-counts of which provides the information about possible sequencing errors. *k*-mers with high *k*-counts (*solid k*-mers) are usually correct, while *k*-mers with low *k*-counts (*weak k*-mers) more likely contain errors (Figure 15.4a). The basic scheme of *k*-mers-based algorithms can be described as follows:

(1) Calculate *k*-mers s and their *k*-counts $kc(s)$. A straightforward calculation and storage of *k*-mers and *k*-counts may be inefficient due to usually large sizes of NGS data sets. This can be resolved by storing *k*-mers in a hash map, where each key is a *k*-mer s and the corresponding value is the array $v(s) = ((r, i) : s$ is a *k*-mer of the read r starting at position $i)$ (15). *k*-counts can be adjusted based on GC-content (34).

(2) Determine a threshold *k*-count (*error threshold*) t_e separating solid and weak *k*-mers.

Algorithm kGEM Error Correction Algorithm (29, 35)

1: Find initial k generalized haplotypes using the following steps. First, iteratively identify k reads $s^1,\ldots,s^k$ as follows: at each iteration, pick the most frequent read, which have not been picked before and such that the minimum Hamming distance to the previously selected reads is maximal. Then define the initial set of generalized haplotypes $\mathcal{G}^1 = \{G^1,\ldots,G^k\}$:
$$G^i_{e,l} \leftarrow \begin{cases} 1 - 4\varepsilon, & \text{if } s^i_l = e; \\ \varepsilon, & \text{otherwise.} \end{cases}$$

2: calculate $h_{j,i} \leftarrow \Pr(r_i|G^j)$ using (15.9); set the iterations count $t \leftarrow 1$; $\mathcal{G}^0 = \emptyset$;

3: **while** $\mathcal{G}^t \neq \mathcal{G}^{t-1}$ **do**

4:      Estimate frequencies $f_1,\ldots,f_k$ and portions $p_{1,1},\ldots,p_{k,N}$ as follows. Starting from the initial solution $f_j^{(0)} \leftarrow \frac{1}{k}, j = 1,\ldots,k$, iteratively perform the following steps:

5:      **while** the Euclidean distance between current and previous solution is greater than predefined precision δ **do**

6:          For each read r_i and generalized haplotype $G^j \in \mathcal{G}^t$ calculate $p_{j,i}$:
$$p_{j,i} \leftarrow \frac{f_j \cdot h_{j,i}}{\sum_{u=1}^{k} f_u \cdot h_{u,i}}$$

7:          Calculate the expected number of reads emitted by G^j that match r_i:
$$e_{j,i} \leftarrow o_i \cdot p_{j,i}$$

8:          Calculate the updated multiplicity and frequency of each generalized haplotype $G^j \in \mathcal{G}^t$:
$$m_j \leftarrow \sum_{i=1}^{N} e_{j,i}, f_j \leftarrow \frac{m_j}{\sum_{u=1}^{k} m_u};$$

9:      **end while**

10:     Update the set of generalized haplotypes and recalculate probabilities as follows: $\mathcal{G}^{t+1} = \{G^1,\ldots,G^k\}$, where
$$G^j_{e,l} \leftarrow \begin{cases} 1 - 4\varepsilon, & \text{if } e = \arg\max_{e' \in \{a,c,t,g,d\}} \sum_{i \in \{1,\ldots,N\}: r_{i,l}=e'} p_{j,i}; \\ \varepsilon, & \text{otherwise.} \end{cases}$$

11:     collapse identical haplotypes and drop haplotypes with a high error probability p^j_{err} (see (15.13)).

12:     $h_{j,i} \leftarrow \Pr(r|G^i)$; $t \leftarrow t + 1$;

13: **end while**

14: round the obtained generalized haplotypes i.e. find the set of sequences $\mathcal{H} = \{hp^1,\ldots,hp^k\}$, where the base at each position v of a haplotype hp^j has the highest frequency in the column u of the generalized haplotype G^j.

Ensure: the set of haplotypes $\mathcal{H}$.

---

(3) For every read r find error regions, that is, the positions segments $[i_e, j_e]$ such that for every $p \in [i_e, j_e]$ the k-mer s_p starting at the position p is weak (or considered as such). If such error regions exist, either correct (15, 31, 36, 37) or remove (34) errors.

All heterogeneous viral genomes contain extremely variable regions, which allow for distinguishing among individual intrahost variants. Consequently, k-mers derived

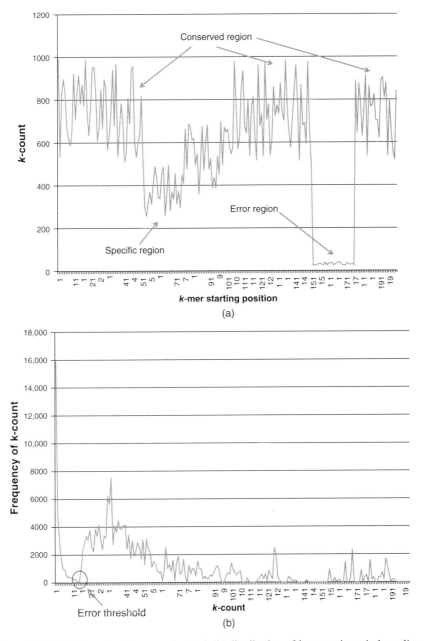

Figure 15.4   (a) *k*-counts of *k*-mers of a read; (b) distribution of *k*-counts in a viral amplicon data set.

from these regions usually have "intermediate" k-counts (Figure 15.4a). Presence of such regions and requirement to preserve even low-frequency haplotypes make the problem of determination of the error threshold daunting. The error threshold can be estimated using the k-counts distribution dk of the data set, the idea which was first proposed in Reference 32 and further developed in References 15, 33, 34. It was noticed that k-counts of weak and solid k-mers follow different distributions (Figure 15.4b), and different models describing these distributions were proposed (the exponential distribution and the series of Poisson distributions (15, 34), Poisson and Gaussian distribution (32)). Based on this observation, the error threshold can be estimated as the minimal k-count separating two different distributions. In References 32, 34, 38, the error threshold is calculated as the first local minimum of the distribution dk, which usually gives satisfactory separation of distributions for a shotgun data. In Reference 15, it was noticed that this method is not always applicable to amplicon data because of the discrete structure of dk. Thus, for amplicon data, the k-mer threshold was defined by the low end of the first long segment of the consecutive zeros of the distribution dk. The subsequent version of the error correction algorithm (15) also includes a more complex algorithm for the error-threshold estimation based on the fitting Poisson distribution to the distribution dk.

Possible errors in reads are localized in error-prone regions, which can be identified by clustering genome positions using the variable bandwidth mean-shift method (39, 40) according to their k-counts (15, 34). In particular, KEC (15) estimates error regions by (i) successively finding maximal segments $[i_e, j_e]$ such that for every $p \in [i_e, j_e]$ we have $kc(s_p) \le t_e$; (ii) clustering positions; (iii) extending the obtained segments in both directions by adding consecutive positions q such that k-mer s_q belongs to the same cluster, as k-mer s_p for some $p \in [i_e, j_e]$, and (iv) joining the overlapping segments. EDAR (34) first clusters positions, partitions clusters into disjoint segments, and considers the segments with the mean k-counts less than t_e as error regions.

Once errors are localized, different algorithms treat errors differently. EDAR removes potentially erroneous nucleotides and split reads, while Reptile (36), KEC (15), and EULER-SR (37) try to correct the errors by finding nucleotide insertions, deletions, or replacements that convert weak k-mers into strong k-mers. Both approaches have advantages and disadvantages. In particular, correction of errors allows to keep as much genetic variability as possible, thus making the corresponding algorithms especially suitable for detection of minor variants in heterogeneous viral populations. However, in the process of transforming k-mers, this approach may introduce new artificial errors (31); moreover, it is more time- and memory-consuming. Filtering of errors does not introduce new errors, but it is not suitable for amplicon sequencing and produces a risk of loss of rare variants and distortion of variant frequencies. We will consider some of these approaches in more detail in the following sections.

A method for detection and removal of potential sequencing errors implemented in EDAR (34) is based on the following idea. If $[i_e, j_e]$ is an error region of the read r of length l, then all k-mers starting within error region probably contain errors, while k-mers starting at positions $i_e - 1$ and $j_e + 1$ are likely to be error-free. In particular, it implies that if $j_e - i_e + 1 > k$, then the base at position $i_e + k - 1$ is most probably incorrect. Using this and similar observations, EDAR estimates a set of error bases $\mathcal{EB}$ as follows:

$$\mathcal{EB} \leftarrow \begin{cases} \{i_e + k - 1\}, \text{ if } j_e - i_e + 1 \leq k \text{ and } j_e = l - k + 1; \\ \{j_e\}, \text{ if } j_e - i_e + 1 \leq k \text{ and } j_e < l - k + 1; \\ \{i_e + k - 1, j_e\}, \text{ if } k < j_e - i_e + 1 < 2k; \\ \{i_e + k - 1, \ldots, j_e\}, \text{ if } j_e - i_e + 1 \geq 2k; \end{cases} \qquad (15.14)$$

Then error bases are removed and reads with errors are split accordingly.

EULER-SR (37) corrects errors in a read r of length $|r|$ containing weak k-mers by solving the following optimization problem: find a sequence r^*, which contains only solid k-mers (*solid sequence*) such that the edit distance $d(r, r^*)$ is minimal. This problem can be solved using a dynamic programming as follows (37). Let S_k be a set of all solid k-mers. For an integer $i \leq |r|$ and $a \in S_k$, the score function score(i, a) is defined as the minimum edit distance between a prefix of r of length i and all solid strings, which have a as a suffix. This function can be computed recursively using this relation:

$$\text{score}(i, a) = \min_{X \in \{A,C,T,G\}} \left\{ \text{score}(i - 1, X \sqcup a_{k-1}) + \begin{cases} 1, \text{ if } s_i \neq a_k; \\ 0, \text{ if } s_i = a_k; \end{cases} \right.$$

$$\left. \text{score}(i - 1, a) + 1, \text{score}(i, X \sqcup a_{k-1}) + 1 \right\}, \qquad (15.15)$$

where $X \sqcup a_{k-1}$ is concatenation of a character X and the prefix of a of length $k - 1$, and the initial condition is score$(0, a) = 0$ for all $a \in S_k$. Solution of the error correction problem corresponds to the optimal score $\min_{a \in S_k} \text{score}(|r|, a)$. The optimal score and the corresponding solid string can be calculated by finding a shortest path between vertices with in-degree 0 and vertices with out-degree 0 in a relation digraph G_{rel} with vertices corresponding to pairs (i, a) ($i \in \{0, \ldots, |r|\}$, $a \in S_k$) and two vertices (i_1, a_1) and (i_2, a_2) being connected by an arc, if they belong to the same recurrence relation with (i_2, a_2) as the left-hand side.

This approach produces a somewhat exact solution, but it is highly time- and memory-consuming. In Reference 37, several heuristic approaches were used to improve the algorithm performance; however, none is scalable for longer reads (37). KEC (15) uses a different heuristic approach to error correction, which takes into account a possible nature of errors (such as high prevalence of errors in homopolymers) and utilizes information derived from the structure of previously detected error regions. Corrections are based on the analysis of different factors, including the length of an error region, nucleotide sequences at its end, and the frequencies of similar solid k-mers. The procedure of error correction is repeated iteratively, after which reads containing weak k-mers, which were not corrected, are discarded.

Let $h_i(X)$ be a homopolymer consisting of i identical bases $X \in \{A, C, T, G\}$ ($i \geq 0$, for convenience we assume that any single base is also a homopolymer of length 1 and a homopolymer of length 0 is deletion). The error correction in KEC is based on the following lemma:

Lemma 1. (15). *Suppose that a read r contains an error region $[i_e, j_e]$, which has the length $l = j_e - i_e + 1$, ends with the base X, is not a prefix or suffix of r and was caused by a single error E. Then*

(1) *If E is a nucleotide replacement, then $l = k$.*
(2) *If E is an insertion creating a homopolymer of the length p $(1 \leq p \leq k)$, then $l = k - p + 1$ and the error region is followed by a homopolymer $h_{p-1}(X)$.*
(3) *If E is a deletion creating a homopolymer of the length p $(0 \leq p \leq k)$, then $l = k - p - 1$ and the error region is followed by a homopolymer $h_p(Y)$, where $Y \neq X$.*

For error regions with length not exceeding k, Lemma 1 allows to quickly and reliably detect and correct errors. However, it is not applicable to error regions with length greater than k. Such "long" error regions may appear, if two or more errors are separated by less than k positions in the read, and the prevalence of those regions depends on the parameter k. The value of k should be carefully chosen, since too large value creates more long error regions, while too small value hampers the reliable separation of k-counts of weak and solid k-mers. In different studies, the best results are achieved with $k = 25$ (15) and $k = 15, 23$ (34). For example, the number of long error reads was very low at $k = 25$ in HCV samples studied in Reference 15 in comparison to the number of error regions of lengths not exceeding k (Figure 15.5).

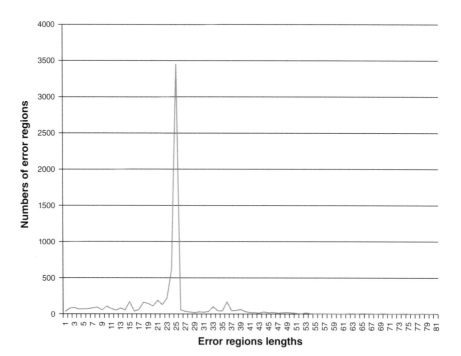

Figure 15.5 Distribution of lengths of error regions for one of samples studied in Reference 15. Source: Skums 2012 15. Reproduced with permission of Springer Science+Business Media.

The basic scheme of k-mer-based error correction implemented in KEC is summarized in the "k-mer-based error correction implemented in KEC algorithm" as follows. It should be noted that although at certain stages of this algorithm, several possible corrections of errors are possible (and the algorithm chooses the correction producing k-mers with highest k-counts), for real viral data in the overwhelming majority of cases (95.9% in average) the correction producing a solid k-mer is unique (15). This reduces the impact of heuristic component of this algorithm and makes the results more reliable.

Algorithm k-Mer-Based Error Correction Implemented in KEC

---

repeat
    for every read $r = (r_1, \ldots, r_{|r|})$ containing error regions **do**
        for every error region $[i_e, j_e]$ of r such, that $l = j_e - i_e + 1 \leq k$ **do**
            if $i_e \neq 1$ and $j_e \neq |r| - k + 1$ **then**
                if $l < k - 1$ **then**
                    Identify the type and position of error E using Lemma 1. In this case, the possible identification is unique. For the insertion remove the base r_{j_e}; for the deletion duplicate r_{j_e} if it introduces a solid k-mer.
                end if
                if $l \in \{k - 1, k\}$ **then**
                    Identify the type and position of error E using Lemma 1. In this case, the error region can be produced by different errors. If $l = k - 1$ and $r_{j_e} = r_{j_e + 1}$, then E could be either insertion of r_{j_e} or deletion of the base $X \neq r_{j_e}$ between r_{j_e} and $r_{j_e + 1}$. If $l = k$, then E can be either replacement of one of the three possible bases or insertion. In the case of ambiguity choose the correction, which produces a solid k-mer with the highest k-count.
                end if
            else
                Remove the prefix of suffix of r corresponding to the error region.
            end if
        end for
    end for
    Recalculate k-mers and error regions.
until all k-mers are solid or the predefined number of iterations is reached
Recalculate k-mers, the error threshold and error regions. Discard all reads containing error regions.

15.2.3 Alignment-Based Algorithms

In this section, we will consider several methods of NGS error correction for viral amplicons, which are based on sequence alignment. The main challenge associated with the application of alignment-based algorithms for viral amplicons is that a high-quality sequence alignment is computationally expensive, and the number of unique NGS reads is usually high due to the extreme heterogeneity of viral populations. Furthermore, viral heterogeneity also leads to the lack of good references, and the quality of alignment may be greatly affected by sequencing errors. However, in addition to a possible use as primary error correction algorithms, such methods can be extremely useful in a pipeline with clustering-based or k-mers-based algorithms as post-processing step for data previously cleaned by these algorithms, when the number of unique reads and sequencing errors is greatly reduced, but a significant number of errors still remain.

Two such alignment-based methods were proposed in Reference 15. The first of these methods is described in "KEC postprocessing using alignment" algorithm. In Reference 15, it is used as a postprocessing step for KEC algorithm described in the previous section. It corrects errors and performs *local haplotypes reconstruction*, that is, identify haplotypes and their relative frequencies.

The second algorithm ET (15) uses empirical frequency thresholds for indels and haplotypes obtained from a parameter calibration based on experimental sequencing results for artificially created single-clone samples with known clone sequences. An indel threshold is defined as the maximum frequency of erroneous indels in single-clone data sets, and a haplotype frequency threshold is estimated as the weighted average frequency of erroneous haplotypes + 9 standard deviations. These thresholds are platform-, virus-, and genomic region-specific. For instance, for 454/Roche GS FLX platform and HCV HVR1 region, the indel and haplotype thresholds were estimated as 5.9% and 0.4%, respectively (15). ET work flow consists of the following steps:

(1) *External references' alignment and lengths filtering.* Each unique read is aligned against a set of external references for the sequenced genomic region. For each read, the best reference of the length L is chosen, and reads with lengths less than $0.9L$ are deleted and reads with lengths greater than $1.L$ are clipped to the size of the corresponding reference.

(2) *Internal references' alignment.* Internal reference set is constructed by selecting 20 most frequent reads with minimal numbers of indels with regard to the corresponding external references. Each read is aligned against the internal reference set and the closest internal reference is chosen.

(3) *Homopolymer errors' correction.* Using alignments to internal references, homopolymers of at least 3 nt long are identified and indels in homopolymers are corrected. Then multiplicities of reads are recalculated.

(4) *PCR or sequencing artifacts' removal.* Using the internal reference, the number of indels in each read is calculated and reads containing at least T indels are discarded, where T is the weighted average of the number of indels in the data set + 4 standard deviations.

(5) *Indel threshold and Ns filtering.* Reads that contain indels with frequencies lower than the indel threshold are discarded. After that reads containing Ns are also discarded. The latter step is performed at one of the latest stages of the algorithm to take advantage of the nucleotide frequencies at positions other than those with N.

(6) *Haplotypes frequencies filtering.* Haplotypes with frequencies lower than the haplotype frequency threshold are removed.

Algorithm KEC Postprocessing Using Alignment (15)

Require: The set of unique reads $R = \{r_1, \ldots, r_n\}$ with multiplicities $m_1, \ldots, m_n$.

1: Set the count of iteration $nIter \leftarrow 1$

2: **repeat**

3:     Calculate the set of maximal reads $R_{max} \subseteq R$, i.e., reads that does not contain other reads as subsequences. Let

$$a_{i,j} \leftarrow \begin{cases} 1, & \text{if } i = j \text{ or the read } r_j \text{ is a subsequence of the read } r_i; \\ 0, & \text{otherwise;} \end{cases} \tag{15.16}$$

$$d_j \leftarrow \sum_{i=1}^{n} a_{i,j}. \tag{15.17}$$

4:     For every read $r_i \in R$, calculate its frequency f_i as $f_i = m_i / \sum_{k=1}^{|R|} m_k$. Recalculate frequencies of maximal reads r_j as follows: $f_j \leftarrow \sum_{k=1}^{|R|} a_{j,k} \frac{f_k}{d_k}$.

5:     **if** $nIter$ is odd **then**

6:         Calculate a multiple alignment of the maximal reads R_{max} and identify homopolymer regions in the alignment. If the position in a homopolymer region contains gap, calculate the total frequencies of reads with and without gap at this position, and, if one of them is significantly greater than the other, correct the position accordingly.

7:     **else**

8:         Calculate the neighbor joining tree T of the set R_{max} using pairwise alignment scores from the previous iteration. Let P be the set of pairs of reads, which have a common parent in the tree T. For every pair $(r_i, r_j) \in P$, calculate a pairwise alignment, identify homopolymer regions containing gaps, and correct those gaps to the more frequent variant if the edit distance between r_i and r_j is 1 or the frequency of one of the reads is significantly greater than the other.

9:     **end if**

10:   $R \leftarrow R_{max}$, $nIter \leftarrow nIter + 1$.

11: **until** no corrections are made or the number of iterations $nIter$ does not exceed the maximal allowed value

Ensure: haplotypes R and their frequencies $f_1, \ldots, f_{|(R)|}$

15.3 ALGORITHM COMPARISON

15.3.1 Benchmark Data

The data from Reference 15 were used for algorithms comparison.[1] For the data generation, a set of 10 plasmid clones C_1–C_{10} with sequences covering a 309 nt segment of the E1/E2 junction region of HCV genome was obtained (GenBank accession number AB047639.1) and 24 samples were created, with samples S_1–S_{14} containing a single clone and samples M_1–M_{10} containing mixtures of clones in

---

[1] Available at http://alan.cs.gsu.edu/NGS/?q=content/pyrosequencing-error-correction-algorithm

TABLE 15.2　Frequencies of 10 Clones in 24 Samples from 454/Roche GS FLX Experiment

	C_1	C_2	C_3	C_4	C_5	C_6	C_7	C_8	C_9	C_{10}
S_1	100	0	0	0	0	0	0	0	0	0
S_2	100	0	0	0	0	0	0	0	0	0
S_3	0	100	0	0	0	0	0	0	0	0
S_4	0	0	100	0	0	0	0	0	0	0
S_5	0	0	0	100	0	0	0	0	0	0
S_6	0	0	0	0	100	0	0	0	0	0
S_7	0	0	0	0	0	100	0	0	0	0
S_8	0	0	0	0	0	0	100	0	0	0
S_9	0	0	0	0	0	0	0	100	0	0
S_{10}	0	0	0	0	0	0	0	0	100	0
S_{11}	0	0	0	0	0	0	0	0	0	100
S_{12}	0	0	0	0	0	0	0	0	0	100
S_{13}	0	0	0	0	0	0	0	0	100	0
S_{14}	0	0	0	0	0	0	0	100	0	0
M_1	2.9	2.9	2.9	2.9	2.9	2.9	2.9	80.0	0	0
M_2	12.5	12.5	12.5	12.5	12.5	12.5	12.5	12.5	0	0
M_3	10.0	10.0	10.0	10.0	10.0	10.0	10.0	30.0	0	0
M_4	10.0	10.0	10.0	10.0	10.0	10.0	20.0	20.0	0	0
M_5	1.0	1.0	1.0	1.0	1.0	1.0	1.0	93.0	0	0
M_6	0	8.7	8.7	8.7	8.7	21.7	21.7	21.7	0	0
M_7	5.0	5.0	5.0	5.0	5.0	5.0	20.0	50.0	0	0
M_8	3.1	3.1	3.1	3.1	3.1	0.0	42.3	42.3	0	0
M_9	80.0	2.9	2.9	2.9	2.9	2.9	2.9	2.9	0	0
M_{10}	2.0	2.0	2.0	2.0	0.0	30.6	30.6	30.6	0	0

different proportions (Table 15.2). The samples were sequenced using 454/Roche GS FLX platform. The entire region was sequenced as a single amplicon, which allowed to compare algorithms' performances using the measures of a qualities of haplotypes reconstruction and their frequency estimations.

15.3.2　Results and Discussion

We compared performances of clustering-based algorithms SHORAH and kGEM, alignment-based algorithm ET, and k-mers-based algorithm KEC with the alignment-based postprocessing step described in "KEC postprocessing using alignment" algorithm. Several modifications of SHORAH clustering hyper-parameter α were tested, and the value $\alpha = 0.1$, which produced the best result, was chosen. For kGEM, the initial value $k = 1000$ was used. For KEC, the k-mers size $k = 25$ was used, and an error threshold was estimated as the end of the first segment of k consecutive zeroes in a k-counts distribution. The algorithms' performances were measured using four parameters: a number of missing true haplo-types (sensitivity), a number of false haplotypes (specificity), root mean square error of frequency estimations of haplotypes, and an average Hamming distance between

TABLE 15.3 Algorithms Comparison for Single-Clone (S) and Mixture (M) samples

	ET				KEC				SHORAH				kGEM			
	MT	FS	RMSE	HD	MT	FS	RMSE	HD	MT	FS	RMSE	HD	MT	FS	RMSE	HD
S1	0	0	0.00	0.00	0	2	1.64	2.50	0	351	29.02	4.83	0	1	17.66	1.00
S2	0	0	0.00	0.00	0	0	0.00	0.00	0	269	30.12	4.44	0	1	17.83	1.00
S3	0	1	1.04	1.00	0	0	0.00	0.00	0	292	23.44	5.31	0	2	15.01	1.50
S4	0	1	0.96	2.00	0	0	0.00	0.00	0	271	44.68	5.37	0	0	0.00	0.00
S5	0	0	0.00	0.00	0	0	0.00	0.00	0	319	9.63	4.47	0	0	0.00	0.00
S6	0	1	0.70	2.00	0	0	0.00	0.00	0	194	18.70	3.90	0	2	11.58	1.00
S7	0	0	0.00	0.00	0	0	0.00	0.00	0	496	21.52	6.70	0	0	0.00	0.00
S8	0	0	0.00	0.00	0	0	0.00	0.00	0	262	14.37	4.58	0	1	4.10	1.00
S9	0	0	0.00	0.00	0	0	0.00	0.00	0	183	6.23	6.97	0	1	1.50	2.00
S10	0	0	0.00	0.00	0	0	0.00	0.00	0	288	7.77	5.11	0	0	0.00	0.00
S11	0	0	0.00	0.00	0	0	0.00	0.00	0	717	24.71	5.03	0	1	5.71	1.00
S12	0	1	0.65	2.00	0	0	0.00	0.00	0	611	25.94	5.52	0	1	5.82	2.00
S13	0	0	0.00	0.00	0	0	0.00	0.00	0	156	5.53	4.93	0	0	0.00	0.00
S14	0	0	0.00	0.00	0	0	0.00	0.00	0	161	6.83	6.60	0	2	4.57	2.50
M1	0	0	1.26	0.00	0	0	0.78	0.00	0	320	1.23	4.51	0	1	8.33	1.00
M2	0	0	1.50	0.00	0	0	1.95	0.00	0	738	3.70	4.44	0	1	2.25	1.00
M3	0	0	2.87	0.00	0	0	4.22	0.00	0	638	3.65	4.25	0	1	4.35	1.00
M4	0	0	2.12	0.00	0	0	3.09	0.00	0	577	2.88	5.20	0	1	3.61	1.00
M5	0	0	0.29	0.00	7	0	7.00	0.00	0	214	0.91	7.37	0	4	6.70	1.75
M6	0	0	2.45	0.00	0	0	1.81	0.00	0	394	3.25	4.50	0	1	4.88	1.00
M7	0	0	1.04	0.00	0	0	2.42	0.00	0	499	2.04	5.00	0	2	7.03	1.00
M8	0	0	0.40	0.00	0	0	2.25	0.00	0	336	3.30	5.48	0	1	6.68	1.00
M9	0	0	2.40	0.00	0	0	1.53	0.00	0	643	6.56	4.49	0	3	3.87	2.33
M10	0	0	3.70	0.00	0	0	4.16	0.00	1	637	6.13	5.30	0	1	5.23	1.00

MT, Missing true haplotypes; FS, false haplotypes; RMSE, root mean square error of frequency estimations of haplotypes; HD, average Hamming distance to the closest true haplotype, averaged over all false haplotypes.

calculated false haplotypes and closest true haplotypes. The algorithms' results for all 24 data sets are shown in Table 15.3, and the results averaged over single-clone samples and mixtures are shown in Table 15.4.

ET was successful in detection of correct haplotypes in all 10 mixtures and in 10 of 14 single-clone samples, but failed to discriminate one false haplotype in four single-clone samples. KEC was correct for 13 of 14 single-clone samples (with 2 false haplotypes not being removed in one sample) and for 9 of 10 mixtures (with the exception of the mixture M5, where it failed to detect low-frequency variants). kGEM detected all true haplotypes in all samples, but failed to discriminate some false haplotypes in 9 of 14 single-clone samples and in all 10 mixtures. SHORAH found all correct haplotypes in all single-clone samples and in 9 of 10 mixtures, producing a large number of false haplotypes and a significant divergence of expected and estimated frequencies of haplotypes.

Thus, all four algorithms successfully found true haplotypes in all single-clone samples and detected most true haplotypes in mixtures, with ET and kGEM being the most sensitive and KEC being the least sensitive. The specificity of SHORAH is lower than of three other algorithms, with KEC being the most specific followed by ET and kGEM. The root mean square error for mixtures is comparable for all four algorithms, with ET being the most accurate followed by KEC, SHORAH and kGEM,

TABLE 15.4   Algorithms Performance on Average

	MT	FS	RMSE	HD
Single-clone samples				
ET	0	0.29	0.24	0.5
KEC	0	0.14	0.12	0.178571
SHORAH	0	326.43	19.18	5.267445
kGEM	0	0.86	5.98	0.928571
Mixture samples				
ET	0	0.00	1.80	0
KEC	0.7	0.00	2.92	0
SHORAH	0.1	499.60	3.37	5.053398
kGEM	0	1.60	5.29	1.208333

MT, Missing true haplotypes; FS, false haplotypes; RMSE, root mean square error of frequency estimations of haplotypes; HD, average Hamming distance to the closest true haplotype, averaged over all false haplotypes.

respectively. The accuracy of frequencies' estimation on single-clone samples was greater for ET and KEC than for kGEM. However, the accuracy of kGEM is higher than of SHORAH. Analysis of Hamming distances between false haplotypes and their closest true neighbors shows that false haplotypes retained by KEC, ET, and kGEM are genetically closer to true haplotypes than the ones obtained by SHORAH.

Thus, all four algorithms are comparable in sensitivity, but showed marked differences in the specificity and accuracy of frequencies' estimation. It should be emphasized that ET is based on experimentally calibrated parameters for the particular genomic region under consideration, while KEC, SHORAH, and kGEM are independent of such estimations and are more universally applicable. However, a solid performance of ET indicates that empirical evaluation of the error profiles for particular amplicons planned to be used in many studies is highly beneficial.

Clustering algorithms assign observed reads to unobserved haplotypes under indirect assumption that reads represent a statistical sample of the underlying viral population in the presence of uniformly distributed sequencing errors, while KEC and ET specifically take into account the presence of systematic homopolymer errors, which have a high prevalence and are not randomly distributed (15). Therefore, for pyrosequencing data, KEC and ET may have a certain advantage over more general clustering algorithms. For a different sequencing platform, such as PacBio or Illumina MiSeq and HiSeq, performance of algorithms may differ. In particular, it can be expected that clustering algorithms are most useful for cleaning Illumina data, which have a low-indel error rate and a more uniform error distribution.

REFERENCES

1. Hajarizadeh B, Grebely J, Dore GJ. Epidemiology and natural history of HCV infection. World J Gastroenterol Nat Rev Gastroenterol Hepatol 2013;10(9):553–562.
2. Kilmarx PH. Global epidemiology of HIV. Curr Opin HIV AIDS 2009;4(4):240–246.

3. Sanjuan R, Nebot MR, Chirico N, Mansky LM, Belshaw R. Viral mutation rates. J Virol 2010;84(19):9733–9748.

4. Domingo E. Biological significance of viral quasispecies. Viral Hepatitis Rev 1996;2:247–261.

5. Duarte EA, Novella IS, Weaver SC, Domingo E, Wain-Hobson S, Clarke DK, Moya A, Elena SF, de la Torre JC, Holland JJ. RNA Virus Quasispecies: significance for viral disease and epidemiology. Infect Agents Dis 1994;3(4):201–214.

6. Archer J et al. Analysis of high-depth sequence data for studying viral diversity: a comparison of next generation sequencing platforms using Segminator II. BMC Bioinformatics 2012;13:47.

7. Huse SM, Huber JA, Morrison HG, Sogin ML, Welch DM. Accuracy and quality of massively parallel DNA pyrosequencing. Genome Biol 2007;8(7):R143.

8. Gilles A, Meglecz E, Pech N, Ferreira S, Malausa T, Martin J. Accuracy and quality assessment of 454 GS-FLX Titanium pyrosequencing. BMC Genomics 2011;19(12):245.

9. Zagordi O, Daumer M, Beisel C, Beerenwinkel N. Read length versus depth of coverage for viral quasispecies reconstruction. PLoS ONE 2012;7(10):e47046.

10. McElroy KE, Luciani F, Thomas T. GemSIM: general, error-model based simulator of next-generation sequencing data. BMC Genomics 2012;13:74.

11. Harismendy O et al. Evaluation of next generation sequencing platforms for population targeted sequencing studies. Genome Biol 2009;10(3):R32.

12. Quail MA et al. A tale of three next generation sequencing platforms: comparison of Ion Torrent, Pacific Biosciences and Illumina MiSeq sequencers. BMC Genomics 2012;13:341.

13. Bragg LM, Stone G, Butler MK, Hugenholtz P, Tyson GW. Shining a light on dark sequencing: characterising errors in ion torrent PGM data. PLoS Comput Biol 2013;9(4):e1003031.

14. Mardis ER. Next-generation DNA sequencing methods. Annu Rev Genomics Hum Genet 2008;9:387–402.

15. Skums P, Dimitrova Z, Campo D, Vaughan G, Rossi L, Forbi J, Yokosawa J, Zelikovsky A, Khudyakov Y. Efficient error correction for next-generation sequencing of viral amplicons. BMC Bioinformatics 2012;13 Suppl 10:S6.

16. Dohm JC, Lottaz C, Borodina T, Himmelbauer H. Substantial biases in ultra-short read data sets from high-throughput DNA sequencing. Nucleic Acids Res 2008;36(16):e105.

17. Campo D, Skums P, Dimitrova D, Vaughan G, Forbi J, Teo C-G, Khudyakov Y, Lau D. Drug-resistance of a viral population and its individual intra-host variants during the first 48 hours of therapy. Clin Pharmacol Ther 2014;95(6):627–635.

18. Luciani F, Bull RA, Lloyd AR. Next generation deep sequencing and vaccine design: today and tomorrow. Trends Biotechnol 2012;30(9):443–452.

19. Luciani F, Alizon S. The evolutionary dynamics of a rapidly mutating virus within and between hosts: the case of hepatitis C virus. PLoS Comput Biol 2009;5(11):e1000565.

20. Tazi L, Imamichi H, Hirschfeld S, Metcalf JA, Orsega S, Perez-Losada M, Posada D, Lane HC, Crandall KA. HIV-1 infected monozygotic twins: a tale of two outcomes. BMC Evol Biol 2011;11:62.

21. Westby M, Lewis M, Whitcomb J, Youle M, Pozniak AL, James IT, Jenkins TM, Perros M, van der Ryst E. Emergence of CXCR4-using human immunodeficiency virus type 1 (HIV-1) variants in a minority of HIV-1- infected patients following treatment with the CCR5 antagonist maraviroc is from a pretreatment CXCR4-using virus reservoir. J Virol 2006;80(10):4909–4920.

22. Skums P, Campo D, Dimitrova Z, Vaughan G, Lau D, Khudyakov Y. Numerical detection, measuring and analysis of differential interferon resistance for individual HCV intra-host variants and its influence on the therapy response. In Silico Biol 2011-2012;11(5-6):263–269.

23. Zagordi O, Klein R, Däumer M, Beerenwinkel N. Error correction of next generation sequencing data and reliable estimation of HIV quasispecies. Nucleic Acids Res 2010;38(21):7400–7409.

24. Zagordi O, Geyrhofer L, Roth V, Beerenwinkel N. Deep sequencing of a genetically heterogeneous sample: local haplotype reconstruction and read error correction. J Comput Biol 2010;17(3):417–428.

25. Zagordi O, Bhattacharya A, Eriksson N, Beerenwinkel N. ShoRAH: estimating the genetic diversity of a mixed sample from next-generation sequencing data. BMC Bioinformatics 2011;26(12):119.

26. Quince C, Lanzen A, Curtis T, Davenport R, Hall N, Head I, Read L, Sloan W. Accurate determination of microbial diversity from 454 pyrosequencing data. Nat Methods 2009;6(9):639–641.

27. Quince C, Lanzen A, Davenport RJ, Turnbaugh PJ. Removing noise from pyrosequenced amplicons. BMC Bioinformatics 2011;28(12):38.

28. Eriksson N, Pachter L, Mitsuya Y, Rhee SY, Wang C, Gharizadeh B, Ronaghi M, Shafer RW, Beerenwinkel N. Viral population estimation using pyrosequencing. PLoS Comput Biol 2008;4(4):e1000074.

29. Artyomenko A et al. kGEM: An EM-based algorithm for local reconstruction of viral quasispecies. Computational Advances in Bio and Medical Sciences (ICCABS), IEEE 3rd International Conference; 2013.

30. Salmela L, Schroder J. Correcting errors in short reads by multiple alignments. Bioinformatics 2011;27(11):1455–1461.

31. Pevzner P, Tang H, Waterman M. An Eulerian path approach to DNA fragment assembly. Proc Natl Acad Sci U S A. 2001;98(17):9748–9753.

32. Chaisson M, Pevzner P. Short read fragment assembly of bacterial genomes. Genome Res 2008;18(2):324–330.

33. Chaisson M, Brinza D, Pevzner P. De novo fragment assembly with short mate-paired reads: does the read length matter? Genome Res 2009;19(2):336–346.

34. Zhao X, Palmer L, Bolanos R, Mircean C, Fasulo D, Wittenberg D. EDAR: an efficient error detection and removal algorithm for next generation sequencing data. J Comput Biol 2010;17(11):1549–1560.

35. Skums P, Artyomenko A, Glebova O, Ramachandran S, Mandoiu I, Campo D, Dimitrova Z, Zelikovsky A, Khudyakov Y. Computational framework for next-generation sequencing of heterogeneous viral populations using combinatorial pooling. Bioinformatics. 2014. DOI: 10.1093/bioinformatics/btu726. [Epub ahead of print].

36. Yang X, Dorman K, Aluru S. Reptile: representative tiling for short read error correction. Bioinformatics 2010;26(20):2526–2533.

37. Chaisson M, Pevzner PA, Tang H. Fragment assembly with short reads. Bioinformatics 2004;20(13):2067–2074.

38. Butler J, Maccallum I, Kleber M et al. ALLPATHS: de novo assembly of whole-genome shotgun microreads. Genome Res 2008;18(5):810–820.

39. Comaniciu D, Ramesh V, Meer P. The variable bandwidth mean shift and data-driven scale selection. Proceedings of the 8th International Conference on Computer Vision; 2001. p 438–445.

40. Comaniciu D, Meer P. Mean shift: a robust approach toward feature space analysis. IEEE Trans Pattern Anal Mach Intell 2002;24:603–619.

41. Georgescu B, Shimshoni I, Meer P. Mean shift based clustering in high dimensions: a texture classification example. Proceedings of the 9th International Conference on Computer Vision; 2003, p 456-463.

42. Mathworks: Matlab. Natick, MA; 2010.

16

PROBABILISTIC VIRAL QUASISPECIES ASSEMBLY

ARMIN TÖPFER AND NIKO BEERENWINKEL

Department of Biosystems Science and Engineering, ETH Zurich, Basel, Switzerland

16.1 INTRA-HOST VIRUS POPULATIONS

Viruses are pathogens that cause infectious diseases (1). A single virus particle, called virion, exhibits its genetic material as a single- or double-stranded linear or circular DNA or RNA molecule in one or multiple copies. A virion alone cannot reproduce itself as cells do. For reproduction, a virus depends on infecting cells of an organism. RNA viruses are of particular interest, as their viral replication enzymes have a low fidelity and produce error-prone copies of the genome, including single point mutations and recombination. This high error rate, among other factors, limits the upper bound of the genome size in the range of 10^3–10^6 base pairs (bp), which is much smaller than bacterial or eukaryotic genomes. Above a certain error rate threshold, mutational meltdown would lead to a virus population that is not able to preserve the genotype with the highest fitness, where fitness is defined as the reproductive capability. Thus, viral error rate and genome size are anticorrelated (2). A viral genome is protected by a protein coat and many viruses are protected by an additional envelope. Receptors on the outside surface are used to attach to and fuse with host cells. The viral life cycle, from infection of a host cell to the release of new virions, is very short and is completed in a few days time. This leads to a vast amount of virions in each infected host. The swarm of virions evolves through mutation and natural selection. It is subject to the host's immune pressure and possibly antiviral therapy. Virions may escape this selective pressure and gain selective advantage by acquiring one or more

Computational Methods for Next Generation Sequencing Data Analysis, First Edition.
Edited by Ion I. Măndoiu and Alexander Zelikovsky.
© 2016 John Wiley & Sons, Inc. Published 2016 by John Wiley & Sons, Inc.
Companion website: www.wiley.com/go/Mandoiu/NextGenerationSequencing

of the following genomic alterations: (i) single-nucleotide variants (SNVs), (ii) loss or gain of one or more amino acids, (iii) large deletions, for example, due to alternative splicing, or (iv) recombination of different strains.

16.1.1 Viral Quasispecies

A single infected host contains millions of virions. This vast amount of virions can be summarized as a set of unique genetically different variants, called haplotypes, together with their respective frequency. Each haplotype is identified by its genome, that is, its pattern of SNVs and structural variations. This distribution of haplotypes is referred to as a quasispecies (Figure 16.1).

Manfred Eigen defined the quasispecies equation as a mathematical model for RNA virus populations that evolve under a mutation–selection process (3). In this

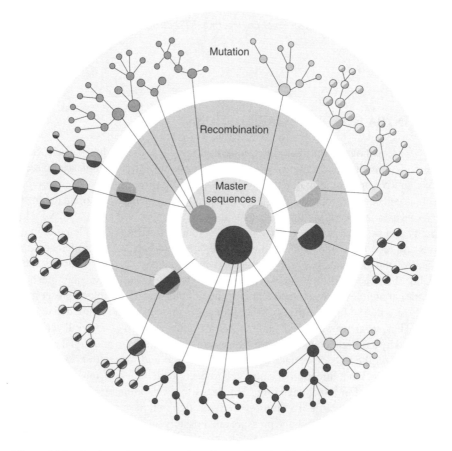

Figure 16.1 A schematic representation of a quasispecies. Haplotypes are represented as dots and the corresponding frequency as size. The quasispecies emerges from a small set of master sequences, inner circle, by recombination, middle ring, and mutation, outer ring. Edges between dots represent the closest distance between haplotypes. Please see www.wiley.com/go/Mandoiu/NextGenerationSequencing for a color version of this figure.

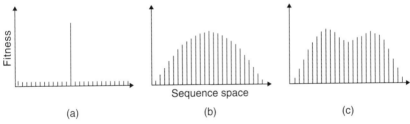

Figure 16.2 For different scenarios, schematic fitness landscapes are visualized. For each scenario, the fitness, reproductive capability, is shown as a function of the discrete sequence space. Under the assumption that only one haplotype has a high fitness (a), mutant variants reproduce with a negligible capability or not at all. In the survival of the flattest scenario, for example, with one (b) or two master sequences (c), mutant variants have only a slightly reduced fitness.

model, mutation can lead to transformation of one viral haplotype into another at the time of replication. Mutations determine strain replication rates that give rise to selection. The mutation process is considered to consist of point mutations only. However, many clinically relevant viruses, including human immunodeficiency virus type 1 (HIV-1) and hepatitis C virus (HCV), also undergo recombination with high frequency. Therefore, the quasispecies model has been extended to account for both mutation and recombination (4). It is assumed that at equilibrium, swarms of low-frequency mutant haplotypes, subject to mutation and recombination, are maintained by few master sequences that have the highest fitness (5, 6). A master sequence can be thought of as the center of one mutant cloud or the mode of the haplotype distribution in high-dimensional sequence space.

16.1.2 Fitness

If a virus has a high error rate, it follows an evolutionarily stable strategy (7) to generate a population of haplotypes with approximately equal fitness instead of one haplotype being the fittest (Figure 16.2). Occupation of such a broad fitness landscape is called survival of the flattest (8). It is a survival strategy. The evolutionary trajectory of the viral infection depends not only on the characteristics of the fittest but also on the whole population.

Natural selection acts on a population of haplotypes, and small variations within this population may improve the chances of survival in changing environments. For example, a low-frequency haplotype that exhibits an appropriate resistance mutation could gain an advantage under treatment. Then, the mass of the quasispecies shifts to the new master sequence and a new haplotype distribution is created. This survival strategy improves chances of adaption to environmental changes, such as treatment switches.

16.1.3 HIV-1 as a Model System

As in many fields of research, fundamental genetic research is performed on model systems, such as *Arabidopsis thaliana* for plants or *Escherichia coli* for bacteria.

In viral quasispecies studies, HIV-1 is a representative viral model. HIV-1 is the cause of a deadly pandemic, the acquired immunodeficiency syndrome (AIDS), and is as such a major focus of viral research. There are 34 million people infected with HIV-1 worldwide. HIV-1 is composed of a small diploid single-strand RNA (ssRNA) genome with ~9.7 kilobases (kb) that evolves fast by point mutation and recombination, leading to a high intrapatient diversity that varies from a single to thousands of haplotypes (9). The genome is divided into three parts, the gag, pol, and env regions (Figure 16.3). The gag region contains genes for the capsid and matrix proteins. The pol region contains genes for the protease, reverse transcriptase, RNase, and integrase. The env region contains genes for the receptor proteins. The genome is surrounded by a shell of capsid proteins, a matrix, and an envelope. Embedded in the envelope, the protein env protrudes the surface. Env consists of a glycoprotein (gp) 120 cap and a gp41 stem. These surface proteins enable the virus to attach to and fuse with host target cells to initiate the infectious cycle.

HIV-1 maintains a high genetic intrapatient diversity that significantly differs from the interpatient diversity (10). High diversity is due to two properties of HIV-1, a high reproductive capability of 10^{10}–10^{12} virions per day and a low-fidelity reverse transcriptase producing 2.5×10^{-5} substitutions per base and reverse transcription (RT) (11). There are at least 10^7–10^8 HIV-1 infected cells per host (12). The effective population size is hard to estimate and has been reviewed in Reference 13. HIV-1 is able to generate every possible mutation along the genome per day. Most of these mutants are not maintained because many of the mutations result in low-fitness or even nonreproductive virions.

Research in the last 30 years has identified the genetic sequences and, more in particular, the positions where drug resistance causing mutations of different HIV-1 sequences are located. Very effective antiretroviral drugs have been developed to target the different steps in the viral life cycle, but as every virus adapts to its host, interpatient diversity is high. Every patient may show different patterns of resistance mutations; therefore, treatment has to be personalized. In this case, genotyping, the identification of resistance mutations, is necessary to tailor the best possible regimen, that is, drug combination, for an individual patient (14, 15).

16.1.4   Recombination

The high mutation rate of HIV-1 is only one of two factors that advances intrapatient diversity. During RT of the RNA genome, the reverse transcriptase is known to switch strands (Figure 16.4). In double infected patients, this can lead to new recombinant DNA molecules, randomized products of existing haplotypes. The rate of recombination events of HIV-1 is assumed to be 10-fold higher, approximately 3×10^{-4} per base and RT, and consequently plays an important role in shaping the quasispecies (16, 17),

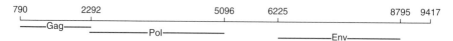

Figure 16.3   Polyprotein coding region of HIV-1 with reference positions of strain HXB2. The three functional regions gag, pol, and env are shown in their respective reading frames.

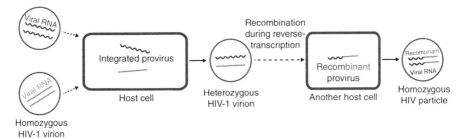

Figure 16.4   Two heterozygous virions of distinct subtypes infect the same host cell and as a subsequent event, virions with recombinant genomes are produces.

but possibly not to disease progression (18). Under the assumed recombination rate, every time a virion infects a host cell there is a 95% chance that recombination occurs at least once.

In the majority of these strand-switching events, there is no consequence since the genome is homozygous and exhibits two clonal copies of the same ssRNA molecule. On the other hand, in a patient infected with many haplotypes, recombination could lead to a much greater genetic diversity. In this case, it may happen that a single host cell is co- or super-infected by two homozygous virions from different HIV-1 sub-types. Both genomes are reverse transcribed and integrated in the host DNA. As a subsequent event in the viral life cycle, new heterozygous virions can be produced that contain both ssRNA molecules from two subtypes. There is physical evidence that virions can contain different strains, proven by simultaneously labeling RNA molecules (19). When such a heterozygous virion infects another host cell, the reverse transcriptase may create a recombinant DNA molecule (Figure 16.4). The effective recombination rate is much lower, as coinfection is required to generate diversity (20), and is found to be 1.4×10^{-4} per base and RT (21). With this process, the virus can create recombinant haplotypes that comprise new mosaics of existing SNV patterns that have an effect on the evolution of drug resistances (22) and intrahost viral diversity (23). Especially in the early stage of an HIV-1 infection, recombination accelerates the acquisition of immune escape mutations from cytotoxic T lymphocytes (24).

16.1.5   Clinical Implications

The quasispecies structure is of clinical importance; thus, viral haplotype sequences and frequency distribution have to be identified. It has been shown to affect virulence (25), as low-frequency genetic variants may harbor resistance mutations that are capable of evolutionary escape from the selective pressure of host immune responses (26) and of medical interventions, such as anti-viral drug treatment (27). In addition, it affects pathogenesis (28); for example, a key element of pathogenesis is a co-receptor switch to escape the immune system (29). Knowing the intrapatient quasispecies structure, one could shed light on critical transmitter–recipient cases (30). In order to determine the quasispecies structure, genotyping is performed.

16.1.6 Genotyping

Genotypic antiretroviral drug resistance testing is performed via sequencing. It has revolutionized the care of HIV-1 infected patients because treatment can be personalized to already existing intrapatient drug-resistance mutations. The sequences for drug-resistance testing have mainly been produced by conventional Sanger bulk sequencing. In contrast to single-cell genomics, viral genomes cannot yet be sequenced individually. Assessing the quasispecies structure from a mixed sample is currently done by SNV calling. This *de facto* standard procedure ignores patterns of co-occurrence among mutations. Even though epistatic interactions are abundant in RNA viruses (31), it is not possible to predict these solely on SNVs. Phasing of haplotypes, the identification of the individual haplotype sequences, is in practice not possible from the traces of Sanger sequencing. Without knowing the underlying mix of haplotypes, viral phenotype prediction is not possible.

Nevertheless, single haplotype sequencing is possible with Sanger sequencing being a labor- and time-intensive approach. Single clones can be isolated by limiting dilution, so that on average only a single molecule is present in the solution. As this approach is very laborious, it has not been adopted in a high-throughput manner. Nevertheless, it has been used to find different haplotypes from patient samples and is nowadays also used to validate results from newer technologies.

16.2 NEXT-GENERATION SEQUENCING FOR VIRAL GENOMICS

Next-generation sequencing (NGS) technologies revolutionized assessing viral genetic diversity experimentally. In a single experiment, the virus population is sampled as short fragments in a highly parallel manner. Employing mathematical models, individual haplotypes can be reconstructed computationally and each read, a sequenced DNA fragment, can be assigned to an individual haplotype. NGS enables ultradeep sequencing of viral populations, that is, every position of the genome is covered with tens of thousands of sequenced fragments. Even though the advantages of NGS outweigh, the experimental approach and the sequencing technology have several pitfalls that lead to difficulties in the statistical analysis of the data (27).

16.2.1 Library Preparation

Prior to sequencing, the plasma sample of an infected individual has to be prepared. The quality of the sample and its preparation is crucial, as accumulated errors are reflected in the reads. If the goal is to assemble the quasispecies, often an amplicon approach is chosen. The preparation includes (i) design of RT-PCR and PCR amplicon primers, (ii) elimination of contaminating DNA or RNA from other cells, (iii) extraction of viral RNA or DNA, (iv) RT of RNA into cDNA, (v) amplification by polymerase chain reaction (PCR), and (vi) ligation of platform specific primers.

Amplicon layout is important, as the amplicon primer choice has a major impact on haplotype assembly and frequency estimation. In practice, these primers are designed to bind to more homogeneous regions. If haplotypes exhibit diversity in the primer binding regions, primers favor to bind to DNA fragments that match better. This bias, amplified during PCR, influences frequency estimation and may lead to large deviations from the true haplotype frequencies. PCR with random primers does not have the bias towards specific strains, but mainly leads to amplification of nonspecific target cDNA. These data sets can only be used to generate a consensus sequence of the quasispecies because the sample does not contain enough target material to reconstruct individual haplotypes.

The RT-PCR step is critical, as it can introduce artificial recombinant cDNAs. This effect needs to be minimized by using special lab protocols. Standard RT-PCR conditions generate 30.6–37.1% recombinants cDNAs, which can be decreased with optimized RT-PCR conditions to 0.9–2.6% (32). Artificial recombinants cannot be distinguished from true biological recombinants and lead to an underestimate of the haplotype frequencies and overestimate of the quasispecies diversity.

16.2.2   Sequencing Approaches

Since the introduction of the first commercial NGS system in 2004, when pyrosequencing was introduced by 454/Roche, the landscape of different sequencing technologies and their performances has changed drastically (Table 16.1). It was the standard platform for sequencing in the last 10 years. In this method, complementary strands of single-stranded templates are produced in an iterative manner. In each round, nucleotides of only one out of the four possible nucleotides, adenine, guanine, cytosine, and thymine, are added. If a tagged nucleotide is incorporated, emitted light is detected by the system. Nowadays the FLX system of 454/Roche produces reads up to 1 kb, but the sample preparation is very laborious. Systematic errors in homopolymeric regions are highly abundant with an error rate of 10%, 30%, and 50% for homopolymers of size 6, 8, and 10 bp, respectively. Another sequencing platform is Ion Torrent semiconductor sequencing, which detects hydrogen ions that are released as the complementary strand of the template is polymerized. Its strengths are speed and costs, but the read length is only up to 400 bp with a maximal throughput of 5.5 million reads that exhibits similar homopolymer errors as 454/Roche. Sequencing by ligation (SOLiD) produces 1.2–1.4 billion very short reads of 50 bp that do not exhibit homopolymeric, but problems with palindromic sequences. A single run takes 1–2 weeks. In the last years, the majority of sequences have been produces with Illumina sequencing machines, because library preparation is much

TABLE 16.1   Comparison of Next- and Third-Generation Sequencing Technologies (33, 34)

Technology	Read Length (bp)	Throughput (Reads)	Run-Time	Most Abundant Errors
454/Roche	1,000	1 million	1 day	Homopolymeric errors
Ion Torrent	400	80 million	2 hours	Homopolymeric errors
SOLiD	50	1.2–1.4 billion	1–2 weeks	Palindromic sequences
Illumina MiSeq	2×300	44–50 million	3 days	Substitutions
PacBio	20,000	55,000	3 hours	Higher indel rate

easier and throughput is much higher. Illumina performs sequencing by synthesis for paired-end reads and provides two major platforms, the MiSeq and HiSeq. HiSeq produces up to 6 billion 2×100 bp reads in approximately 11 days and is used for human genome sequencing. The MiSeq platform is more likely used for viral sequencing, as it produces 44–50 million 2×300 bp reads in 3 days. The latest technology, classified as third-generation sequencing, is single-molecule real-time sequencing by Pacific Biosciences (PacBio). Single molecules can be sequenced with an average read length of 20 kb, but the throughput is only 55,000 reads with higher technical error rates of ~0.3%, ~8%, and ~5% per base for substitutions, insertions, and deletions, respectively.

In viral quasispecies analysis, there are two main goals: (i) the identification of low-frequency variants and (ii) haplotype assembly on a whole-genome scale. The two sequencing platforms, Illumina MiSeq and Pacific Biosciences, have the potential to provide data to solve these questions. Even though 454/Roche can be used for sequencing of viral samples, it will not be further discussed, as its development and sale have been discontinued. The two approaches and their performances are discussed in more detail in the following.

For the Illumina platform, amplicons are randomly fragmented in smaller molecules prior to sequencing. Ligation of Illumina and strand-specific adapters is performed to enable cluster generation. Cluster generation, the amplification of each DNA molecule by PCR on the chip, is performed to increase the signals produced during sequencing. Each strand is sequenced from the $3'$-end (Figure 16.5). With this procedure, Illumina generates paired-end reads, which stem from the same DNA fragment. Depending on the fragment size, the forward and backward reads can overlap or include an unsequenced insert fragment. Even though the single reads are only 300 bp long and are often too short to span homogeneous regions, the additional paired-end information can help. If the insert fragment is large enough, the paired reads can flank homogeneous regions and allow linkage of distant variant sites. In practice, a flat fragment size distribution between 500 and 1500 bp is sought. With a fragment size of 500 bp, overlapping paired-end reads can be merged and used for subsequent analysis. With a three times longer fragment size of 1500 bp, distant co-occurring SNVs can be linked. Another main advantage of Illumina is the low-accumulated substitution rate of approximately 0.3% and the low indel rate of 10^{-6} per base. These rates can be measured by sequencing a single strain, performing pair-wise alignment to the known reference genome, and counting the errors.

PacBio performs single-molecule sequencing. It is the first commercially available platform of its kind that is able to sequence single molecules without amplification. Even though the length of the sequenced molecules outperforms 454/Roche by a factor of at least 20, PacBio currently has several limitations. The PacBio RS machine needs more raw cDNA material for sequencing. This is a major disadvantage in the laboratory preparation step, as many nested PCR runs and length purifications have to be performed. This is critical, as every PCR cycle can introduce additional substitutions, indels, and artificial recombinations. On the other hand, Illumina performs this step on the instrument and customers cannot control or optimize this step. In addition, the throughput is 55,000 reads, limited by the chip design. PacBio's major drawback

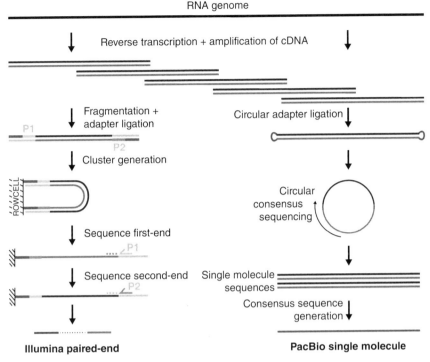

RNA genome

Reverse transcription + amplification of cDNA

Fragmentation + adapter ligation

Circular adapter ligation

P1

P2

Cluster generation

FLOWCELL

Circular consensus sequencing

Sequence first-end

P1

Sequence second-end

P2

Single molecule sequences

Consensus sequence generation

Illumina paired-end

PacBio single molecule

Figure 16.5 Sequencing workflow from the purified RNA genome to the final reads. The workflow is visualized for Illumina paired-end and PacBio single-molecule sequencing.

is a higher insertion and deletion rate. To overcome this limitation and reduce the error rate, PacBio introduced circular consensus sequencing (ccs) (Figure 16.5). A short DNA molecule is circularized and then sequenced multiple times. The long read contains the single-molecule sequences that only differ in sequencing errors as they stem from the same molecule. The single-molecule sequences can be extracted by identifying and cutting the circular adapters. In a postprocessing step, an algorithm integrates over all alignments of the single-molecule sequences and generates a circular consensus sequence. With this approach, PacBio produces reads of much higher accuracy, as their errors are truly random, compared to all other technologies.

16.2.3 Specialized Viral Sequencing Methods

Specialized laboratory methods, complementary to the computational approaches, have been developed to perform error correction, frequency estimation, and low-frequency variant calling (35–39). Each cDNA template is tagged with a unique primer-ID in the RT reaction, also called barcoding. After PCR amplification, in theory every fragment should be amplified in the same order of magnitude. The sequenced templates can be clustered by primer-ID, and the cluster consensus sequence corresponds to the error-corrected sequence. In practice, a high resampling of a small number of templates can be observed that influences the frequency estimation and leads to wrong frequency estimates. The reasons for this are still yet to be determined.

Another orthogonal approach is the circularization of ssDNA templates and rolling circle amplifications to create long fragments that contain multiple copies of the same template. The individual copies of the same template on each read only differ in PCR and sequencing errors that can be corrected by calling the consensus sequence. This method can lower the detection threshold of low-frequency variant calling by several orders of magnitude (40).

16.2.4 Data Preprocessing and Read Alignment

Prior to viral haplotype reconstruction, reads have to be filtered and aligned. The first step of data processing is the trimming of platform-dependent adapters and amplicon primers. For PacBio reads, this step is not trivial, as most of the primers are degenerated due to the high error rate and partially clipped by consensus sequence generation. The next step involves quality clipping of the reads from both ends, as quality drops at the read ends. The viral quasispecies reconstruction approaches described in the following rely on a high-quality alignment of the reads. This alignment can be produced with standard alignment methods such as BWA-mem (41) against a well-known reference strain. Alignment artifacts, such as long soft-clipped read ends and misaligned bases due to a large distance to the reference, can be decreased by a progressive re-alignment strategy. In an iterative procedure, reads are aligned against a reference genome, a consensus sequence is generated, and reads are re-aligned against the consensus sequence. The consensus sequence is generated from the alignment with ambiguous nucleotides and major abundant insertions can be included; the software package ConsensusFixer (42) is capable of performing this step. The alignment quality can be increased by pairwise-sensitive Smith–Waterman alignments of the reads with a reference genome. This process can be accelerated by k-mer matching of the reads with the reference to first find an approximate alignment region (43). Depending on the sequencing platform, different affine gap costs are required. Such sensitive Smith–Waterman alignment to a reference with ambiguous nucleotides, which tests multiple sets of affine gap costs to find an optimal alignment, is implemented in InDelFixer (44). In general, higher quality baseline alignment leads to better reconstruction results.

16.2.5 Spatial Scales of Viral Haplotype Reconstruction

Given the alignment of the reads, diversity estimation can be performed on different spatial scales (Figure 16.6). On the smallest scale, where only a single position of the genome is considered, position-wise diversity estimation means detecting SNVs; SNV calling assumes independence of genomic positions, which precludes detection of co-occurring SNVs. Local reconstruction focuses on a genomic region that can be covered with a single read. For some applications, local reconstruction is sufficient, for example, the protease gene of HIV-1 with length 297 bp can be covered by Illumina reads. If a region of interest is larger than the average read, the problem is

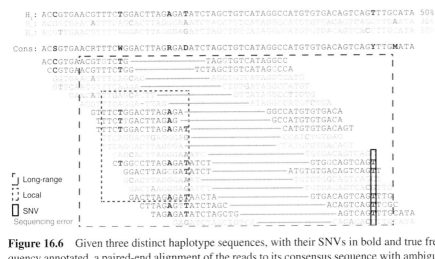

Figure 16.6 Given three distinct haplotype sequences, with their SNVs in bold and true frequency annotated, a paired-end alignment of the reads to its consensus sequence with ambiguous bases is illustrated. Sequencing errors are indicated as pink characters. Reconstruction can be performed on three different spatial scales: position-wise SNV calling (solid box), local with a maximal length of a read (dotted box), and long-range (dashed box). Please see www.wiley.com/go/Mandoiu/NextGenerationSequencing for a color version of this figure.

known as global, or long-range haplotype reconstruction, or quasispecies assembly. Since at least the left- and rightmost reads do not overlap, a jigsaw puzzle of reads has to be solved to phase overlapping and nonoverlapping reads.

16.2.6 Quasispecies Assembly Performance

Performance of quasispecies assembly depends on the many factors of the input sample and the resulting data set. The amplicon layout has to be designed for the underlying assembly problem. Sufficient overlap of neighboring amplicons is necessary to enable read assembly across amplicons.

The heterogeneity of the underlying population including the haplotype distribution and the spatial heterogeneity distribution along the genome, the read length, and the error rate of the sequencing technology are closely coupled in the haplotype reconstruction. Haplotypes that differ by an amount that is close to the error rate of the sequencing technology are challenging to reconstruct. Even if there is enough diversity to distinguish between haplotypes, the distribution over the genome has to be approximately random. If homogeneous regions are longer than the average read length, haplotype phasing is limited to regions with heterogeneity (Figure 16.7).

Solutions to overcome homogeneous regions are either very long PacBio (Figure 16.7d) or paired-end reads (Figure 16.7c) that contain a sufficiently large insert fragment size to bridge conserved regions. Another strategy is to find matching haplotype frequencies of the flanking regions. In general, too short reads cannot be compensated for by a higher coverage (45). Reconstruction of low-frequency haplotypes demands an ultradeep sequencing approach with a uniform coverage.

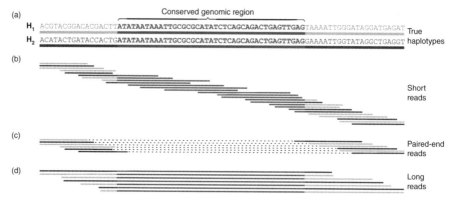

Figure 16.7 Schematic overview for a haplotype structure with a conserved region. (a) The underlying two true haplotypes, which are identical in the red colored region. (b) The multiple sequence alignment of reads sampled from the two haplotypes. (c) Paired-end reads with an insert size longer than the conserved region. (d) Very long reads that are longer than the conserved region. Please see www.wiley.com/go/Mandoiu/NextGenerationSequencing for a color version of this figure.

16.3 PROBABILISTIC RECONSTRUCTION METHODS

The concept of haplotype reconstruction has its roots in human genetics. In this section, human haplotyping and its relationship to probabilistic viral haplotype reconstruction methods are elaborated.

16.3.1 From Human to Viral Haplotype Reconstruction

Our human diploid genome consists of 23 chromosome pairs and an approximate combined length of 3 billion bp. Each pair, one copy is from our mother and the other copy from our father.

The sequence along one chromosome is called a human haplotype and is defined as an element of $\{A, C, G, T\}^L$, with L being the length of the chromosome. A genotype is the mating of two haplotypes for an individual. As the majority of the positions are conserved, a haplotype can be reduced to the variant sites. Sequencing errors can be easily corrected, as variants are bi-allelic, that is, a variant occurs with a 50% frequency. All low-frequency variants are identified as sequencing errors and corrected to the dominant nucleotide. Variant sites are bi-allelic, they can only take one of two possible values, and a haplotype can be represented as an element of $\{0, 1\}^{L'}$, with L' being the number of heterogeneous positions. Current technologies are not able to sequence chromosomes individually and only the genotype, the combined information of a chromosome pair, is available. In the example of Figure 16.8, the goal is to reconstruct the individual haplotypes from a sequence of binary pairs. The haplotype structure is of importance, as some genetic diseases are associated with linkage of specific mutant alleles on one haplotype, for example, Crohn's disease (46).

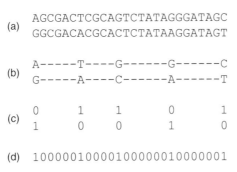

(a)
```
AGCGACTCGCAGTCTATAGGGATAGC
GGCGACACGCACTCTATAAGGATAGT
```

(b)
```
A-----T----G------G------C
G-----A----C------A------T
```

(c)
```
0     1     1     0     1
1     0     0     1     0
```

(d) `10000010000100000010000001`

Figure 16.8   (a) Two human haplotype DNA sequences of length $L = 26$, (b) only the $L' = 5$ variant sites, (c) its binary representation with 1 for the major allele, and (d) the genotype.

16.3.1.1   *Existing Human Phasing Algorithms*   Apart from probabilistic modeling, one important approach to human haplotype inference has been formulated with respect to the evolutionary history of a sampled set of haplotypes, the population-genetic concept of a coalescent (47). Given a single haplotype site of one individual, its history can be traced back in an acyclic graph to the site of one parental haplotype. This means that the history of two haplotypes at one site merges at one haplotype of their most recent common ancestor. Under the infinite-sites assumption and without recombination, the coalescent model says that $2n$ haplotypes from $2n$ individuals can be displayed as a tree with $2n$ leaves with m edges representing the individual sites. If haplotypes are binary and the root is labeled as a binary sequence, the tree displaying the evolution of the haplotypes is called perfect phylogeny. The perfect phylogeny problem is to find the set of explaining haplotypes that defines a perfect phylogeny for a given set of genotypes. Various perfect phylogeny methods have been developed (48–50) and it has been first introduced and solved in Reference 51.

The first probabilistic models in human haplotyping belong to the "gene counting" models (52–56). These models compute the maximum likelihood estimate of the haplotype frequencies from multilocus, unphased genotype count data via an expectation maximization (EM) algorithm . This approach provides haplotype frequencies that can be used to compute the most likely pairing of haplotypes for a given genotype.

The maximum likelihood approach can be regarded as a special case of Bayesian inference with a flat prior for the haplotype frequencies, that is, all haplotypes are equally likely. An informative prior for a guided inference process has the potential to favor solutions that respect an underlying genealogy and to lead to estimates of the haplotype frequencies closer to the true ones. This approach has been implemented in the software PHASE (57), which provides better results than its predecessors, but the speed of convergence is slow for whole-genome data sets. To achieve the run-time improvement in Reference 58, the partition–ligation strategy (54) has been adopted. A genomic region is split into multiple smaller regions. Haplotype inference is performed on adjacent regions individually. Using the current parameter estimates, adjacent haplotypes are allowed to merge into a larger haplotype. This pairwise merging procedure is performed until all regions are merged into haplotypes that do not grow longer in size. The partition–ligation approach has gained wider adoption (59–62) and has also been implemented as a new Bayesian algorithm called HAPLOTYPER (54). Here, the haplotype frequency prior has been replaced by a Dirichlet prior. Even though the Dirichlet prior is computationally more convenient, it cannot model the mutational and recombinational relationship between haplotypes. Such an advanced prior has been proposed and implemented in MC-VL (63) and ELB (64). It has the potential to lead to more accurate results.

The successor of the Bayesian algorithm PHASE is the software package Fast-PHASE (65). It employs a hidden Markov model (HMM) to model each haplotype as a puzzle of smaller fragments. Instead of modeling each haplotype as a hidden state, FastPHASE has a fixed state space size and clusters similar haplotypes. In fact, each haplotype is not limited to stem from a single cluster, but can emerge from a mosaic of clusters to allow for a recombinant structure. Finally, FastPHASE takes genotype data to estimate the haplotype sequences and frequencies.

A similar approach, called Beagle (66), allows a varying number of states across the genome that depends on the empirical linkage disequilibrium. A probabilistic long-range haplotype reconstruction algorithm has been introduced with the new phasing algorithm called emphases (67).

16.3.1.2 *Human versus Viral Genomics*

Human and viral genomics are related, but face different difficulties. In human genetics, reconstruction of the two human haplotypes within one individual and the identification of minor variants on a population scale have to be distinguished. First, the relationship of human haplotyping within one individual and the viral quasispecies assembly is discussed. In the setting of viral haplotype reconstruction, variant sites may exhibit all four nucleotides. Information compression by binarization, as utilized in the human context, is not possible. Relevant resistance mutations may occur with low abundance and, as such, naïve error correction, calling SNVs over a certain threshold, is not sufficient. Viral haplotype phasing cannot be restricted to the variant sites, since low-frequency mutations and sequencing errors cannot be distinguished easily without the risk of information loss. Compared to a diploid human genome, a viral quasispecies consists of an unknown and possible large number of haplotypes with an unknown frequency distribution. In general, viral haplotype reconstruction (i.e., sequence assembly), error correction, and frequency distribution estimation are closely coupled and should ideally be addressed at the same time. Removing noise prior to reconstruction may lead to miscorrections that cannot be improved in subsequent steps. The solutions to the quasispecies reconstruction problem, discussed in this chapter, employ probabilistic models. These models allow the combination of the reconstruction of haplotypes, error correction, and frequency estimation.

In human genetics, the identification of genome-wide variation on a population level is also of interest, and its relationship to a viral population is obvious. In this approach, DNA pooling of multiple samples is performed (68–71). To distinguish reads of different individuals, it is possible to barcode each sample, but workload for sample preparation is too high. Instead, equimolar amounts of DNA of multiple individuals are mixed into one sample prior amplification and sequencing. The bioinformatics challenge is to discriminate low-frequency variants from sequencing errors and not the phasing of haplotypes. This approach is very similar to the *de facto* standard SNV calling for viral populations.

16.3.2 Viral Haplotype Inference Methods Overview

Viral haplotype reconstruction is a fairly new field of research. The first ideas were published in 2003 (72). Wildenberg et al. proposed a method to deconvolute sequence variation from a mixed DNA sample that had been sequenced with Sanger technology. This method, tested in simulation studies for substitutions, has not been adopted in practice, as common single indels make the problem NP-complete and thus computationally infeasible. The earliest approaches to estimate the structure of virus populations with the help of NGS have been proposed in 2008 by Eriksson et al. (73) and Westbrooks et al. (74). Distinctive computational methods have been proposed from various domains to solve the viral haplotype reconstruction problem; each specialized for different NGS technologies, experimental designs, and quasispecies structures. For local reconstruction, probabilistic clustering methods (73, 75–77) and *k*-mer statistics (78) have been developed. For the more complex quasispecies assembly problem HMMs (79, 80), probabilistic mixture models (81), sampling schemes (82), combinatorial approaches based on analyzing the read overlap graph (73, 74, 83–85), coloring of overlap and conflict graphs by constraint programming (86), and max-clique enumeration (87) approaches have been developed. In the following, the probabilistic methods for viral haplotype inference are discussed.

16.3.3 Local Viral Haplotype Inference Approaches

Local haplotype inference methods employ probabilistic clustering approaches (73, 77). Their main assumption is that erroneous reads from the same haplotype are more likely to cluster together than reads from different haplotypes. Thus, the consensus sequence of each cluster represents the error-corrected local haplotype.

In Eriksson et al. (73), the reconstruction task is split into read clustering given a fixed cluster number and a method to determine the number of clusters. For read clustering, a *k*-means algorithm is employed. As the number of clusters is unknown, a statistical test is used to determine the number k of haplotypes. The limitation of this approach is that a fixed error rate has to be assumed for the hypothesis test.

A unified read clustering method has been proposed by Zagordi et al. (77) and implemented in the software ShoRAH. They developed a generative probabilistic model for clustering reads based on the Dirichlet process mixture (DPM). The DPM defines a prior distribution in a Bayesian manner to capture the uncertainty in the number of haplotypes. The model assumes that reads are sampled from a Bayesian multinomial mixture model, each mixture component representing a haplotype. The mixture proportions are an estimate of the haplotype frequencies. The model parameters include the DNA sequences of the haplotypes, the assignment of reads to haplotypes, and the error rate of the sequencing process. In general, Gibbs sampling is performed if sampling from the joint distribution is too hard or not possible, but sampling from the conditional distribution is feasible. Here, Gibbs sampling is used to sample the posterior distribution. The probability of creating new clusters is governed by a single hyperparameter. This allows to fine-tune the performance of the Gibbs sampler.

16.3.4 Quasispecies Assembly

The quasispecies assembly problem refers to the reconstruction of haplotypes that are longer than the original reads. For this, a jigsaw puzzle of short reads has to be solved. Similar to the partition–ligation strategy for human haplotyping (54), ShoRAH computes overlapping local viral haplotypes and solves the minimal path cover, that is, searches for a minimal set of long-range haplotypes that explain all local haplotypes employing a read overlap graph.

The software PredictHaplo (88), developed by Prabhakaran et al., can be seen as a global extension to the local DPM of ShoRAH. PredictHaplo combines local and long-range haplotype reconstruction in one probabilistic model.

In a first step, reconstruction is performed locally. In a second step, parameter estimates of the local reconstruction are used as a prior for extending a local window. As windows are progressively increased, results are used to update the prior information about cluster probabilities for the subsequent window. This procedure enables long-range haplotype reconstruction from perfectly overlapping reads. In order to speed up computation, reads are compressed to variant positions above a certain entropy threshold. This might lead to information loss, but PredictHaplo has proven to be able to reconstruct low-frequency variants below 1% for a coverage above 10,000×.

16.3.5 Recombinant Quasispecies Assembly

In a viral quasispecies, it is assumed that there are a few master sequences and each is surrounded by a cloud of low-frequency variants. These low-frequency variants are very similar to their master sequences and only differ in mutations of single bases. In addition, there are haplotypes that emerge by recombination of master sequences.

QuasiRecomb (80) is a generative model that incorporates the mutational and recombinational relationships of haplotypes to their master sequences. Unlike ShoRAH and PredictHaplo, which aim at enumerating all haplotypes given the data, QuasiRecomb estimates a parametric haplotype distribution of the underlying quasispecies. It is similar to the human haplotype reconstruction method FastPHASE (65). Both models are HMMs, have a fixed number of clusters and estimate jointly recombination between and the DNA composition of the clusters. QuasiRecomb models the master sequences as sequence profiles, called generators. Generators are represented by position-wise probability tables over $\mathcal{A} = \{A, C, G, T, -\}$ to account for low-frequency mutations at each position. Additional position-wise error rates are inferred to represent sequencing errors. In order to learn the recombinational relationship of haplotypes, each haplotype may emerge from different generators in a mosaic structure. The workflow of QuasiRecomb is (i) to infer the generators from an ultradeep sequencing data set and (ii) to predict the haplotype distribution (Figure 16.9). If a recombinant quasispecies structure is present, QuasiRecomb compares favorably to existing methods.

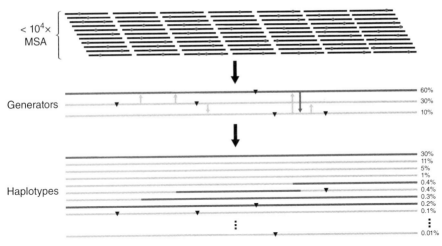

Figure 16.9 Schematic workflow of `QuasiRecomb`. The input data is an alignment of short erroneous reads from an ultradeep sequencing experiment. Sequencing errors are depicted as dots. Given the alignment, sequence profile generators and their frequencies are inferred. Mutations and recombination patterns between generators are illustrated as triangles and arrows, respectively. Given the estimated sequence profile generators, the haplotype distribution is sampled from the model.

16.3.5.1 Model of QuasiRecomb

`QuasiRecomb` is a probabilistic model that generates an estimate of the underlying quasispecies structure by switching among K different generators, each of length L. It employs an HMM, in which different generators are hidden states. Here, the Markov chain is a statistical process that allows state transitions on a state space to explain the generation of haplotypes from a set of master sequences and subsequently reads from these haplotypes. The different profile sequence generators are denoted by P_k with $k \in [1, \ldots, K]$. The number of generators K is unknown and yet to be determined by model selection. Let Z_j be the hidden random variable with state space $\{P_1, \ldots, P_K\}$ and j be the position in the generator (Figure 16.10). The probability to begin a read with generator P_k at any sequence position j is $P(Z_k) = \pi_k$. A recombination event between two generators occurs with transition probability $P(Z_j = P_k \mid Z_{j-1} = P_l) = \rho_{jkl}$, for $k \neq l$; if $k = l$ then recombination did not occur. Profile sequence generators P_k indicate the parental sequence generating viral haplotypes $H \in \mathcal{A}^L$. The hidden state H_j indicates the haplotype character at position j that is produced with emission probability $P(H_j = v \mid Z_j = P_k) = \mu_{jkv}$, for each $v \in \mathcal{A}$. Each observed read R, subjected to sequencing errors, with bases R_j is obtained from an haplotype H with probability $P(R_j = b \mid H_j = v) = 1 - (n - 1)\epsilon_j$ if $r = v$, and ϵ_j otherwise. The model parameters are summarized as $\theta = (\pi, \rho, \mu, \epsilon)$.

The likelihood $P(R \mid \theta)$ of the model factorizes into the product over independent reads, and for each read i, it can be computed efficiently using the Markov property,

$$P(R \mid \theta) = \prod_i \sum_{Z^i, H^i} \prod_j P(R_j^i \mid H_j^i) P(H_j^i \mid Z_j^i) P(Z_j^i \mid Z_{j-1}^i).$$

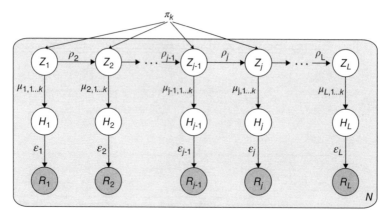

Figure 16.10 Graphical representation of the HMM of `QuasiRecomb`. Only one observation i is depicted; for the full model, the graph is replicated for $i = 1, \ldots, N$ indicated by the plate notation. Variable Z_j represents the master sequence, H_j the nucleotide of the emitted haplotype, and R_j the nucleotide of the observed read at position j.

Parameter Inference For parameter inference, a prior distribution for the model parameters is defined and the maximum a posteriori (MAP) estimate of the parameters is computed by maximizing the posterior probability

$$P(\theta \mid R) \propto P(R \mid \theta)P(\theta).$$

Employing Dirichlet priors leads to sparse MAP solutions if the Dirichlet hyperparameters approach zero. The regularized model is also a biological plausible assumption. Most of the genomic regions in real RNA virus populations are conserved or nearly conserved. This is caused by selective pressure on certain regions to ensure reproductive capability. The virus is defined by these regions. In these regions, the different generating sequences cannot be distinguished because there is no or little diversity.

Despite regularization of the transition probabilities, identical regions between generators are a problem (Figure 16.7). As reads of the conserved region cannot be assigned uniquely to one of the generators and there is no information about the linkage of the flanking regions, a recombinant structure is estimated. This can be avoided by requiring the recombination probability to be less than 0.5.

Model Selection The described model uses an unknown set of sequence profile generators to explain the underlying quasispecies structure. To determine the unknown number of generators K, for model selection the Bayesian information criterion (BIC) is used. Empirical studies have shown that in practice, model selection works up to at least $K = 5$ (80). This is sufficient for practical purposes, as the number of master sequences in the quasispecies model is assumed to be small.

Prediction Given the MAP estimate of the parameters, there are three objects of interest. First, the SNV posterior probability can be computed. This method can detect low-frequency mutations, as it takes co-occurring mutations into account and therefore has more statistical power to distinguish sequencing errors from low-frequency SNVs. Second, recombination spots can be identified that gave rise to the underlying quasispecies with respect to the master sequences. Last, the main object of interest is to derive the viral quasispecies structure. Computing the probability of each haplotype gives rise to the estimated haplotype distribution $P(H)$, but enumeration is computationally infeasible as there are 4^L possible haplotypes. However, as most of the emission and transition probability tables are concentrated on one entry, that is, all other probabilities are exactly or close to zero, the mass is centered around a few haplotypes. Having this in mind, the MAP estimate of the parameters are used for sampling from $P(H)$.

16.3.5.2 *Extension of QuasiRecomb*

The model of `QuasiRecomb`, as described earlier, can only be used for local haplotype reconstruction (Figure 16.11). In this case, the length of the genomic region of interest has to be the read length. A single read can emerge from a set of profile sequence generators.

Quasispecies Assembly The local reconstruction model can be extended to support the reconstruction of long-range haplotypes. In this case, reads are shorter than the sought haplotypes and therefore shorter than the profile generators (Figure 16.12a). Silent begin (B) and end (E) states are introduced, which do not generate haplotype characters, but allow for an arbitrary read placement (Figure 16.12).

A read may start at the first position in one of the profile generators P_k with probability π_k or in the begin state B. If it started in B, the Markov chain may continue in B or jump to any P_k with probability π_k. Once the chain is in the state P_k, it is allowed to jump between the different P_k to explain the read or jump into E to end the read, but it is not allowed to jump back to B. If the chain is in state E, it has to

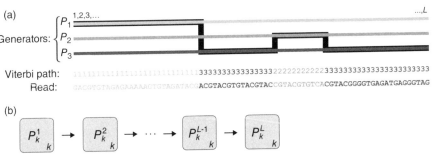

Figure 16.11 (a) Schematic illustration of a local reconstruction model with three profile generators P_1, P_2, and P_3. Given a read that covers the full length of interest, the Viterbi path is shown in black as a mosaic of the profile generators. (b) Transition state diagram in plate notation. Transition between plates indicates all possible transitions between profile sequence generators. Please see www.wiley.com/go/Mandoiu/NextGenerationSequencing for a color version of this figure.

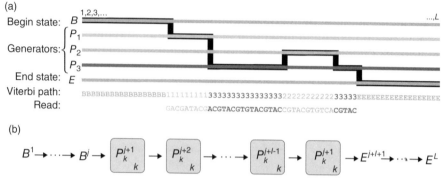

Figure 16.12 (a) Schematic illustration of a long-range reconstruction model with three profile generators P_1, P_2, and P_3 and silent begin (B) and end (E) states. The read is shorter than the region of interest and can arbitrarily be located in the region of interest. (b) Transition state diagram in plate notation. Transition between plates indicates all possible transitions between profile sequence generators. Silent states B^1 to B^j and E^{j+l+1} to E^L enable read placement. Please see www.wiley.com/go/Mandoiu/NextGenerationSequencing for a color version of this figure.

continue in E (Figure 16.12b). In practice, the transitions from and to the silent states are not parameters that need to be inferred, as the read placement is given by the input alignment. If generators are multiple times longer than the reads, it can happen that the optimization procedure gets stuck in a local optima and many EM restarts are needed. To handle this, the partition–ligation strategy (54) can be used to overcome this limitation, as parameter inference of shorter regions is more robust.

Paired-End Quasispecies Assembly The model can also support paired-end reads (Figure 16.13a). For paired-end reads, the length of the unsequenced fragment between the paired sequences is given by the alignment, and more importantly, both reads stem from the same haplotype as they stem from the same fragment. Paired-end reads can help to phase haplotypes that contain conserved regions if read pairs exist that overlap the flanking regions, exhibiting sufficient diversity to distinguish between haplotypes.

Without introducing additional parameters, the existing probabilistic model can be extended to gain advantage of this new data. The earlier introduced generators, which embody sequence profiles to represent the observed haplotypes, are enhanced to model the inserts between the paired-end reads. For this, insertion states I_k represent the silent, noncharacter emitting, version of the existing profile generators P_k (Figure 16.13). The state space is extended to $\{B, P_1, \ldots, P_K, I_1, \ldots, I_K, E\}$. These insertion states are needed because it should be possible to switch between generators in the unsequenced region of the paired-end reads. This is due to the fact that the read on the second pair may emerge from another generator though it is from the same haplotype. The information for the recombination between generators I_k and I_l is inferred from other observed reads, which cover this region. Indeed, this information is already available by ρ_j, which contains the transition probabilities between P_k

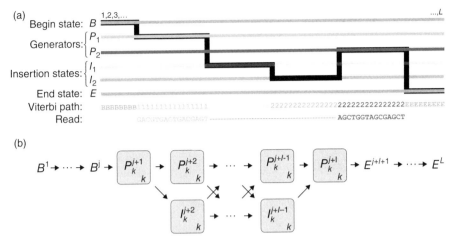

Figure 16.13 Schematic illustration of a paired-end long-range reconstruction model with two profile generators P_1 and P_2, their respective silent states I_1 and I_2, and silent begin (B) and end (E) states. The unsequenced fragment between reads is shown as a dashed line between read pairs. The additional silent insertion states allow recombination to occur between these two observed read pairs. (B) Transition state diagram in plate notation. Transition between plates indicates all possible transitions between profile and silent sequence generators. Silent states B^1 to B^j and E^{j+l+1} to E^L enable read placement. Please see www.wiley.com/go/Mandoiu/NextGenerationSequencing for a color version of this figure.

and P_l. Therefore, transition matrices of the insertion states are identical to those of sequence generators. This is computationally convenient as the number of parameters does not increase. The transition probabilities between P_k and I_l at the first position of the unsequenced fragment, and I_k and P_l at the first position of the second read pair are also in ρ_j. The positions of the transitions from and to the insertion states are given by the alignment. With this extension, accuracy of long-range haplotype reconstruction increases for quasispecies with low diversity and/or conserved regions, as there is more statistical power to distinguish haplotypes.

16.4 CONCLUSION

Probabilistic haplotype reconstruction is well studied and established in human genetics. In recent years, these methods gained increasing interest in the field of viral haplotype reconstruction, as NGS enabled the study of complex viral quasispecies. Even though both fields face different problems, they share and address many problems similarly. Probabilistic models can address the major questions of viral quasispecies assembly combined in a single model: (i) determine the number of haplotypes, (ii) assemble individual haplotype sequences, (iii) estimate the frequency distribution, and (iv) distinguish between true biological variation and sequencing errors. Viral quasispecies assembly has the potential to replace the current *de facto* diversity estimation by SNV calling. The study of co-occurring mutations might give more insights into the dynamics and possibly the evolutionary trajectory of intra-patient viral populations. With advances in library preparation, increasing sensitivity of sequencing platforms, and more sophisticated models, it might be possible to detect all or most viral strains in a single individual. Such valueable information could be used to improve treatment outcome by adjusting regimens.

REFERENCES

1. Harper D. Online Etymology Dictionary; 2001. Etymology Online www.etymonline.com.
2. Gago S, Elena SF, Flores R, Sanjuán R. Extremely high mutation rate of a hammerhead viroid. Science 2009;323(5919):1308–1308.
3. Eigen M, McCaskill J, Schuster P. The molecular quasi-species. Adv Chem Phys 1989;75:149–263.
4. Boerlijst MC, Bonhoeffer S, Nowak MA. Viral quasi-species and recombination. Proc R Soc London, Ser B 1996;263:1577–1584.
5. Domingo E, Escarmis C, Sevilla N, Moya A, Elena SF, Quer J, Novella IS, Holland JJ. Basic concepts in RNA virus evolution. FASEB J 1996;10(8):859–864.
6. Domingo E, Sheldon J, Perales C. Viral quasispecies evolution. Microbiol Mol Biol Rev 2012;76(2):159–216.
7. Calsina A, Cuadrado S. Small mutation rate and evolutionarily stable strategies in infinite dimensional adaptive dynamics. J Math Biol 2004;48(2):135–159.
8. Sardanyés J, Elena SF, Solé RV. Simple quasispecies models for the survival-of-the-flattest effect: the role of space. J Theor Biol 2008;250(3):560–568.
9. Rouzine IM, Coffin JM, Weinberger LS. Fifteen years later: hard and soft selection sweeps confirm a large population number for HIV in vivo. PLoS Genet 2014;10(2):e1004179.
10. Alizon S, Fraser C. Within-host and between-host evolutionary rates across the HIV-1 genome. Retrovirology 2013;10(1):49.
11. Mansky LM. The mutation rate of human immunodeficiency virus type 1 is influenced by the *VPR* gene. Virology 1996;222(2):391–400.
12. Haase AT, Henry K, Zupancic M, Sedgewick G, Faust RA, Melroe H, Cavert W, Gebhard K, Staskus K, Zhang ZQ, Dailey PJ, Balfour HH Jr., Erice A, Perelson AS. Quantitative image analysis of HIV-1 infection in lymphoid tissue. Science 1996;274(5289):985–989.
13. Kouyos RD, Althaus CL, Bonhoeffer S. Stochastic or deterministic: what is the effective population size of HIV-1? Trends Microbiol 2006;14(12):507–511.
14. Thompson MA, Aberg JA, Cahn P, Montaner JSG, Rizzardini G, Telenti A, Gatell JM, Günthard HF, Hammer SM, Hirsch MS, Jacobsen DM, Reiss P, Richman DD, Volberding PA, Yeni P, Schooley RT, International AIDS Society-USA. Antiretroviral treatment of adult HIV infection: 2010 recommendations of the International Aids Society-USA panel. JAMA 2010;304(3):321–333.
15. Beerenwinkel N, Montazeri H, Schuhmacher H, Knupfer P, von Wyl V, Furrer H, Battegay M, Hirschel B, Cavassini M, Vernazza P et al. The individualized genetic barrier predicts treatment response in a large cohort of HIV-1 infected patients. PLoS Comput Biol 2013;9(8):e1003203.
16. Zhuang J, Jetzt AE, Sun G, Yu H, Klarmann G, Ron Y, Preston BD, Dougherty JP. Human immunodeficiency virus type 1 recombination: rate, fidelity, and putative hot spots. J Virol 2002;76(22):11273–11282.

17. Jetzt AE, Yu H, Klarmann GJ, Ron Y, Preston BD, Dougherty JP. High rate of recombination throughout the human immunodeficiency virus type 1 genome. J Virol 2000;74(3):1234–1240.

18. Bretscher MT, Althaus CL, Müller V, Bonhoeffer S. Recombination in HIV and the evolution of drug resistance: for better or for worse? Bioessays 2004;26(2):180–188.

19. Chen J, Nikolaitchik O, Singh J, Wright A, Bencsics CE, Coffin JM, Ni N, Lockett S, Pathak VK, Hu W-S. High efficiency of HIV-1 genomic RNA packaging and heterozygote formation revealed by single virion analysis. Proc Natl Acad Sci U S A 2009;106(32):13535–13540.

20. Levy DN, Aldrovandi GM, Kutsch O, Shaw GM. Dynamics of HIV-1 recombination in its natural target cells. Proc Natl Acad Sci U S A 2004;101(12):4204–4209.

21. Neher RA, Leitner T. Recombination rate and selection strength in HIV intra-patient evolution. PLoS Comput Biol 2010;6(1):e1000660.

22. Althaus CL, Bonhoeffer S. Stochastic interplay between mutation and recombination during the acquisition of drug resistance mutations in human immunodeficiency virus type 1. J Virol 2005;79(21):13572–13578.

23. Charpentier C, Nora T, Tenaillon O, Clavel F, Hance AJ. Extensive recombination among human immunodeficiency virus type 1 quasispecies makes an important contribution to viral diversity in individual patients. J Virol 2006;80(5):2472–2482.

24. Mostowy R, Kouyos RD, Fouchet D, Bonhoeffer S. The role of recombination for the coevolutionary dynamics of HIV and the immune response. PLoS ONE 2011;6(2):e16052.

25. Töpfer A, Höper D, Blome S, Beer M, Beerenwinkel N, Ruggli N, Leifer I. Sequencing approach to analyze the role of quasispecies for classical swine fever. Virology 2013;438(1):14–19, 3.

26. Henn MR, Boutwell CL, Charlebois P, Lennon NJ, Power KA, Macalalad AR, Berlin AM, Malboeuf CM, Ryan EM, Gnerre S et al. Whole genome deep sequencing of HIV-1 reveals the impact of early minor variants upon immune recognition during acute infection. PLoS Pathog 2012;8(3):e1002529.

27. Beerenwinkel N, Günthard HF, Roth V, Metzner KJ. Challenges and opportunities in estimating viral genetic diversity from next-generation sequencing data. Front Microbiol 2012;3:329.

28. Vignuzzi M, Stone JK, Arnold JJ, Cameron CE, Andino R. Quasispecies diversity determines pathogenesis through cooperative interactions in a viral population. Nature 2006;439(7074):344–348.

29. Berger EA, Murphy PM, Farber JM. Chemokine receptors as HIV-1 coreceptors: roles in viral entry, tropism, and disease. Annu Rev Immunol 1999;17(1):657–700.

30. Wolfs TFW, Zwart G, Bakker M, Goudsmit J. HIV-1 genomic RNA diversification following sexual and parenteral virus transmission. Virology 1992;189(1):103–110.

31. Hinkley T, Martins J, Chappey C, Haddad M, Stawiski E, Whitcomb JM, Petropoulos CJ, Bonhoeffer S. A systems analysis of mutational effects in HIV-1 protease and reverse transcriptase. Nat Genet 2011;43(5):487–489.

32. Di Giallonardo F, Zagordi O, Duport Y, Leemann C, Joos B, Künzli-Gontarczyk M, Bruggmann R, Beerenwinkel N, Günthard HF, Metzner KJ. Next-generation sequencing of HIV-1 RNA genomes: determination of error rates and minimizing artificial recombination. PLoS ONE 2013;8(9):e74249.

33. Quail MA, Smith M, Coupland P, Otto TD, Harris SR, Connor TR, Bertoni A, Swerdlow HP, Gu Y. A tale of three next generation sequencing platforms: comparison of ion torrent, pacific biosciences and illumina miseq sequencers. BMC Genomics 2012;13(1):341.

34. Liu L, Li Y, Li S, Hu N, He Y, Pong R, Lin D, Lu L, Law M. Comparison of next-generation sequencing systems. J Biomed Biotechnol 2012;2012:251364.

35. Jabara CB, Jones CD, Roach J, Anderson JA, Swanstrom R. Accurate sampling and deep sequencing of the HIV-1 protease gene using a primer ID. Proc Natl Acad Sci U S A 2011;108(50):20166–20171.

36. Kinde I, Wu J, Papadopoulos N, Kinzler KW, Vogelstein B. Detection and quantification of rare mutations with massively parallel sequencing. Proc Natl Acad Sci U S A 2011;108(23):9530–9535.

37. Schmitt MW, Kennedy SR, Salk JJ, Fox EJ, Hiatt JB, Loeb LA. Detection of ultra-rare mutations by next-generation sequencing. Proc Natl Acad Sci U S A 2012;109(36): 14508–14513.

38. Acevedo A, Brodsky L, Andino R. Mutational and fitness landscapes of an RNA virus revealed through population sequencing. Nature 2014;505(7485):686–690.

39. Wu NC, De La Cruz J, Al-Mawsawi LQ, Olson CA, Qi H, Luan HH, Nguyen N, Du Y, Le S, Wu T-T et al. HIV-1 quasispecies delineation by tag linkage deep sequencing. PLoS ONE 2014;9(5):e97505.

40. Lou DI, Hussmann JA, McBee RM, Acevedo A, Andino R, Press WH, Sawyer SL. High-throughput dna sequencing errors are reduced by orders of magnitude using circle sequencing. Proc Natl Acad Sci U S A 2013;110(49):19872–19877.

41. Li H. Aligning sequence reads, clone sequences and assembly contigs with BWA-MEM; 2013. arXiv preprint arXiv:1303.3997.

42. Töpfer A. ConsensusFixer; 2014. Available at https://github.com/armintoepfer/ConsensusFixer. Accessed 2014 Jan 31.

43. Archer J, Rambaut A, Taillon BE, Harrigan PR, Lewis M, Robertson DL. The evolutionary analysis of emerging low frequency HIV-1 CXCR4 using variants through time–an ultra-deep approach. PLoS Comput Biol 2010;6(12):e1001022.

44. Töpfer A. Indelfixer; 2014. Available at https://github.com/armintoepfer/InDelFixer. Accessed 2014 Jan 31.

45. Zagordi O, Däumer M, Beisel C, Beerenwinkel N. Read length versus depth of coverage for viral quasispecies reconstruction. PLoS ONE 2012;7(10):e47046.

46. Croucher PJP, Mascheretti S, Hampe J, Huse K, Frenzel H, Stoll M, Lu T, Nikolaus S, Yang S-K, Krawczak M et al. Haplotype structure and association to Crohn's disease of CARD15 mutations in two ethnically divergent populations. Eur J Hum Genet 2003;11(1):6–16.

47. Hudson RR et al. Gene genealogies and the coalescent process. Oxford Surv Evol Biol 1990;7(1):44.

48. Bonizzoni P, Della Vedova G, Dondi R, Li J. The Haplotyping problem: an overview of computational models and solutions. J Comput Sci Technol 2003;18(6):675–688.

49. Gusfield D, Eddhu S, Langley C. Optimal, efficient reconstruction of phylogenetic networks with constrained recombination. J Bioinform Comput Biol 2004;2(1):173–213.

50. Halldorsson BV, Bafna V, Edwards N, Lippert R, Yooseph S, Istrail S. A survey of computational methods for determining haplotypes. In: Istrail S, Waterman MS, Clark AG, editors. *Computational Methods for SNPs and Haplotype Inference*. Volume 2983 of Lecture Notes in Computer Science. Springer-Verlag; 2004. p 26–47.

51. Gusfield D. Haplotyping as perfect phylogeny: conceptual framework and efficient solutions. Proceedings of the 6th Annual International Conference on Computational Biology. ACM; 2002. p 166–175.

52. Excoffier L, Slatkin M. Maximum-likelihood estimation of molecular haplotype frequencies in a diploid population. Mol Biol Evol 1995;12(5):921–927.

53. Hawley ME, Kidd KK. HAPLO: a program using the EM algorithm to estimate the frequencies of multi-site haplotypes. J Hered 1995;86(5):409–411.

54. Niu T, Qin ZS, Xu X, Liu JS. Bayesian haplotype inference for multiple linked single-nucleotide polymorphisms. Am J Hum Genet 2002;70(1):157–169.

55. Zhao JH, Curtis D, Sham PC. Model-free analysis and permutation tests for allelic associations. Hum Hered 1999;50(2):133–139.

56. Zhao JH, Lissarrague S, Essioux L, Sham PC. Genecounting: haplotype analysis with missing genotypes. Bioinformatics 2002;18(12):1694–1695.

57. Stephens M, Smith NJ, Donnelly P. A new statistical method for haplotype reconstruction from population data. Am J Hum Genet 2001;68(4):978–989.

58. Stephens M, Donnelly P. A comparison of bayesian methods for haplotype reconstruction from population genotype data. Am J Hum Genet 2003;73(5):1162–1169.

59. Li SS, Khalid N, Carlson C, Zhao LP. Estimating haplotype frequencies and standard errors for multiple single nucleotide polymorphisms. Biostatistics 2003;4(4):513–522.

60. Zhang P, Sheng H, Morabia A, Gilliam TC. Optimal step length em algorithm (OSLEM) for the estimation of haplotype frequency and its application in lipoprotein lipase genotyping. BMC Bioinformatics 2003;4(1):3.

61. Thomas A. GCHap: fast MLEs for haplotype frequencies by gene counting. Bioinformatics 2003;19(15):2002–2003.

62. Qin ZS, Niu T, Liu JS. Partition-ligation–expectation-maximization algorithm for haplotype inference with single-nucleotide polymorphisms. Am J Hum Genet 2002;71(5):1242.

63. Eronen L, Geerts F, Toivonen H. A Markov chain approach to reconstruction of long haplotypes. In: Altman RB, Dunker AK, Hunter L, Jung TA, Klein TE, editors. *Pacific Symposium on Biocomputing*. World Scientific; 2004. p 104–115.

64. Excoffier L, Laval G, Balding D et al. Gametic phase estimation over large genomic regions using an adaptive window approach. Hum Genomics 2003;1(1):7–19.

65. Scheet P, Stephens M. A fast and flexible statistical model for large-scale population genotype data: applications to inferring missing genotypes and haplotypic phase. Am J Hum Genet 2006;78(4):629–644.

66. Browning BL, Browning SR. A unified approach to genotype imputation and haplotype-phase inference for large data sets of trios and unrelated individuals. Am J Hum Genet 2009;84(2):210–223.

67. Johnson NA, London SJ, Romieu I, Wong WH, Tang H. Accurate construction of long range haplotypes in ulrelated individuals. Stat Sin 2013;23:1441–1461.

68. Kofler R, Pandey RV, Schlötterer C. PoPoolation2: identifying differentiation between populations using sequencing of pooled DNA samples (Pool-Seq). Bioinformatics 2011;27(24):3435–3436.

69. Ferretti L, Ramos-Onsins SE, Pérez-Enciso M. Population genomics from pool sequencing. Mol Ecol 2013;22(22):5561–5576.

70. Catchen J, Hohenlohe PA, Bassham S, Amores A, Cresko WA. Stacks: an analysis tool set for population genomics. Mol Ecol 2013;22(11):3124–3140.

71. Gautier M, Foucaud J, Gharbi K, Cézard T, Galan M, Loiseau A, Thomson M, Pudlo P, Kerdelhué C, Estoup A. Estimation of population allele frequencies from next-generation sequencing data: pool-versus individual-based genotyping. Mol Ecol 2013;22(14):3766–3779.

72. Wildenberg A, Skiena S, Sumazin P. Deconvolving sequence variation in mixed DNA populations. J Comput Biol 2003;10(3-4):635–652.

73. Eriksson N, Pachter L, Mitsuya Y, Rhee S-Y, Wang C, Gharizadeh B, Ronaghi M, Shafer RW, Beerenwinkel N. Viral population estimation using pyrosequencing. PLoS Comput Biol 2008;4(4):e1000074.

74. Westbrooks K, Astrovskaya I, Campo D, Khudyakov Y, Berman P, Zelikovsky A. HCV quasispecies assembly using network flows. In: *Bioinformatics Research and Applications, 4th International Symposium, ISBRA 2008, Atlanta (GA), USA, May 6-9, 2008. Proceedings.* Volume 4983 of Lecture Notes in Computer Science. Springer-Veralg; 2008. p 159–170.

75. Macalalad AR, Zody MC, Charlebois P, Lennon NJ, Newman RM, Malboeuf CM, Ryan EM, Boutwell CL, Power KA, Brackney DE, Pesko KN, Levin JZ, Ebel GD, Allen TM, Birren BW, Henn MR. Highly sensitive and specific detection of rare variants in mixed viral populations from massively parallel sequence data. PLoS Comput Biol 2012;8(3): e1002417.

76. Quince C, Lanzen A, Davenport RJ, Turnbaugh PJ. Removing noise from pyrosequenced amplicons. BMC Bioinformatics 2011;12:38.

77. Zagordi O, Geyrhofer L, Roth V, Beerenwinkel N. Deep sequencing of a genetically heterogeneous sample: local haplotype reconstruction and read error correction. J Comput Biol 2010;17(3):417–428.

78. Skums P, Dimitrova Z, Campo DS, Vaughan G, Rossi L, Forbi JC, Yokosawa J, Zelikovsky A, Khudyakov Y. Efficient error correction for next-generation sequencing of viral amplicons. BMC Bioinformatics 2012;13:(Suppl 10):S6.

79. Jojic V, Hertz T, Jojic N. Population sequencing using short reads: HIV as a case study. In: Altman RB, Dunker AK, Hunter L, Murray T, Klein TE, editors. *Pacific Symposium on Biocomputing.* World Scientific; 2008. p 114–125.

80. Töpfer A, Zagordi O, Prabhakaran S, Roth V, Halperin E, Beerenwinkel N. Probabilistic inference of viral quasispecies subject to recombination. J Comput Biol 2013;20(2):113–123.

81. Prabhakaran S, Rey M, Zagordi O, Beerenwinkel N, Roth V. HIV haplotype inference using a constraint-based dirichlet process mixture model. NIPS Workshop on Machine Learning in Computational Biology; 2010.

82. Prosperi MCF, Salemi M. QuRe: software for viral quasispecies reconstruction from next-generation sequencing data. Bioinformatics 2012;28(1):132–133.

83. Astrovskaya I, Tork B, Mangul S, Westbrooks K, Măndoiu I, Balfe P, Zelikovsky A. Inferring viral quasispecies spectra from 454 pyrosequencing reads. BMC Bioinformatics 2011;12 Suppl 6:S1.

84. Mancuso N, Tork B, Skums P, Mandoiu I, Zelikovsky A. Viral quasispecies reconstruction from amplicon 454 pyrosequencing reads. Bioinformatics and Biomedicine Workshops (BIBMW), 2011 IEEE International Conference on. IEEE; 2011. p 94–101.

85. O'Neil ST, Emrich SJ. Haplotype and minimum-chimerism consesus determination using short sequence data. BMC Genomics 2012;13 Suppl 2:S4.

86. Huang A, Kantor R, DeLong A, Schreier L, Istrail S. Qcolors: an algorithm for conservative viral quasispecies reconstruction from short and non-contiguous next generation sequencing reads. IEEE International Conference on Bioinformatics and Biomedicine Workshops; 2011. p 130–136.

87. Töpfer A, Marschall T, Bull RA, Luciani F, Schönhuth A, Beerenwinkel N. Viral quasispecies assembly via maximal clique enumeration. PLoS Comput Biol 2014;10(3): e1003515.

88. Prabhakaran S, Rey M, Zagordi O, Beerenwinkel N, Roth V. HIV haplotype inference using a propagating dirichlet process mixture model. IEEE/ACM Trans Comput Biol Bioinf 2014;11(1):182–191.

17

RECONSTRUCTION OF INFECTIOUS BRONCHITIS VIRUS QUASISPECIES FROM NGS DATA

BASSAM TORK*, EKATERINA NENASTYEVA*, ALEXANDER ARTYOMENKO, NICHOLAS MANCUSO

Department of Computer Science, Georgia State University, Atlanta, GA, USA

MAZHAR I. KHAN

Department of Pathobiology and Veterinary Science, University of Connecticut, Storrs, CT, USA

RACHEL O'NEILL

Department of Molecular and Cell Biology, University of Connecticut, Storrs, CT, USA

ION I. MĂNDOIU

Department of Computer Science and Engineering, University of Connecticut, Storrs, CT, USA

ALEXANDER ZELIKOVSKY

Department of Computer Science, Georgia State University, Atlanta, GA, USA

17.1 INTRODUCTION

Poultry farms are susceptible to viral infections that cause significant economic losses worldwide in terms of impaired growth, reduced egg production and quality, and even mortality. In the United States where infections with virulent strains of Newcastle disease and highly pathogenic avian influenza are not common, the infectious bronchitis

*Contributed equally.

Computational Methods for Next Generation Sequencing Data Analysis, First Edition.
Edited by Ion I. Măndoiu and Alexander Zelikovsky.
© 2016 John Wiley & Sons, Inc. Published 2016 by John Wiley & Sons, Inc.
Companion website: www.wiley.com/go/Mandoiu/NextGenerationSequencing

virus (IBV) is the biggest single cause of economic loss. Viral quasispecies sequences reconstruction by analyzing high-throughput sequencing (HTS) data contributes to understanding of the roles and interactions of host animal and viral genomes. This understanding is necessary for improving animal health, well-being, and production efficiency, which is one of the goals included in Blueprint for USDA Efforts in Agricultural Animal Genomics (2008-2017).

17.2 BACKGROUND

17.2.1 Infectious Bronchitis Virus

RNA viruses are causing a significant burden on the health and productivity of agriculturally important animals since the rapid evolution of RNA viruses within infected hosts is coupled in this context with frequent transmissions between animals, due to the typically high animal density in production environments. Poultry farms are particularly susceptible to viral infections, which cause significant economic losses worldwide in terms of impaired growth, reduced egg production and quality, and even mortality.

In the United States where infections with virulent strains of Newcastle disease and highly pathogenic avian influenza are not common, the IBV is considered the biggest single cause of economic loss. It infects the domestic fowl (1). Apart from causing respiratory disease, it may also replicate at many non-respiratory epithelial surfaces. First described in the United States in 1930, IBV has undergone exponential population growth (2) and is now distributed worldwide, with dozens of circulating serotypes that may differ by as much as 25% in the amino acid sequence of the hypervariable region of the spike glycoprotein.

As serotypes cross-protect poorly, chicken may get infected multiple times during their lifetime. IBV infection of broilers retards growth while in layers it drops down egg production. Young chicken may die directly from IBV infection, but a greater number die due to secondary bacterial infections. Vaccination, most commonly with live attenuated vaccines, is broadly used to control IBV disease but protection is short-lived and layers have to be revaccinated multiple times during their lifespan, sometimes with different serotypes.

Several IBV serotypes can co-circulate in a region, creating conditions for recombination between strains and resulting in complex evolution and epidemiology (3). Cloning and sequencing have been used to show that IBV variants exist in infected poultry (4) as well as in several widely used commercial live attenuated vaccines (5). Furthermore, McKinley et al. (5) has shown that attenuated live vaccines undergo in vivo selection following vaccination and can infect contact-exposed chicken. Coupled with significant field persistence of Arkansas-type vaccine viruses (6), this reinforces the evidence that infectious IBV strains can emerge from the evolution of attenuated live vaccines in commercial flocks (7).

17.2.2 High-Throughput Sequencing

Recent advances in HTS technologies such as the Roche 454 FLX Titanium, Illumina Genome Analyzer, and ABI SOLiD have led to orders of magnitude of higher throughput compared to classic Sanger sequencing (8). Indeed, at full capacity, each one of these HTS sequencers is capable of producing millions/billions of sequenced bases per day. Coupled with continuously decreasing prices, HTS has profoundly transformed genomics research, enabling a host of novel HTS applications ranging from transcription analysis (9, 10) and small noncoding RNAs (11) to detection of epigenetic changes, such as DNA methylation (12) and histone modifications (13), and from individual genome resequencing (14, 15) to metagenomics (16, 17) and paleogenomics (18, 19). HTS is also emerging as a key technology for quasispecies analysis since it eliminates the cloning step and allows sampling a much larger fraction of the quasispecies (20). Initial studies have focused on using HTS to identify extremely low-frequency drug-resistant variants in human patients chronically infected with HIV (21–25). Very recently, HTS has been used to monitor the effects of an siRNA therapy for the treatment of HCV infection (26) and to characterize the quasispecies of the Pandemic H1N1 Influenza A virus (27). Just as PCR and automated Sanger sequencing revolutionized molecular biology, HTS is expected to transform research on phylodynamics of RNA viruses (28).

However, fully realizing the potential of HTS technologies requires the development of novel analysis methods. Analysis is challenging due to the huge amount of data generated by HTS technologies, on the one hand, and to the short read lengths and high error rates, on the other. As a consequence, many tools developed for Sanger reads do not work at all or have impractical run-times when applied to HTS data. Even newly developed algorithms for *de novo* genome assembly from HTS data (29–32) are tuned for reconstruction of haploid genomes, and work poorly when the sequenced sample contains a large number of closely related sequences, as is the case in viral quasispecies. Methods designed to reconstruct diploid haplotypes from shotgun genome sequencing data (33–39) cannot be extended to reconstruct the full spectrum of sequence variants in a quasispecies either, as they critically assume uniform abundance of each haplotype in the sample.

Viral quasispecies sequences reconstruction by analyzing HTS/Next Generation Sequencing (NGS) data will contribute to understanding the roles and interactions of infected animal and viral variants for improving animal health and productivity. The contributions in this chapter are as follows:

– Using different reconstruction methods with different methods and different parameter settings for accurate reconstruction of viral quasispecies sequences and their frequencies from NGS data.
– Comparison with other methods.
– Experimental validation of used methods.

17.3   METHODS

Many tools developed for Sanger reads do not work at all or have impractical run-times when applied to NGS data. Even newly developed algorithms for *de novo* genome assembly from NGS data are tuned for reconstruction of haploid genomes and work poorly when the sequenced sample contains a large number of closely related sequences, as is the case in viral quasispecies. To address these shortcomings, we introduce evaluated reconstruction flows for accurate reconstruction of viral quasispecies sequences and estimate their frequencies from HTS data where it incorporates different NGS reads error correction methods, aligners, and genome assemblers (ViSpA (40), and ShoRAH (41, 42)), using different tuning parameters. It should be noted some other recent genome reconstruction tools such as QuasiRe-comb (43) (another program from ShoRAH's development team), BIOA:VirA (44), QuRe (45), and PredictHaplo (46). In the current work, we considered only ShoRAH and ViSpA. We applied experiments on IBV 454 shotgun reads and collected from commercial poultry farms. For method validation, we use IBV sanger clones (as ground truth).

The proposed reconstruction flows consist of the following stages (see Figure 17.1):

1. Read Error Correction. This step is necessary since the reads produced by 454 Life Sciences system are prone to errors, and it is important to distinguish read-ing errors from rare viral variants. As mentioned earlier, 454 Life Sciences can erroneously sequence 1 bp per 1000 bp (47). The error rate is strongly related to the presence and size of homopolymers (48), that is, genome regions, consisting of consecutive repetition of a single base (e.g., TTTTT).

   We use KEC, SAET, and ShoRAH programs to do error correction prior to assembly. These error correction algorithms involve clustering of reads. While ShoRAH clusters the reads in Bayesian manner using the Dirichlet process mixture (41), and KEC clusters reads based on k-mers (49) , SAET uses reads of quality scores for error correction (50).

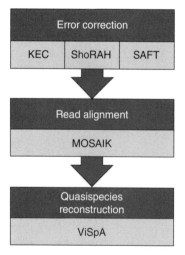
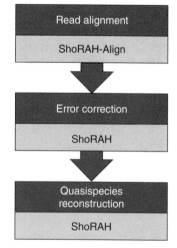

Figure 17.1   Evaluated reconstruction flows. The evaluated reconstruction flows consists of three steps: (i) read error correction (ii) read alignment, and (iii) reconstruction of viral qua-sispecies.

2. Read Alignment. In this step, we use independent alignment program to map reads against a reference viral sequence (51), this aligner can be easily replaced with another one.
3. Reconstruction of Viral Quasispecies. In this step, we use two assembly programs ViSpA (40) and ShoRAH (41) to reconstruct variants from aligned reads and estimate their relative frequencies.

ViSpA (40) executes the following steps and outputs the quasispecies spectrum (i.e. variant sequences and their relative frequencies):

- Preprocess Aligned Reads. ViSpA uses placeholders I and D for aligned reads containing insertions and deletions; in this process, it does a simplistic error correction. Deletions supported by a single read are replaced either with the allele present in all the other reads in the same position if they are the same or with N (unknown base pair), and it removes insertions supported by a single read.
- Construct the Read Graph. In the read graph, each vertex corresponds to a read and each directed edge connects two overlapping reads. ViSpA differentiates between two types of reads: super-read and sub-read, which is a substring of the super-read. The read graph consists only of super-reads.
- Assemble Candidate Quasispecies Sequences. Each candidate variant corresponds to a path in the read graph. ViSpA uses what is the so-called max-bandwidth paths for assembly.
- Estimate Frequency of Haplotype Sequences. In this step, ViSpA uses expectation maximization algorithm to estimate the frequency of each reconstructed sequence using both super-reads and sub-reads.

ShoRAH (41, 42) executes the following steps and outputs the quasispecies spectrum:

- Align Reads. The first step for ShoRAH is producing a Multiple Sequence Alignment (MSA) of reads. We used the version 0.5 of the ShoRAH. This version had its own aligner that was used in the study. It aligns all reads to the reference and from the set of pairwise alignments it builds an MSA.
- Correct Reads from Genotyping Errors (Local Haplotype Reconstruction). While ViSpA uses independent error correction programs, ShoRAH uses its own error correction method. Sequencing errors are corrected by a Bayesian inference algorithm, which estimates the quality of the reconstruction, although only the maximum likelihood estimate is passed on to subsequent steps. ShoRAH implements a specific probabilistic clustering method based on the Dirichlet process mixture for correcting technical errors in deep sequencing reads and for highlighting the biological variation in a genetically heterogeneous sample.

- Reconstruct Global Haplotype. This step is similar to assembly of candidate quasispecies sequences in ViSpA.
- Estimate Frequency. In this step, ShoRAH estimates the frequency of each candidate sequence.

17.3.1 Compared Methods

We vary different parameter values, the following tuning parameters are used for ViSpA quasispecies reconstruction:

- n: number of mismatches between super-reads and sub-reads
- m: number of mismatches in the overlap between two super-reads
- t: mutation rate

For ShoRAH, we used the default parameters for quasispecies reconstruction.

17.4 RESULTS AND DISCUSSION

17.4.1 Data Sets

Read samples were collected from IBV-infected chickens, and quasispecies variants were sequenced using life sciences 454 shotgun sequencing, followed by Sanger sequencing of individual variants. Reads coverage profile is represented on Figure 17.2. The initial number of reads was 21,040. After SAET error correction, the number of reads did not change. It decreased to 19,439 after ShoRAH correction and to 17,122 after KEC correction. Ten Sanger clones (true variants) were used for validation. These clones are considered as the golden standards or the ground truth for parameter calibration and comparison between different methods (see Figure 17.3).

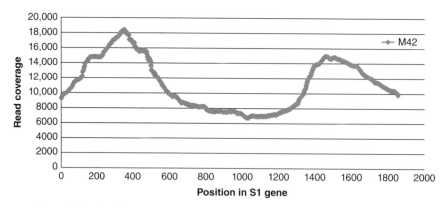

Figure 17.2   Read coverage. Number of reads covered every position in S1 gene.

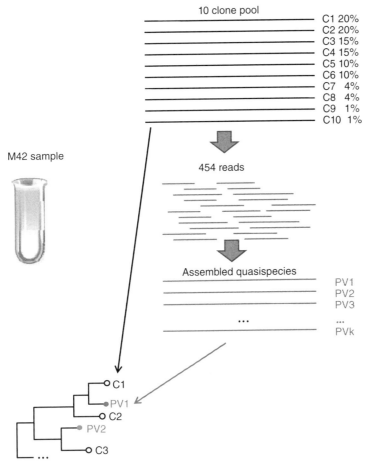

Figure 17.3 Schematic representation of calibration, validation experiments based on Sanger clones

17.4.2   Validation of Error Correction Methods

Experiment 1.

We mapped the 454 IBV reads to each of the 10 Sanger clones using Mosaik Aligner and computed the edit distance for each 454 read (or mismatch), and for each Sanger clone we categorized reads according to 0, 1, 2, ..., 15 mismatches to this Sanger clone. We computed the average coverage of each read set/category to this Sanger clone by summing up the coverage count of each position in the clone sequence then dividing by its length.

After the previous procedures, we computed the average coverage for uncorrected reads, reads corrected by SAET, reads corrected by KEC, and reads corrected by ShoRAH. Sanger clones got the highest average coverage when mismatches are less than 5 for uncorrected and SAET corrected reads, and less than 4 mismatches for KEC and ShoRAH corrected reads. The variant 42V1H3_G01_04 (the suffix _04

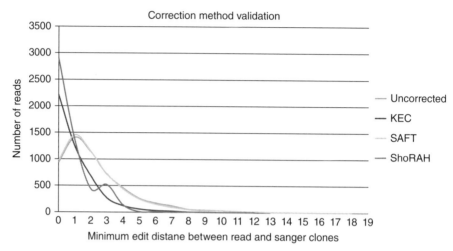

Figure 17.4 The distribution of 454 IBV reads categories (edit distance to the closest Sanger clone) for different correction methods.

means that the frequency of this Sanger clone is 4%) had low average coverage along all mismatches values and we considered it as an outlier.

Experiment 2. We mapped each of 454 IBV reads to all Sanger clones using Mosaik Aligner and computed the minimum edit distance of the read to some Sanger clones. Figure 17.4 shows the number of reads per category (i.e., minimum edit distance to some clones) for used correction methods.

From Figure 17.4, we see that for 0 and 1 mismatches (minimum edit distance category), ShoRAH outputs more reads than KEC, which comes in the second place. Then SAET comes in the third place, we notice that SAET is too conservative, that is, it covers nearly the same number as uncorrected reads.

17.4.3 Tuning, Comparison, and Validation of Methods for Quasispecies Reconstruction

We measured the pairwise edit distance for the 10 Sanger clones as follows. We ran pairwise alignment for the Sanger clones using ClustalW, then cut the sticking out ends, and computed the pairwise edit distance (Levenshtein Distance) for all clones in the overlapped region (see Table 17.1).

To validate the reconstruction of the quasispecies using different methods, we computed the pairwise distances between Sanger clones and reconstructed variants. We measured the distances as follows. We computed pairwise alignment between Sanger clones and reconstructed variants using ClustalW, then cut sticking out ends, and got overlapped region between each clone and each variant, the reason behind that was to avoid calculating edit distance for uncovered fragments (all overlapped regions happened to be more than 500 bp long, which is close to Sanger clones average length (=550 base pairs)).

As a result of the previous step, we got several groups of identical overlapped regions of reconstructed variants and collapsed them and summed up their

TABLE 17.1   Pairwise Edit Distance between the 10 Sanger Clones

Frequencies of Reconstructed Variants	Sanger Clones									
	42E9_A08_20	42V1H7_C02_20	42E3_C07_15	42A6_F04_15	42E10_B08_10	H1_E11_10	42V1B10_B04_04	42V1H3_G01_04	42H5_A12_01	42V1C6_B08_01
42E9_A08_20	0	0	1	1	1	1	3	3	2	3
42V1H7_C02_20		0	1	1	1	1	3	3	2	3
42E3_C07_15			0	2	2	2	4	4	3	4
42A6_F04_15				0	0	2	2	2	3	2
42E10_B08_10					0	2	2	2	3	2
42H1_E11_10						0	3	2	1	2
42V1B10_B04_04							0	2	4	3
42V1H3_G01_04								0	3	2
42H5_A12_01									0	3
42V1C6_B08_01										0

The suffix number of the clone id is the clone frequency, for example, 42E9_A08_20 means that the frequency of the clone 42E9_A08 is 20%

frequencies (the suffix of the variant sequences ID's represent the collapsed variant abundance, for example, the frequency of the variant 42E9_A08_40 is 40%).

Identical Sanger clone fragments were also collapsed (e.g., the two most frequent clones 42E9_A08_20 and 42V1H7_C02_20 are collapsed to 42E9_A08_40). Then we computed edit distance (Levenshtein distance) for every overlapped region.

Table 17.2 shows edit distance values between distinct Sanger clones and reconstructed variants for one of the dominating methods using ViSpA where the

TABLE 17.2   Edit Distance between Collapsed Sanger Clones and ViSpA Reconstructed Variants Using Parameters 1, 2, 5 (Number of Mismatches between Sub-Reads and Super-Reads, Number of Mismatches between Two Overlapped Reads and Mutation Rate, Respectively), Threshold=0.005 on KEC Corrected Reads

Frequencies of Reconstructed Variants	Sanger Clones							
	42E9_A08_40	42A6_F04_25	42E3_C07_15	42H1_E11_10	42V1B10_B04_04	42V1H3_G01_04	42H5_A12_01	42V1C6_B08_01
0.6647	0	1	1	1	3	3	2	3
0.1212	1	0	2	2	2	2	3	2
0.0399	1	2	0	2	4	4	3	4
0.1054	2	3	3	3	5	5	4	5
0.0461	2	1	3	3	3	3	4	3
0.0228	1	2	2	2	4	4	3	4

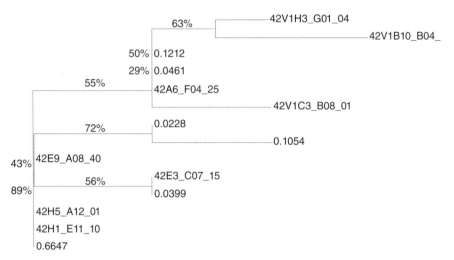

Figure 17.5 Phylogenetic tree over collapsed Sanger clones and collapsed reconstructed variants inferred from the method with parameters 1_2_5 on KEC corrected reads using ViSpA.

number of mismatches between super-reads and sub-reads (n) is 1, the number of mismatches in the overlapping region between two reads (m) is 2, and the mutation rate (t) is 5. We considered only reconstructed variants with frequencies more than 0.005 (threshold=0.005). Reads are corrected using KEC program.

As we see from Table 17.2, the method correctly reconstructed three of the most frequent variants. Thus, in the overlapped region, there was edit distance 0 between the most frequent Sanger clone with id 42E9_A08_40 and frequency 0.40 and the reconstructed variant with the frequency 0.6647. In addition, 42A6_F04_25 clone was identical to the variant with frequency 0.1212 and 42E3_C07_15 clone was identical to the variant with frequency 0.0399. For the most popular variant the method overestimated the frequency assigning it to 0.6647 instead of true 0.40. On the other hand, it underestimated the second and the third popular variants—obtained frequencies 0.1212 instead of 0.25 and 0.0399 instead of 0.15. On the whole, the validation with distinct Sanger clones show that existing reconstruction methods can give adequate representation of IBV variant population.

To compare how well the reconstructed variant sequences of the previous method were compared to Sanger clones, we reconstructed the phylogenetic tree, see Figure 17.5 (it shows only the frequencies for the reconstructed variants, while a Sanger clone starts with the number 42). As we can see from the phylogenetic tree, the reconstructed variants by ViSpA are close to Sanger clones.

To compare between different methods (with different parameter settings), we use the following two measures:

- average distance to clones ADC$= \sum_i d(c_i) \cdot f(c_i)$
- average prediction error APE$= \sum_j d(q_j) \cdot f(q_j)$,

where $c_1, \cdots, c_{10}$ are Sanger clones, $q_1, \cdots, q_j$ are reconstructed variants, $f(c_i)$ and $f(q_j)$ are frequencies of ith clone and jth variant, and $d(c_i)$ and $d(q_j)$ are edit distance from ith clone to the closest reconstructed variant and edit distance from jth reconstructed variant to the closest clone.

Average Distance to Clones (ADC) and Average Prediction Error (APE) are analogous to sensitivity and ppv, respectively. The difference is that ADC and APE indicate better quality whenever they are closer to 0, while sensitivity and ppv indicate better quality when they are closer to 1.

In addition to the previous method. We ran many other experiments using different methods (see Tables 17.3–17.5 and Figures 17.6–17.8) and calculated the values of ADC) and APE for each method as shown in Table 17.6 and Table 17.7, respectively

TABLE 17.3 Edit Distance between Collapsed Sanger Clones and ViSpA Reconstructed Variants Using Parameters 2, 2, 10 (the Number of Mismatches between Sub-Reads and Super-Reads, the Number of Mismatches between Two Overlapped Reads, and Mutation Rate, Respectively), Threshold=0.005 on KEC Corrected Reads, where 85% of Reconstructed Variants Have Perfect Match with 65% of the Sanger Clones

	Sanger Clones							
Frequencies of Reconstructed Variants	42E9_A08_40	42A6_F04_25	42E3_C07_15	42H1_E11_10	42V1B10_B04_04	42V1H3_G01_04	42H5_A12_01	42V1C6_B08_01
0.6703	**0**	1	1	1	3	3	2	3
0.1845	1	**0**	2	2	2	2	3	2
0.0054	4	3	5	5	5	5	6	5

TABLE 17.4 Edit Distance between Collapsed Sanger Clones and ViSpA Reconstructed Variants Using Parameters 1, 2, 0 (the Number of Mismatches between Sub-Reads and Super-Reads, the Number of Mismatches between Two Overlapped Reads, and Mutation Rate, Respectively), Threshold=0.005 on SAET Corrected Reads

	Sanger Clones							
Frequencies of Reconstructed Variants	42E9_A08_40	42A6_F04_25	42E3_C07_15	42H1_E11_10	42V1B10_B04_04	42V1H3_G01_04	42H5_A12_01	42V1C6_B08_01
0.1448	**0**	1	1	1	3	3	2	3
0.0714	1	**0**	2	2	2	2	3	2
0.0067	1	2	**0**	2	4	4	3	4
0.1923	1	2	2	2	4	4	1	4
0.0469	2	3	3	3	5	5	4	5
0.0061	4	5	5	5	7	7	6	7

TABLE 17.5 Edit Distance between Collapsed Sanger Clones and ShoRAH Reconstructed Variants Using Default Parameters, Threshold=0.005 on Uncorrected Reads

Frequencies of Reconstructed Variants	42E9_A08_40	42A6_F04_25	42E3_C07_15	42H1_E11_10	42V1B10_B04_04	42V1H3_G01_04	42H5_A12_01	42V1C6_B08_01
0.1277	0	1	1	1	3	3	2	3
0.0663	1	0	2	2	2	2	3	2
0.0386	1	2	0	2	4	4	3	4
0.0535	4	5	5	5	7	7	6	7
0.0526	2	3	3	3	5	5	4	5
0.0434	8	9	9	9	11	11	10	11
0.0289	3	4	4	4	6	6	5	6
0.0268	9	10	10	10	12	12	11	12
0.0263	5	6	6	6	8	8	7	8
0.0259	4	5	5	5	7	7	6	7
0.0243	3	2	4	4	4	4	5	4
0.0191	6	7	5	7	9	9	8	9
0.0189	5	6	4	6	8	8	7	8
0.0172	3	4	4	4	6	6	5	6
0.01718	1	2	2	2	4	4	3	4
0.0167	9	10	8	10	12	12	11	12
0.0159	2	3	1	3	5	5	4	5
0.0156	3	2	4	4	4	4	5	4
0.0127	10	9	11	11	11	11	12	11
0.0110	4	5	5	5	7	7	6	7
0.0097	1	2	2	2	4	4	3	4
0.0089	12	13	13	13	15	15	14	15
0.0088	1	2	2	2	4	4	3	4
0.0084	5	4	6	6	6	6	7	6

(ViSpA 1_2_5 means that the variants are reconstructed by ViSpA using parameter values $n = 1, m = 2, t = 5$). Table 17.5 shows that the method is able to recall or reconstruct 12.7% of the variants with 0 edit distance with the most frequent Sanger clones (40%), 6.6% of the variants with 0 edit distance with 25% frequent Sanger clones, 3.8% of the variants with 0 edit distance with 15% frequent Sanger clones; therefore, the used method is able to reconstruct or recall 23% ($0.127 + 0.066 + 0.038$) of the variants with 0 edit distance with 80% ($0.40 + 0.25 + 0.15$) of Sanger clones with relatively small precision since it reconstructs more false negatives.

We say that the method A dominates the method B if both ADC and APE values of A are at most the corresponding ADC and APE values of B. Figure 17.9 (a pictorial diagram of Tables 17.6 and 17.7) shows that methods V125KEC (V: ViSpA

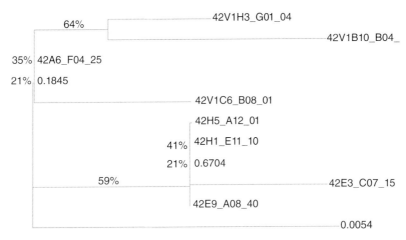

Figure 17.6 Phylogenetic tree over collapsed Sanger clones and collapsed reconstructed variants inferred from one of the dominating methods with parameters 2_2_10 on KEC corrected reads using ViSpA.

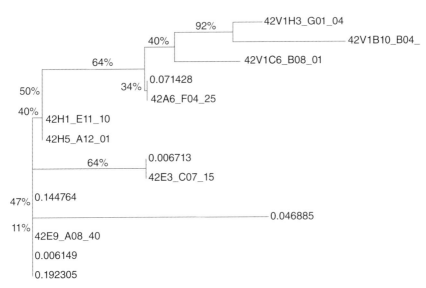

Figure 17.7 Phylogenetic tree over collapsed Sanger clones and collapsed reconstructed variants inferred from one of the dominating methods with parameters 1_2_0 on SAET corrected reads using ViSpA.

assembler, 1:*n*, 2:*m*, 5:*t*, KEC:correction method), V2210KEC, and V120SAET dominate all other methods, that is, have the best values in terms of ADC and APE. Our results suggest that using different methods with different parameter calibration and parameter settings can improve the solution and its predictive power for the quasispecies inference problem in terms of recall and precision.

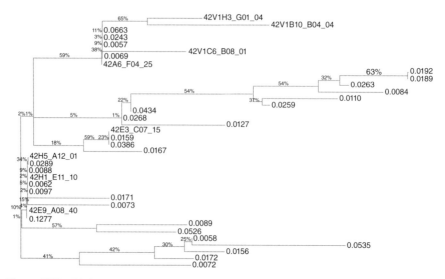

Figure 17.8 Phylogenetic tree over collapsed Sanger clones and collapsed reconstructed variants inferred from one of the methods with default parameters on Uncorrected reads using ShoRAH (close to the dominating methods).

TABLE 17.6   Average Distance to Clones (ADC) for the Reconstructed Variants Using Different Methods

	Error Correction Method		
	Uncorrected	KEC	SAET
ViSpA 1_2_0	0.45	0.79	0.29
ViSpA 1_2_5	0.3	0.3	0.45
ViSpA 1_2_10	0.45	0.45	0.3
ViSpA 2_2_0	0.45	0.79	0.3
ViSpA 2_2_5	0.45	0.3	0.3
ViSpA 2_2_10	0.3	0.45	0.3
ShoRAH	0.3	0.3	0.3

TABLE 17.7   Average Prediction Error (APE) for the Reconstructed Variants Using Different Methods

	Error Correction Method		
	Uncorrected	KEC	SAET
ViSpA 1_2_0	0.2	0.16	0.66
ViSpA 1_2_5	0.63	0.28	0.58
ViSpA 1_2_10	0.53	0.18	0.45
ViSpA 2_2_0	1.35	0.26	3.12
ViSpA 2_2_5	0.2	0.29	0.31
ViSpA 2_2_10	0.41	0.02	0.41
ShoRAH	2.84	2.89	2.45

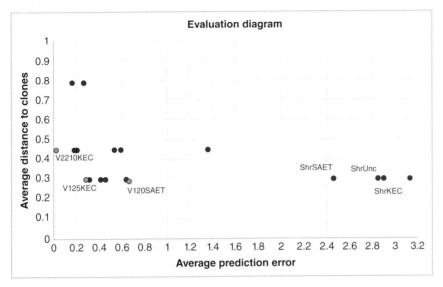

Figure 17.9 Evaluation diagram for average prediction error (APE) and average distance to clones (ADC) values for different methods. Each point corresponds to a method and the dominant solutions correspond to red points.

ACKNOWLEDGMENTS

This work has been partially supported by NSF award IIS-0916401, NSF award IIS-0916948, Agriculture and Food Research Initiative Competitive Grant No. 201167016-30331 from the USDA National Institute of Food and GSU Molecular Basis of Disease Fellowship.

REFERENCES

1. Cavanagh D. Coronavirus avian infectious bronchitis virus. Vet Res 2007;38(2):281–297.

2. Vijayykrishna D, Smith G, Zhang J, Peiris J, Chen H, Guan Y. Evolutionary insights into the ecology of coronaviruses. J Virol 2007;81(8):4012–4020.

3. Liu S, Zhang X, Wang Y, Li C, Han Z, Shao Y, Li H, Kong X. Molecular characterization and pathogenicity of infectious bronchitis coronaviruses: complicated evolution and epidemiology in china caused by cocirculation of multiple types of infectious bronchitis coronaviruses. Intervirology 2009;52(4):223–234.

4. Jackwood M, Hilt D, Callison S. Detection of infectious bronchitis virus by real-time reverse transcriptase-polymerase chain reaction and identification of a quasispecies in the beaudette strain. Avian Dis 2003;47(3):718–724.

5. McKinley ET, Hilt DA, Jackwood MW. Avian coronavirus infectious bronchitis attenuated live vaccines undergo selection of subpopulations and mutations following vaccination. Vaccine 2008;26(10):1274–1284.

6. Jackwood MW, Hilt DA, McCall AW, Polizzi CN, McKinley ET, Williams SM. Infectious bronchitis virus field vaccination coverage and persistence of Arkansas-type viruses in commercial broilers [cobertura de la vacunación a nivel de campo contra el virus de la bronquitis infecciosa y persistencia de virus del tipo arkansas en pollos de engorde comerciales]. Avian Dis 2009;53(2):175–183. cited By (since 1996) 1.

7. Nix W, Troeber D, Kingham B, Keeler C Jr., Gelb J Jr. Emergence of subtype strains of the Arkansas serotype of infectious bronchitis virus in Delmarva broiler chickens. Avian Dis 2000;44(3):568–581.

8. Holt RA, Jones SJ. The new paradigm of flow cell sequencing. Genome Res 2008;18(6):839–846. DOI: 10.1101/gr.073262.107.

9. Marioni JC, Mason CE, Mane SM, Stephens M, Gilad Y. RNA-seq: an assessment of technical reproducibility and comparison with gene expression arrays. Genome Res 2008;18:1509–1517. DOI: 10.1101/gr.079558.108.

10. Mortazavi A, Williams BAA, McCue K, Schaeffer L, Wold B. Mapping and quantifying mammalian transcriptomes by RNA-Seq. Nat Methods 2008;5(7):621–628. DOI: 10.1038/nmeth.1226.

11. Sittka A, Lucchini S, Papenfort K, Sharma CM, Rolle K, Binnewies TT, Hinton JC, Vogel J. Deep sequencing analysis of small noncoding RNA and mRNA targets of the global post-transcriptional regulator, HFQ. PLoS Genet 2008;4(8):1000163. DOI: 10.1371/journal.pgen.1000163.

12. Pomraning KR, Smith KM, Freitag M. Genome-wide high throughput analysis of DNA methylation in eukaryotes. Methods 2009;47(3):142–150.

13. Barski A, Cuddapah S, Cui K, Roh TY, Schones DE, Wang Z, Wei G, Chepelev I, Zhao K. High-resolution profiling of histone methylations in the human genome. Cell 2007;129(4):823–837. DOI: 10.1016/j.cell.2007.05.009.

14. Bentley DR et al. Accurate whole human genome sequencing using reversible terminator chemistry. Nature 2008;456(7218):53–59. DOI: 10.1038/nature07517.

15. Wang J et al. The diploid genome sequence of an Asian individual. Nature 2008; 456(7218):60–65. DOI: 10.1038/nature07484.

16. Cox-Foster DL, Conlan S, Holmes EC, Palacios G, Evans JD et al. A metagenomic survey of microbes in honey bee colony collapse disorder. Science 2007;318(5848):283–287. DOI: 10.1126/science.1146498.

17. Dinsdale E, Edwards R, Hall D, Angly F, Breitbart M, Brulc J et al. Functional metagenomic profiling of nine biomes. Nature 2008;452(7187):629–632. DOI: 10.1038/nature06810.

18. Briggs AW, Stenzel U, Johnson PL, Green RE, Kelso J, Prufer K, Meyer M, Krause J, Ronan M, Lachmann M, Paabo S. Patterns of damage in genomic DNA sequences from a neandertal. Proc Natl Acad Sci U S A 2007;104(37):14616–14621. DOI: 10.1073/pnas.0704665104.

19. Gilbert M, Tomsho LP, Rendulic S, Packard M, Drautz DI, Sher A et al. Whole-genome shotgun sequencing of mitochondria from ancient hair shafts. Science 2007;317(5846):1927–1930. DOI: 10.1126/science.1146971.

20. Eriksson N, Pachter L, Mitsuya Y, Rhee SY, Wang C, Gharizadeh B, Ronaghi M, Shafer RW, Beerenwinkel N. Viral population estimation using pyrosequencing. PLoS Comput Biol 2008;4(4):e1000074. DOI: 10.1371/journal.pcbi.1000074.

21. Archer J, Braverman MS, Taillon BE, Desany B, James I, Harrigan PR, Lewis M, Robertson DL. Detection of low-frequency pretherapy chemokine (CXC motif) receptor 4 (CXCR4)-using HIV-1 with ultra-deep pyrosequencing. AIDS 2009;23(10):1209–1218.

22. Hoffmann C, Minkah N, Leipzig J, Wang G, Arens MQ, Tebas P, Bushman FD. DNA bar coding and pyrosequencing to identify rare HIV drug resistance mutations. Nucleic Acids Res 2007;35(13):91.

23. Simons JF, Egholm M, Lanza J, Desany B, Turenchalk G et al. Ultradeep sequencing of HIV from drug resistant patients. Antivir Ther 2005;10:157.

24. Tsibris AMN, Russ C, Lee W, Paredes R, Arnaout R et al. Detection and quantification of minority HIV-1 env V3 loop sequences by ultra-deep sequencing: preliminary results. Antivir Ther 2006;11:74.

25. Wang C, Mitsuya Y, Gharizadeh B, Ronaghi M, Shafer RW. Characterization of mutation spectra with ultra-deep pyrosequencing: application to HIV-1 drug resistance. Genome Res 2007;17(8):1195–1201.

26. Lanford RE, Hildebrandt-Eriksen ES, Petri A, Persson R, Lindow M, Munk ME, Kauppinen S, Rum H. Therapeutic silencing of microRNA-122 in primates with chronic hepatitis C virus infection. Science 2010;327(5962):198–201.

27. Kuroda M, Katano H, Nakajima N, Tobiume M, Ainai A, Sekizuka T, Hasegawa H, Tashiro M, Sasaki Y, Arakawa Y, Hata S, Watanabe M, Sata T. Characterization of quasispecies of pandemic 2009 influenza A Virus (A/H1N1/2009) by *De Novo* sequencing using a next-generation DNA sequencer. PLoS ONE 2010;5(4):10256. DOI: 10.1371/journal.pone.0010256.

28. Holmes EC, Grenfell BT. Discovering the phylodynamics of RNA viruses. PLoS Comput Biol 2009;5(10):e1000505.

29. Butler J, MacCallum I, Kleber M, Shlyakhter IA, Belmonte MK, Lander ES, Nusbaum C, Jaffe DB. ALLPATHS: De novo assembly of whole-genome shotgun microreads. Genome Res 2008;18:810–820.

30. Chaisson MJ, Pevzner PA. Short read fragment assembly of bacterial genomes. Genome Res 2008;18(2):324–330. doi: 10.1101/gr.7088808.

31. Medvedev P, Brudno M. Ab initio whole genome shotgun assembly with mated short reads. Proceedings of RECOMB 2008; 2008. p 50–64.

32. Zerbino DR, Birney E. Velvet: algorithms for de novo short read assembly using de Bruijn graphs. Genome Res 2008;18:821–829.

33. Bansal V, Bafna V. HapCUT: an efficient and accurate algorithm for the haplotype assembly problem. Bioinformatics 2008;24:153–159.

34. Chen Z, Fu B, Schweller R, Yang B, Zhao Z, Zhu B. Linear time probabilistic algorithms for the singular haplotype reconstruction problem from SNP fragments. J Comput Biol 2008;15(5):535–546. DOI: 10.1089/cmb.2008.0003.

35. Genovese LM, Geraci F, Pellegrini M. A fast and accurate heuristic for the single individual SNP haplotyping problem with many gaps, high reading error rate and low coverage. In: *Algorithms in Bioinformatics*. Berlin Heidelberg: Springer-Verlag; 2007. p 49–60.

36. Li LM, Kim JH, Waterman MS. Haplotype reconstruction from SNP alignment. J Comput Biol 2004;11(2-3):505–516.

37. Lippert R, Schwartz R, Lancia G, Istrail S. Algorithmic strategies for the single nucleotide polymorphism haplotype assembly problem. Brief Bioinform 2002;3(1):23–31.

38. Panconesi A. Fast Hare: a fast heuristic for single individual SNP haplotype reconstruction. Lect Notes Comput Sci 2004;3240:266–277.

39. Xie M, Wang J, Chen J. A model of higher accuracy for the individual haplotyping problem based on weighted SNP fragments and genotype with errors. Bioinformatics 2008;24(13):105–113.

40. Astrovskaya I, Tork B, Mangul S, Westbrooks K, Măndoiu I, Balfe P, Zelikovsky A. Inferring viral quasispecies spectra from 454 pyrosequencing reads. BMC Bioinformatics 2011;12 Suppl 6:1. DOI: 10.1186/1471-2105-12-S6-S1.

41. Zagordi O, Geyrhofer L, Roth V, Beerenwinkel N. Deep sequencing of a genetically heterogeneous sample: local haplotype reconstruction and read error correction. J Comput Biol 2010;17(3):417–428. DOI: 10.1089/cmb.2009.0164.

42. Zagordi O, Bhattacharya A, Eriksson N, Beerenwinkel N. ShoRAH: estimating the genetic diversity of a mixed sample from next-generation sequencing data. BMC Bioinformatics 2011;12:119.

43. Töpfer A, Zagordi O, Prabhakaran S, Roth V, Halperin E, Beerenwinkel N. Probabilistic inference of viral quasispecies subject to recombination. J Comput Biol 2013;20(2):113–123.

44. Mancuso N, Tork B, Skums P, Ganova-Raeva L, Măndoiu I, Zelikovsky A. Reconstructing viral quasispecies from NGS amplicon reads. In Silico Biol 2011;11(5):237–249.

45. Prosperi MC, Salemi M. Qure: software for viral quasispecies reconstruction from next-generation sequencing data. Bioinformatics 2012;28(1):132–133.

46. Prabhakaran S, Rey M, Zagordi O, Beerenwinkel N, Roth V. HIV haplotype inference using a propagating dirichlet process mixture model. IEEE/ACM Trans Comput Biol Bioinform 2014;11(1):182–191. DOI: 10.1109/TCBB.2013.145.

47. 454 Life Sciences. Available at http://www.454.com. Accessed 2016 19 Mar.

48. Brockman W, Alvarez P, Young S, Garber M, Giannoukos G, Lee WL, Russ C, Lander ES, Nusbaum C, Jaffe DB. Quality scores and snp detection in sequencing-by-synthesis systems. Genome Res 2008;18(5):763–770. DOI: 10.1101/gr.070227.107.

49. Skums P, Dimitrova Z, Campo DS, Vaughan G, Rossi L, Forbi JC, Yokosawa J, Zelikovsky A, Khudyakov Y. Efficient error correction for next-generation sequencing of viral amplicons. BMC Bioinformatics 2012;13 Suppl 10:6.

50. Brinza D, Hyland F. Error correction methods in next generation sequencing. Proceedings of IEEE International Conference on Computational Advances in Bio and Medical Sciences (ICCABS 2011); 2011. p 268.

51. Strömberg M. Mosaik Aligner. Available at https://code.google.com/p/mosaik-aligner. Accessed 2016 Mar 19.

18

MICROBIOME ANALYSIS: STATE OF THE ART AND FUTURE TRENDS

MITCH FERNANDEZ, VANESSA AGUIAR-PULIDO, JUAN RIVEROS, AND WENRUI HUANG

Bioinformatics Research Group (BioRG), School of Computing and Information Sciences, Florida International University, Miami, FL, USA

JONATHAN SEGAL

Herbert Wertheim College of Medicine, Florida International University, Miami, FL, USA

ERLIANG ZENG

Department of Computer Science and Engineering, University of Notre Dame, Notre Dame, IN, USA

MICHAEL CAMPOS

Miller School of Medicine, University of Miami, Miami, FL, USA

KALAI MATHEE

Herbert Wertheim College of Medicine, Florida International University, Miami, FL, USA

GIRI NARASIMHAN

Bioinformatics Research Group (BioRG), School of Computing and Information Sciences, Florida International University, Miami, FL, USA

18.1 INTRODUCTION

Microbes form complex heterogeneous interacting communities, whether in the environment or in specific niches within humans and other host organisms (1). *Metagenomics* approaches have been used to study the composition and dynamics of such

Computational Methods for Next Generation Sequencing Data Analysis, First Edition.
Edited by Ion I. Măndoiu and Alexander Zelikovsky.
© 2016 John Wiley & Sons, Inc. Published 2016 by John Wiley & Sons, Inc.
Companion website: www.wiley.com/go/Mandoiu/NextGenerationSequencing

microbial communities. Distinct communities of bacteria are present at different sites of the human body, and changes in their structure have strong implications for human health. The Human Microbiome Project (HMP) focuses on the study of microbial communities that inhabit the healthy human body (2, 3). It is conjectured that 90% of the cells in the human body are bacterial cells and that bacterial communities play such critical roles as aiding in the digestion of food, synthesizing essential vitamins, and assisting the immune system in fending off pathogenic invaders. Human microbiome studies have revealed that diseases and disorders are strongly correlated with changes in microbial community profiles (4–6). These studies have also demonstrated that microbial community structure in five niches of the human body (gut, mouth, airways, urogenital, and skin) are quite distinct and appear to transcend gender, age, and ethnicity (7).

It is possible to employ classical microbiological methods for the analysis of microbiomes, by which one could culture individual taxa and then use specific protocols developed to identify them. However, classical approaches will always be limited by (i) the ability to culture—it is broadly believed that most bacterial taxa are not readily culturable by traditional culturing techniques (although more extensive cultivation approaches have yielded considerably higher percentages of cultured taxa, as shown by the work of M. Surette and colleagues (8)); (ii) time it takes to culture—it could take several weeks for a culture to grow; and (iii) the potentially staggering number and diversity of microbial taxa present in the sample.

Molecular approaches to microbiome studies involve extracting microbial DNA from a sample followed by a process of determining the profile of the microbial community present. A reasonable approach is that of PCR amplifications of marker genes, such as the gene for 16S rRNA, followed by one of a variety of approaches to investigate length heterogeneity of variable regions of the microbial genomes (9). Here, the limitations are the result of the fact that length heterogeneity of specific regions of certain marker genes in the bacterial world is considerably smaller than the number of taxa.

Exploiting sequence heterogeneity is clearly a more informative approach than using length heterogeneity. More recent methods involve the use of next-generation sequencing. The DNA is first extracted and a portion of the marker gene is amplified with appropriate PCR primers. Then, next-generation sequencing techniques generate a large number of reads from the amplicons. Bioinformatics tools are finally used to classify the reads by the different taxa from which they have arisen.

Although this will be discussed in detail later, the limitations of these methods are the following: (i) the extent of our databases and the sequences cataloged in them limits the bacterial taxa that can be successfully identified; (ii) next-generation sequencing techniques produce short reads that may or may not be from distinguishable regions of the individual taxon; and (iii) the software classifier used to classify individual reads that may produce inaccurate results, in part, because the independence assumptions inherent in existing classifiers may not be biologically valid.

Microbiome studies have largely been applied on phylogenetic marker genes, most often the gene for 16S rRNA (e.g., see References 10–15), but others include 23S rRNA, *recA*, *gyrA*, and many more (16–19). There is no consensus on the ideal marker gene, and the popularity of 16S rRNA is simply due to its use in standard reference manuals (19). However, recent studies have taken a different approach by employing whole genome sequencing (7, 20, 21). The advantage of marker genes is the availability of large databases (22–24) to aid in classification of reads. The advantage of the whole genome approach is that it becomes possible to obtain functional profiles of the microbial community by investigating the functional annotations of the genes to which the reads map. We discuss later why this is potentially of greater interest than the marker gene approach.

Regardless of the method of identifying bacterial taxa or the functional elements present, the next step is to leverage the information to understand the difference between samples. Such measures include α- and β-diversity indices, which have been shown to be useful in distinguishing between samples (25, 26). Specific indices for calculating these types of diversity include the Shannon diversity index, popular for its dealing with the uncertainty in predicting unobserved species (27); the Jaccard index, which measures similarity between samples by the proportion of shared clusters or species between them (28); and the popular inverse-Simpson index, which estimates diversity by considering the probability that a randomly selected individual from a sample has already been observed (29). It should be noted that there are situations where well-established diversity indices are not sufficient to differentiate between groups of samples. In recent studies, it has been observed that the diversity indices for the airways microbiome of smoking and nonsmoking subjects are not distinguishable, even though these studies have readily identified individual bacterial genera that tend to favor the airways of smokers over nonsmokers or former smokers (30, 31).

More advanced techniques to understand and differentiate microbial community profiles have been investigated. It is clear that metagenomic studies have to go further and dig deeper to uncover interesting features of microbial communities. The next set of promising investigations have to focus on understanding the structure of microbial communities and their interactions in any environmental niche. One of the greatest challenges in understanding human health is uncovering the large number of complex interactions that occur within the microbial community, and between the community and the human host. There is a great need to interpret the results in a way that is useful to both research scientists and clinicians. Studying the structure of microbial communities will shed light on the nature of bacterial "social networks" and their consequences.

18.2   THE METAGENOMICS ANALYSIS PIPELINE

No single approach is sufficient to fully characterize a source environment due to the complexity of the microbial communities in metagenomic samples. Thus, any metagenomics analysis pipeline will involve a series of sequential steps. This begins with data preprocessing to filter reads by quality and length, remove contaminants, remove chimeric sequences generated during PCR amplification, and prepare data for subsequent analysis. A classification and clustering step is used to map each read to a sequence in a reference database and group reads by taxonomy or sequence similarity. This will be discussed in more detail later. A single-sample analysis step employs standard measures, such as making estimates of the richness and diversity of taxa in each sample; when whole genome sequencing is used, the analysis involves studying the functional pathways and protein families that are present and/or overrepresented. Finally, multiple sample comparisons are used to identify patterns in related groups of samples. Every step in the pipeline has the potential of introducing errors and biases, which will be carried forward to the rest of the analyses (15). It is essential that parameters be selected with care, understanding the limitations of the data and the biases introduced during collection, amplification, and sequencing.

The marker gene approach is unable to directly provide information about the functional elements present in a sample due to its limited ability to resolve taxonomic identity. The gene for 16S rRNA is present in every bacterial genome. This gene contains a mixture of highly conserved and hypervariable regions, the former making it an easy target for amplification and the latter for mapping reads to taxa (32). However, it is generally not possible to use the 16S rRNA gene to distinguish between strains or species, especially with short reads, so identity is normally limited to the genus level or above (33). This can pose a problem for our purposes, since members of the same genera can behave very differently. For example, *Campylobacter hominis* is considered a member of the normal flora of the gut, whereas *Campylobacter jejuni* is known to be pathogenic (34). Furthermore, closely related bacterial species and strains are often competitors for the same environmental niches (35). It is desirable to differentiate between any species and strains present in order to better understand the dynamics of the communities being studied.

The limited resolution of marker gene methods does not mean that reads cannot be intelligently assigned to distinct groups, but rather that those groups cannot be easily mapped to specific taxa. A commonly used approach is to first cluster reads based on sequence similarity into *operational taxonomic units* (OTUs) and then assign the best available taxonomic identity to each OTU. This typically results in several OTUs mapping to the same taxon, with each OTU being roughly analogous to a strain or species at or above a specified sequence similarity. A commonly used threshold for the 16S rRNA marker gene is 97% similarity. It is sometimes useful to approximate higher taxonomic levels, for example, to conduct an analysis at the phylum level. This can be accomplished by adjusting the similarity threshold. Labeling each OTU with reference to its taxon is a convenience, but should be done with caution. There is experimental evidence that even the best 16S classifiers exhibit some taxon-specific biases (36). The use of OTUs may help minimize the effects of these biases.

There are other technical limitations with marker gene approaches. The choice of primers used for PCR can have an appreciable effect on amplification efficiency and resulting coverage (37). Primers are required for the process of amplification and are used by DNA polymerases to identify start sites for copying the complementary strand, as well as to serve as a scaffold on which to build the new strand. Another advantage of primers is that they can be designed to amplify the specific marker genes of our choosing by matching the start of the genomic sequences that interest us. The process of evolution will inevitably produce minor differences in DNA sequences as species diverge, even in the highly conserved regions of marker genes. Ideally designed primers must be able to bind to a maximum number of different target species while capturing genomic regions that maximize their distinguishability.

The most commonly used primers were developed many years ago when 16S rRNA databases were still relatively sparse (17, 38, 39). More recently designed degenerate primers, which contain certain "wildcard" bases, have been shown to be a vast improvement over these and can capture substantially more taxa (40). However, one can never know what has been missed by this approach. Another confounding problem with marker genes is the issue of copy number. All bacteria usually have multiple copies of 16S rRNA genes. Differences in copy number might give an inflated estimate of the relative abundance of certain taxa (41). More troubling is the problem of noise, the presence of a large number of reads that form very low membership OTUs, resulting in thousands of OTUs with as little as a single read. It can be difficult to determine if these OTUs should properly be merged with some other larger OTU (by adjusting the cutoff for similarity), if they represent amplification or sequencing artifacts, or if they are indicative of truly low abundance OTUs present in the sample (42). All of these limitations beg for intelligent ways of filtering and normalizing the data, and a standard for dealing with sequencing efficiency, copy number, and noise has yet to emerge. For now, it is important to recognize that every study is based on a blurry temporal snapshot of a microbial community potentially filled with large, gaping holes. This is obviously a challenge for making inferences about interactions between bacteria when trying to describe the microbial social network, but not an insurmountable one.

A large number of open-source analytical tools for metagenomics have emerged in recent years. These include MG-RAST, MOTHUR, QIIME, CloVR, VAMPS, and others (43–48). These can automatically step through a basic analytical pipeline for metagenomics. However, this is just a starting point for determining how these billions of cells are interacting and the implications for the health of the host. Are there groups of bacteria that tend to work together? Do they appear in healthy as well as diseased tissue? If these groups are present, do they actively compete with other groups for nutrition in the environmental niche, and just how fierce is this competition? What are the implications of the presence of specific groups on the health status of the human host? Can we classify certain bacteria as implicitly pathogenic based on their group membership? Can we infer causal relationships between these groups and disease states or the prognosis of the subject? These more interesting and challenging questions require reflection and are not always answered using cookie-cutter tools.

18.3 DATA LIMITATIONS AND SOURCES OF ERRORS

As stated earlier, every step in the metagenomics pipeline introduces new errors and biases, which will be carried forward to the rest of the analyses. Following are the sources of error in the analyses that are listed.

1. Contamination is a widespread problem, especially during sample collection and DNA extraction. However, propagation of contamination issues can be mitigated by incorporating appropriate controls to subtract out in the downstream analysis.

2. Partial or complete inhibition of amplification from different compounds or chemicals present in the sample, despite efforts to remove them in DNA extraction.

3. PCR amplification of some regions of bacterial genomes may not work efficiently due to issues such as inappropriate PCR amplicon size (outside the range of ideal amplicon size), low copy number, unfavorable nucleotide content (AT-rich or repetitive sequences), faulty universal primers, improperly optimized PCR conditions (temperature, reagent concentrations, or cycling parameters), or stochastic effects related to DNA concentration.

4. Multiple sources of error exist in the sequencing phase, resulting in many "noisy" reads. One way to eliminate these reads is to set a cutoff or threshold for including OTUs in the analysis. However, an incorrectly chosen threshold can be another source of error.

5. Clustering and classification of reads is a major source of error. If reads are clustered, then it uses evidence from a collection of reads in order to decide on a classification, thus reducing the reliance on the classification of individual reads. On the other hand, if the classification is incorrect, it could misclassify entire clusters.

6. When dealing with whole genome sequencing data, often more than one cluster is mapped to the same taxon, which is typically inferred to be caused by the presence of subcategories of that taxon (e.g., species or strains of the same bacterial group). Depending on the region used for the classification, this inference may be incorrect. It is possible that members of a different bacterial group acquire critical genes by *horizontal gene transfer* (HGT) and end up being misclassified.

7. Normalization is necessary, but can be fraught with errors because of the assumptions made. The choice of normalization method remains controversial (49, 50).

8. Data imputations, if needed, can be another source of error.

9. It is essential that parameters be selected with care and awareness of known limitations in how data was collected, amplified, and sequenced.

10. Any multistep inference process can be erroneous because of potential compounding of errors.

11. There are a multitude of external factors, which may influence the structure and activity of microbial communities but which are not considered by either amplicon or whole genome sequencing. These factors might be highly dynamic and can include temporary influxes of invading taxa from environmental exposure, the presence of non-microbial biological entities, changes in the availability of nutrients or the build-up of toxins due to host behavior or cyclical patterns (i.e., inconsistent sleep patterns, dietary changes, hormonal fluctuations), and other unknowns which may profoundly affect any analysis.

18.3.1 Designing Degenerate Primers for Microbiome Work

Metagenomic studies require an efficient PCR amplification of the DNA. This is currently achieved using "universal" forward and reverse primers that amplify appropriate regions of DNA from all targeted microbes. Jaric et al. (40) argue that currently used universal primers for the 16S rRNA region were designed many years ago, are not as efficient, and fail to bind to recently cataloged species. Their analysis shows that 22 of the most widely used primer pairs are far from optimal in the sense that they do not abide by primer design rules and fail to produce amplicons in a "virtual PCR" experiment, because they fail to hybridize to a large number of 16S sequences in the current database.

Jaric et al. (40) focus on the optimal design of degenerate primer pairs so that (i) the number of 16S sequences "virtually" amplified by the primer pair is maximized and (ii) the number of 16S sequences that produce distinguishable amplicons is maximized. Jaric et al. proposed an automated general method for designing PCR primer pairs that abide by primer design rules and uses current sequence databases as input. Since the method is automated, primers can be designed for targeted microbial species or updated as species are added or deleted from the database.

The designed primers were shown (in "virtual PCR" experiments) to achieve the goals mentioned earlier. They take the work one step further to design sets of primer pairs that extend the number of distinguishable amplicons for 16S sequences in the database. Wet lab experiments lend further credence to their claims on the effectiveness of the designed primer pairs.

18.4 DIVERSITY AND RICHNESS MEASURES

The early steps in an analytical pipeline produce a noisy snapshot; however, there are a number of preliminary measures that can subsequently be employed to gain some insight into the environment being studied, including making estimates about the richness and diversity of the community (51). *Richness* simply indicates the number of different OTUs present in a sample given that it is not possible to make an exact count. It assumes that not every OTU present has been directly observed. *Diversity*, like entropy, measures the evenness of the distribution of the abundance of the OTUs. If each OTU has a similar number of members, then the diversity of the community is relatively high, whereas if a few OTUs make up the bulk of all the cells present, then diversity is considered very low.

This, of course, tells you very little about how the members of these communities are interacting. Richness and diversity measures are most useful when compared between different communities, such as the communities housed by different human subjects at the same body site, or between different body sites in the same individual. This is where the coarsest patterns may begin to emerge. Consider samples collected from two groups of subjects, one of healthy controls and the other suffering from some illness. You observe that the number of distinct OTUs present in both groups is about the same, but that diversity is much higher in your healthy subjects. This suggests that a small number of OTUs in your diseased subjects have found an environment that favors them, and they are using up the bulk of the resources that are available at the expense of the other OTUs present. This would be a nice, simple pattern and you could claim to have found a biomarker for this disease. Unfortunately, in most studies to date, patterns like this have failed to emerge. There is simply too much variability between humans, and it is often found that richness and diversity tell you almost nothing about an individual's health status. (See Figure 18.1, where richness and diversity say little that is useful about the three groups being compared.) What has been observed is that an individual's diversity levels tend to be consistent across body sites (7). In other words, someone who has high diversity at body site A is likely to also have high diversity at body site B. In addition, these relative diversity levels seem to remain constant over time. It is not known if diversity levels change when an individual changes status, say from healthy to diseased. Although these broad standard measures merit further tracking, they are probably mainly useful as indicators of the complexity of the community being studied. It is generally more interesting to know which particular OTUs are present and how their abundance levels are changing at different stages of disease progression. This is the true starting point for constructing a network of interactions.

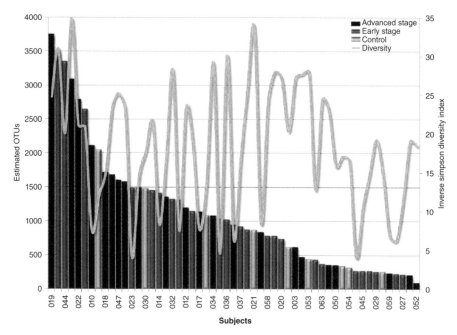

Figure 18.1 Plot of the standard measures of richness and diversity for a collection of samples. Each bar represents the richness (i.e., estimated number of OTUs present in the sample), and the line graph indicates the diversity of the sample. Bars are color-coded (grey shades) by clinical category and are arranged in descending order. However, no apparent pattern is discernible between the categories that are being compared, either in richness or in diversity (unpublished data). Please see www.wiley.com/go/Mandoiu/NextGenerationSequencing for a color version of this figure.

18.5 CORRELATIONS AND ASSOCIATION RULES

It is important to know which OTUs tend to occur together and which do not. More precisely, it is interesting to identify instances when high relative abundance levels for one OTU tend to coincide with high abundance levels for a second OTU. Alternatively, it is also interesting to identify when high relative abundance levels for one OTU coincide with low abundance levels for a second OTU. In addition, there could be other factors that might be considered along with OTU abundance levels. For biomedical studies, demographics of the subject from which samples were taken, or other known environmental conditions (such as acidity or aerobicity of the niche) can be included.

Finding these types of correlations may be the first step in deducing interactions between OTUs in a community. If two OTUs are positively correlated, this could suggest that they are not competing for the same resources or niche in the community and could even indicate some kind of dependency between them. For example, the waste products of one OTU may be modifying the pH levels in the environment, making it more agreeable to a second OTU. Likewise, if the change in pH is toxic to another OTU, we would observe a negative correlation between them. However, there are many alternative interpretations for correlations in the abundance of OTUs. Consider a situation in which three OTUs are competing for the same limited resource. If one OTU is especially efficient at utilizing this resource, its numbers would likely increase while the abundance of its competitors would decline. So there should be a negative correlation between this exceptionally competitive OTU and its two rivals. The abundance of the rivals, however, would tend to change concordantly, assuming they are equally fit. The presence and success of the efficient OTU would cause the number of members of the two less efficient OTUs to decline, concordantly, resulting in a positive correlation between them despite the fact that they are competitors for the same limited resource. Although it is not usually straightforward to understand the reason for the existence of a correlation, establishing the connections can help build a picture of what is going on in a community. We will explore ways of visualizing a network of correlations shortly.

Of greater interest than just pairs of correlated OTUs is to determine sets of OTUs that tend to co-occur, and whether the presence of one OTU is predictive of the presence of another for a subset of the subjects. For this, we can turn to statistical data mining tools for generating association rules, such as the Apriori algorithm for basket data (52). This works by first identifying specific sets (combinations) of bacterial taxa or other items that are frequently observed together in a large number of subjects, and then generating a rule for each item based on the other items in the set. The validity of the rule can be quantified for the given data by computing quantities such as *support*, *confidence*, and *lift* (53). Apriori can be used to essentially cluster OTUs by how often they occur together and suggest which OTUs we should be seeing based on "guilt-by-association." Although association rules are of interest, Apriori can also be used as a classifier to predict disease states. If certain item sets are frequently found in a group of diseased individuals but are absent from otherwise healthy individuals, the item sets can act as markers for the disease.

Several limitations hamper the use of the Apriori algorithm. Basically, it requires discrete values as inputs, and the data resulting from the metagenomics pipeline after normalization consists of relative abundance values for each OTU, which are continuous values. Numerous discretization methods are available to categorize abundance levels. One could discretize the values as "present" or "absent," or as "high," "medium," or "low," or into even finer categories, but these tend to flatten the data, and finding appropriate cutoffs for each category can be fairly arbitrary. There has been some effort toward methods that find frequent item sets and association rules without discretization, but relying on correlations (54). In addition, the number of item sets and association rules generated is normally very high, with many of the item sets showing intersections and similarities. Effective and compact visualizations of rules can be difficult (see Figure 18.2), although graph-based visualizations have great promise (see Section 18.7). Furthermore, determining statistical significance remains an open problem (55). Still, if properly wielded, the use of Apriori can lead to the discovery of interesting associations and interactions between bacterial taxa and help identify indicators of critical changes in human health.

18.6   MICROBIAL FUNCTIONAL PROFILES

A recent study has revealed that the gene content of the bacterial community is more constant than the phylogenetic content (56). Therefore, there is a compelling impetus to move beyond species composition-centric studies of microbial communities and toward functional composition analysis. As discussed earlier, a key preliminary step in metagenomic analysis is to decipher the microbial community structure of a given niche by categorizing the microbes residing therein and understanding their diversity. In the context of microbial communities, the term "species" refers to a fundamental and distinct rank of taxonomic hierarchy. There are limits to what taxonomic profiling can tell us about a community. Although interesting, community membership does little to describe the role being played by the taxa present. It is therefore much more important to investigate the functional profile of microbial communities. The *functional profile* of a sample can be defined as the set of expressed genes and active metabolic pathways in a community.

Consider a situation in which two species have very similar characteristics and functions, both of them being very efficient at synthesizing vitamin B12, for example. These species would be competitors in the same niche and their presence is likely to be mutually exclusive. Imagine doing a comparison of two communities, one in which species A is present in high abundance, and the other in which species B is present. Based on the taxonomic profile, it would seem that these two communities are different, but in fact since the species play the same role, all other things being equal, the communities are actually similar. There is evidence to support situations like the one described. The Human Microbiome Project has observed that although taxonomic membership is highly variable between human subjects at particular body sites, the functional profiles at those sites vary little (7).

Differences in functional profiles of the microbiota can be predictive of the health of a host. Knights et al. (57) reviewed supervised classification methods that exploit

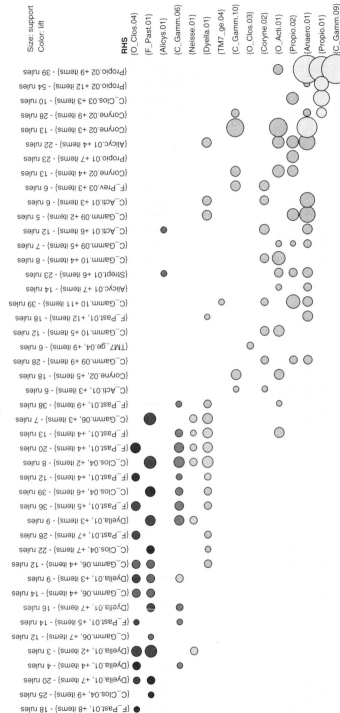

Figure 18.2 Example of a bubble plot for visualizing association rules. Columns consist of overlapping antecedents (item sets) leading to the consequents (the rules) in the rows. Bubble size indicates support and color the strength of the interest measure. Please see www.wiley.com/go/Mandoiu/NextGenerationSequencing for a color version of this figure.

411

this fact. It is therefore of paramount importance to understand the functional profiles of microbial communities in various sites and their dynamics. Can the dynamics of the profile shed light on the health of the host and will it also enlighten us on the etiology, progression, and prognosis?

As previously discussed, two approaches have been adopted for characterizing the diversity of metagenomes. The first approach focuses on the sequencing of phylogenetic marker genes, such as the gene for 16S rRNA (10–15). The second approach is based on whole genome sequencing. Whole genome sequencing is superior to 16S rRNA approaches primarily because it is able to directly characterize the functional profile of the community. Typically, sequences are compared to public databases and genes predicted by homology. Taxonomic identification can be helpful, but is not essential. The major limitation here is that functional annotations are still far from complete, and accurately ascribing functions to newly predicted gene sequences will take many years. Ontology databases will not catch up to the available data anytime soon.

As mentioned earlier, the most popular approaches continue to be those based on 16S rRNA (10–15). The functional analysis of 16S rRNA metagenomic data focuses on what "species" are present. In cases where the roles of the bacterial species are well understood, taxonomic identity can be an indicator of the more active functions in the community. Methods also exist for imputing functional profiles based on 16S data. These involve reconstruction of the ancestral state of taxa to predict the presence of gene families in descendants (58). Although these methods have been tested with good results, problems remain. The 16S approaches cannot resolve taxonomy below the genus level, and it is known that different species or strains in the same genus can behave very differently. The specific changes leading to these behaviors cannot be predicted from ancestry. Furthermore, taxonomic identification is limited by the completeness of marker gene databases. The number of sequences that cannot even be classified down to the genus level can be very high in some studies (M. Fernandez and G. Narasimhan. Unpublished results.). Additionally, when analyzing whole genome sequencing data, the problem of HGT needs to be considered. There are indications that the HGT rate can be as high as 8% in some bacterial communities (59). Although there is merit to attempts to create functional profiles based on taxonomy, these methods may be most useful for identifying environments, which would strongly benefit from whole genome sequencing.

Zhang et al. (60–63) have developed a novel approach to analyze the whole genome data of a microbiome. Their strategy is to compute a matrix, where the rows correspond to features such as genes, gene families, functional groups, or even pathways, and the columns correspond to OTUs. Thus, instead of mapping OTUs to bacterial taxa, the matrix maps them to *functional entities* that enable us to ascribe functional characteristics. The matrix can be seen as a collection of columns (OTUs with a functional profile) or as a collection of rows (Clusters of Orthologous Groups (COGs) with measures of their contributions to each OTU). COGs was used as functional entities to create an edge-weighted network of COGs where the edge weights are the correlations between the row vectors. It is possible to construct a single COG correlation network for each sample or set of samples. If dense subgraphs of this network are identified as clusters, then such a cluster corresponds to COGs that have a similar abundance profile across all OTUs. The authors extrapolated that these COGs have similar functions and are involved in the same pathways. Using GO annotations of these COGs, functional enrichment analysis was pursued. Given two networks, one for each of two samples or sets of samples, differential analysis can highlight the functional differences between them. Another possibility is to investigate which network motifs are conserved or changed between the samples. Differential modules can be mapped to KEGG pathways using iPath, an interactive pathway explorer (64). An example of such a mapped module is shown in Figure 18.3. As can be seen, the pathways of genes in this module include functional categories such as energy metabolism; glycan biosynthesis and metabolism; metabolism of terpenoids and polyketides; and metabolism of cofactors and vitamins. An added advantage of using functional entities is that it helps reduce the number of dimensions nearly 10-fold. OrthoMCL (65), a BLAST-driven method, can also be used to identify putative COGs.[1]

18.7 MICROBIAL SOCIAL INTERACTIONS AND VISUALIZATIONS

Bacterial communities are intricate collections of bacterial species that each provide functions which contribute to the stability of the community. In most natural environments, microbes do not live in isolation but form a complex ecological interaction web. Recent research shows that complex social behaviors are commonly observed not only in animals but also in bacterial species (67, 68). Those social behaviors involve complex systems of cooperation, communication, and synchronization. Thus, the communities are dynamic consortia of microbial species populations.

Because of the complexity of interactions and the sheer size of the communities being studied, efficient ways of analyzing and visually summarizing the analysis are needed. The use of network diagrams is appropriate for this purpose. These consist of nodes to represent each OTU, with edges connecting those nodes that have some kind of relationship. Nodes may have one, multiple, or no edges between them. An example of a complex network diagram is shown in Figure 18.4.

Here we see some of the more abundant OTUs found in the lungs of active smokers (M. Campos, M. Fernandez, A. Wanner, G. Holt, E. Donna, E. Mendes, M. Jaric, E. Silva-Herzog, L. Schneper, J. Segal, D. Moraga, J. D. Riveros, V. Aguiar-Pulido, M. Salathe, K. Mathee, and G. Narasimhan. unpublished data). Each OTU is labeled with the best taxon to which it maps. In the case of multiple OTUs mapping to the same taxon, an arbitrary number is appended to the label to distinguish one from the other. Edges between the nodes indicate correlations between the populations of the OTUs present. The size of the node is adjusted to indicate the size of the population. Correlations are a measure of co-occurrence of the OTUs in the subjects being considered. A green edge is used for a positive correlation, and a red edge is used for a negative correlation. The thickness of the edge increases with the strength of the correlation.

---

[1] Note that in this work, COGs refer to orthologous gene groups obtained directly from results of the computational tool OrthoMCL (65) on genes from the genomes of interest. It should be noted that these COGs are different from the NCBI COG database (66), which contains clusters from 66 unicellular organisms.

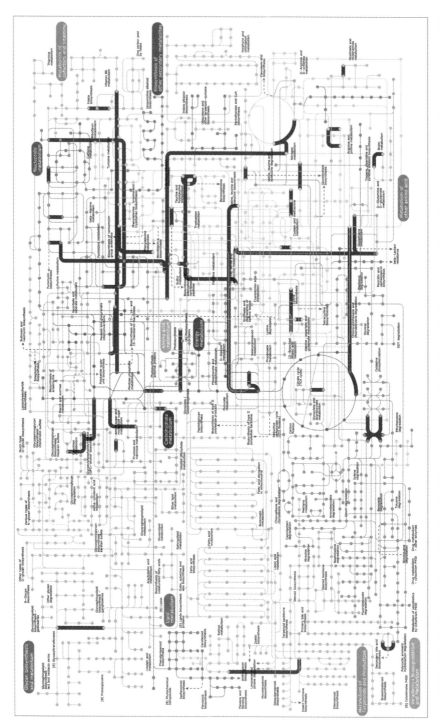

Figure 18.3 Metabolic map of a module identified from a gene network. Nodes symbolize compounds, and lines connecting nodes are enzymes. All enzymes (lines) corresponding to a single KEGG map have the same color. All enzymes (lines) corresponding to a single module are highlighted and colored with module color. Please see www.wiley.com/go/Mandoiu/NextGenerationSequencing for a color version of this figure.

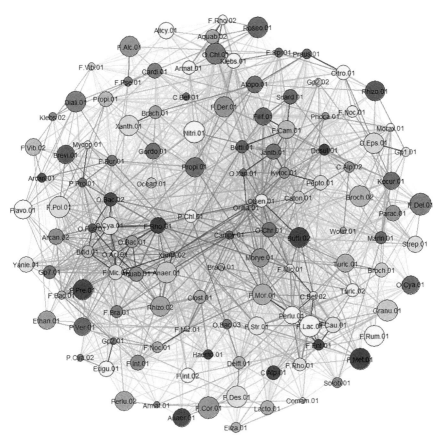

Figure 18.4 Basic network diagram where nodes represent OTUs and edges represent co-occurrence in subjects. Edge color indicates positive (green) or negative (red) correlations. Node size is adjusted to reflect relative abundance using a log scale. A force-directed layout using the Fruchterman–Reingold algorithm is used. The position of each node is dependent on the strength of its interactions with all other nodes in the system. A heatscale has been used to assign a color to each node based on differential abundance between two groups of subjects. The greater the significance in the difference, the hotter (redder) the color of the node. Please see www.wiley.com/go/Mandoiu/NextGenerationSequencing for a color version of this figure.

The size of each node is adjusted to express the relative abundance of the OTU. The nodes are colored according to a heat scale. The less significant the difference in abundance between two groups, the cooler (bluer) the color. The more significant the difference, the hotter (redder) the color. Finally, each OTU is positioned in space according to the Fruchterman–Reingold algorithm (69), which works as follows imagine that a spring exists between every pair of nodes; the strength of the springs varies depending on the size of the node and the strength of the correlation between them. Initially, each node is placed at an arbitrary position in space, and the overall energy of the system due to the pull of the springs is calculated. Two strongly (positively or negatively) correlated nodes will tend to attract each other, but there may be many other interactions acting to pull them apart. Springs that are "stretched" store energy. The positions of the nodes are iteratively readjusted according to these combinations of forces, with the overall goal of minimizing the total energy of the system. This iterative process stops when the total energy of the system remains unchanged, that is, a local minimum is reached.

The above-mentioned diagram contains a large amount of information in a very compact way. However, the density of diagrams like these requires careful study to properly interpret them. Some relationships can be easily picked out visually, while others are much less clear. As should be expected, two strongly positively (or negatively) correlated OTUs tend to be located in close proximity. However, the converse is not true. Two OTUs located in close proximity may have no detectable correlation. There is typically no clear delineation between groups of positively and negatively correlated OTUs.

The network diagrams shown earlier indicates relationships similar to those found in social networks (70). Cooperation and competition between bacteria has been well studied in the field of bacterial ecology. It is therefore natural to ask whether these network diagrams reveal interactions between bacterial taxa. In this context, it makes sense to ask if there are "clusters" in the network graphs and if these clusters are different for different groups of subjects.

We informally define a *club* as a cluster of bacterial taxa such that every member of the group has a stronger correlation with the rest of the group than it does with members not in the group. The concept of clustering has been well studied in the statistics and computing literature and has been defined and computed in a multitude of ways (71). Different clustering methods have used different distance and similarity measures or modeled the data using statistical distributions. Others have defined different objective functions using density and tightness measures, variance ratios, weakest links, relative margins, and other statistical measures, resulting in a wide range of definitions of clusters and algorithms for clustering.

Clubs with positive correlations between each other are likely to be indications of "cooperation" between the members of the group, while negative correlations are likely to be evidence of "competition." The former group of interacting OTUs may represent taxa that complement each other in a given environment, and their composition could indicate a core group of functions needed to thrive.

A substantially more interesting structure to be found in the network diagrams turns out to be "competing groups" of bacterial taxa. We informally define *rival clubs* to be a pair of clubs such that every member of one club has a negative correlation with some or all of the members of the "rival" group. As with the definition of a club, this concept can be formalized and computed in a variety of different ways. Rival clubs are likely to indicate "competing" groups requiring the same scarce resources in a niche. In the example shown in Figure 18.5, Club-A, Club-B, and Club-C form a trio of rival clubs (72).

Interpreting a cluster of this kind raises many unanswered questions. What can be inferred about a cluster of OTUs? Do these OTUs compete for different nutrients

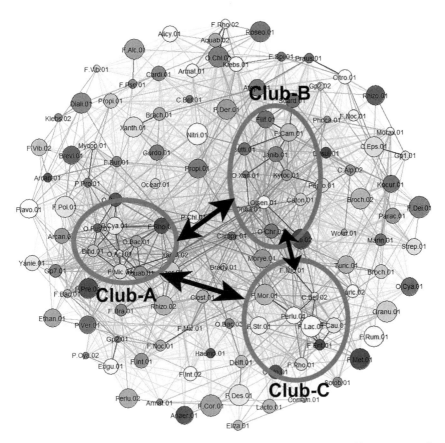

Figure 18.5 Once clubs have been identified, rival clubs are characterized by many negative edges between them. Please see www.wiley.com/go/Mandoiu/NextGeneration Sequencing for a color version of this figure.

in the environment? Are they critical to each other? Does the presence of one help to recruit others to the same environment? What is the effect of the elimination of one of the members of a cluster? Can anything be said about their pathogenicity? Are they all pathogenic or all non-pathogenic? Or can a mix occur? What functional genes do their genomes represent? Do they collectively represent a useful functional profile that helps them to thrive? Are there differences in gene expression? Do they collectively represent a useful gene expression profile? How do they communicate, and what chemicals and metabolites are present in the niche? What can be said about the level of HGT within the members of a cluster? Are there quorum sensing genes or other communication genes strongly expressed within a cluster? Is a cluster meaningful in posting a response to environmental stimulus or stress, such as an antibiotic?

There are some limitations to these network diagrams herein. We are clearly extrapolating interactions based on information on co-occurrence. While positive interactions do suggest co-occurrence, the converse need not be true.

18.8 BAYESIAN INFERENCES

Although the correlation networks described earlier can produce meaningful insights, there are other network structures that can be learned from data which can be helpful in answering interesting questions about the communities being studied. The use of probabilistic graphical models (PGMs) for analyzing microbial communities is still uncommon, but they can serve as useful complements by helping to infer dependencies between OTUs and to possibly identify causal reasons for disease states. A PGM is a directed acyclic graph in which each node represents a variable of interest and each directed arc indicates a conditional dependence relationship between nodes (73). A probability distribution is associated with each node, and through the application of Bayes' rule, predictions can be made about the presence of a node given information about other connected nodes. As a simple example, consider the network represented in Figure 18.6.

Here we see that the probability of the lawn being wet depends only on the probability of it having rained. In contrast, the probability of the neighbor's lawn being wet depends on the probability of rain as well as the probability of the neighbor turning on the sprinkler system. Given the condition that it has rained, the probability of the lawn being wet changes and can be easily calculated. However, given that the neighbor turned on the sprinkler system gives no additional information about whether or not the lawn is wet. In essence, to determine the state of the lawn, the relevant node is only R since the state of M does not depend on any of the other nodes in the network. In a large network, a structure like this dramatically reduces the number of calculations required for determining the probability of the state of a node of interest.

Figure 18.7 gives an example of a more complex PGM constructed from metagenomic data. In this case, most nodes represent OTUs and also included are nodes that represent disease states, host demographics, and other aspects of host phenotype. Once a model is constructed, it can be used to predict the likelihood of a subject developing the disease when any combination of dependent nodes is observed. According to Bayesian theories, certain substructures in these networks can also be used to infer causation.

The use of force-directed layouts in constructing the graphical representations provides similar insights as in the correlation networks mentioned earlier. However,

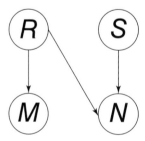

Figure 18.6   The likelihood of event M ("lawn is wet") depends only on the probability of event R ("it rained"). In contrast, the likelihood of event N ("neighbor's lawn is wet") depends on event R and the probability of the event S ("neighbor turned on the sprinkler").

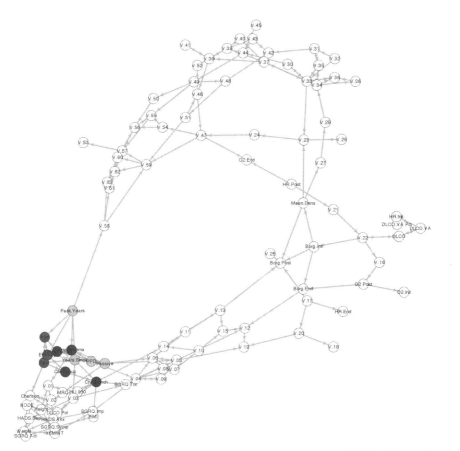

Figure 18.7 A complex PGM. Please see www.wiley.com/go/Mandoiu/NextGeneration Sequencing for a color version of this figure.

learning the structure of the network can be much more computationally challenging. Even a modestly-sized network constructed using Markov Chain Monte Carlo sampling will normally require parallelization on a high-performance computing cluster. The joint probability distribution may include prior knowledge about relationships or can be learned on the fly. Although the use of priors will not affect the structure of the network, having sound probabilities is essential for any inferences being made.

18.9 CONCLUSION

When we talk about the structure of a microbial community, we refer to all of the elements present and the connections between them: a network. Too few microbiome studies attempt to analyze their subject from this perspective. Those that do, tend to focus on taxa, or sometimes the functional elements present, without much regard to other possible nodes. To be sure, a great deal of information can be inferred about these networks by measuring changes in the abundance of either taxa or genes in different environments. These noisy snapshots can clue us in many of the interactions that are occurring. Network diagrams are powerful tools for examining the complex relationships in microbial communities. A large number of connections can be studied at once, and they allow for the discovery of unexpected associations between OTUs, both cooperative and competitive. Yet this is still an incomplete picture. There are too many elements that are not captured by most current methods, and in many cases it is likely that we lack sufficient information to correctly deduce the direction of individual interactions or the topology of the networks being studied. More effort needs to be focused on ways of discovering the structure of microbial networks.

Studying these communities in humans is particularly challenging. We all have different experiences, eat different foods, live in different places, and participate in different social networks. All of these differences contribute to the high variability in the environment that our bodies provide to our bacterial residents. Yet some markers may exist that can distinguish between the microbial communities present at different body sites. Our intuition tells us that there should be a difference between the community residing in healthy tissue versus diseased tissue. Something must be different about the network of bacteria in the lungs of someone with emphysema as compared to someone with no respiratory problems at all. Is there a signature that we have yet to discover? If there is validity to the concept of an enterotype, a classification of individuals based on the bacteria present in their gut, is there also a *"respirotype"*? A *"dermatype"*? If so, it is likely not sufficient to simply define these types based on absence or presence of specific bacteria. What role does host genotype play? Age? Life history? It is the interactions between functional elements, taxon-specific characteristics, host tissue properties, the presence of metabolites, toxins, sensing molecules, and exogenous influences that will produce distinct signatures between healthy and diseased environments. Discovery of these highly heterogeneous subnetworks will be challenging, but their value for rapidly assessing state of health, predicting disease progression, and developing intervention strategies cannot be understated.

REFERENCES

1. Costello EK, Lauber CL, Hamady M, Fierer N, Gordon JI, Knight R. Bacterial community variation in human body habitats across space and time. Science 2009;326(5960):1694–1697.

2. Peterson J et al. The NIH human microbiome project. Genome Res 2009;19(12):2317–2323.

3. Turnbaugh PJ, Ley RE, Hamady M, Fraser-Liggett CM, Knight R, Gordon JI. The human microbiome project. Nature 2007;449(7164):804–810.

4. Turnbaugh PJ, Gordon JI. The core gut microbiome, energy balance and obesity. J. Physiol. (London) 2009;587(Pt 17):4153–4158.

5. Marrazzo JM, Martin DH, Watts DH, Schulte J, Sobel JD, Hillier SL, Deal C, Fredricks DN. Bacterial vaginosis: identifying research gaps proceedings of a workshop sponsored by DHHS/NIH/NIAID. Sex Transm Dis 2010;37(12):732–744.

6. Qin J et al. A human gut microbial gene catalogue established by metagenomic sequencing. Nature 2010;464(7285):59–65.

7. Huttenhower C et al. Structure, function and diversity of the healthy human microbiome. Nature 2012;486(7402):207–214.

8. Sibley CD, Grinwis ME, Field TR, Eshaghurshan CS, Faria MM, Dowd SE, Parkins MD, Rabin HR, Surette MG. Culture enriched molecular profiling of the cystic fibrosis airway microbiome. PLoS ONE 2011;6(7):e22702.

9. Doud MS, Light M, Gonzalez G, Narasimhan G, Mathee K. Combination of 16S rRNA variable regions provides a detailed analysis of bacterial community dynamics in the lungs of cystic fibrosis patients. Hum Genomics 2010;4(3):147–169.

10. Chaffron S, Rehrauer H, Pernthaler J, von Mering C. A global network of coexisting microbes from environmental and whole-genome sequence data. Genome Res 2010;20(7):947–959.

11. Gonzalez A, Knight R. Advancing analytical algorithms and pipelines for billions of microbial sequences. Curr Opin Biotechnol 2012;23(1):64–71.

12. Barberan A, Bates ST, Casamayor EO, Fierer N. Using network analysis to explore co-occurrence patterns in soil microbial communities. ISME J 2011;6:343–351.

13. Freilich S, Kreimer A, Meilijson I, Gophna U, Sharan R, Ruppin E. The large-scale organization of the bacterial network of ecological co-occurrence interactions. Nucleic Acids Res 2010;38(12):3857–3868.

14. Kuczynski J, Liu Z, Lozupone C, McDonald D, Fierer N, Knight R. Microbial community resemblance methods differ in their ability to detect biologically relevant patterns. Nat Methods 2010;7(10):813–819.

15. Faust K, Sathirapongsasuti JF, Izard J, Segata N, Gevers D, Raes J, Huttenhower C. Microbial co-occurrence relationships in the human microbiome. PLoS Comput Biol 2012;8(7):e1002606.

16. Sogin SJ, Sogin ML, Woese CR. Phylogenetic measurement in procaryotes by primary structural characterization. J Mol Evol 1972;1(2):173–184.

17. Lane DJ. 16S/23S rRNA sequencing. In: Stackebrandt E, Goodfellow M, editors. *Nucleic Acid Techniques in Bacterial Systematics*. New York: John Wiley and Sons, Inc.; 1991. p 125–175.

18. Zeigler DR. Gene sequences useful for predicting relatedness of whole genomes in bacteria. Int J Syst Evol Microbiol 2003;53(6):1893–1900.

19. Konstantinidis KT, Tiedje JM. Towards a genome-based taxonomy for prokaryotes. J Bacteriol 2005;187(18):6258–6264.

20. Nelson KE et al. A catalog of reference genomes from the human microbiome. Science 2010;328(5981):994–999.

21. Chain PS et al. Genomics. Genome project standards in a new era of sequencing. Science 2009;326(5950):236–237.

22. Cole JR, Wang Q, Fish JA, Chai B, McGarrell DM, Sun Y, Brown CT, Porras-Alfaro A, Kuske CR, Tiedje JM. Ribosomal Database Project: data and tools for high throughput rRNA analysis. Nucleic Acids Res 2014;42(Database issue):D633–D642.

23. Pruesse E, Quast C, Knittel K, Fuchs BM, Ludwig W, Peplies J, Glockner FO. SILVA: a comprehensive online resource for quality checked and aligned ribosomal RNA sequence data compatible with ARB. Nucleic Acids Res 2007;35(21):7188–7196.

24. DeSantis TZ, Hugenholtz P, Larsen N, Rojas M, Brodie EL, Keller K, Huber T, Dalevi D, Hu P, Andersen GL. Greengenes, a chimera-checked 16S rRNA gene database and workbench compatible with ARB. Appl Environ Microbiol 2006;72(7):5069–5072.

25. Eckburg PB, Bik EM, Bernstein CN, Purdom E, Dethlefsen L, Sargent M, Gill SR, Nelson KE, Relman DA. Diversity of the human intestinal microbial flora. Science 2005;308(5728):1635–1638.

26. McKenna P, Hoffmann C, Minkah N, Aye PP, Lackner A, Liu Z, Lozupone CA, Hamady M, Knight R, Bushman FD. The macaque gut microbiome in health, lentiviral infection, and chronic enterocolitis. PLoS Pathog 2008;4(2):e20.

27. Shannon CE, Shannon CE. The mathematical theory of communication. 1963. MD Comput 1997;14(4):306–317.

28. Jaccard P. The distribution of the flora in the alpine zone. New Phytol 1912;11(2):37–50.

29. Simpson EH. Measurement of diversity. Nature 1949;163(4148):688.

30. Erb-Downward JR et al. Analysis of the lung microbiome in the "healthy" smoker and in COPD. PLoS ONE 2011;6(2):e16384.

31. Morris A et al. Comparison of the respiratory microbiome in healthy nonsmokers and smokers. Am J Respir Crit Care Med 2013;187(10):1067–1075.

32. Woese CR, Gutell R, Gupta R, Noller HF. Detailed analysis of the higher-order structure of 16S-like ribosomal ribonucleic acids. Microbiol Rev 1983;47(4):621–669.

33. Fox GE, Wisotzkey JD, Jurtshuk P, Fox GE. How close is close: 16S rRNA sequence identity may not be sufficient to guarantee species identity. Int J Syst Bacteriol 1992;42(1):166–170.

34. Zaneveld JR, Lozupone C, Gordon JI, Knight R. Ribosomal RNA diversity predicts genome diversity in gut bacteria and their relatives. Nucleic Acids Res 2010;38(12):3869–3879.

35. Hibbing ME, Fuqua C, Parsek MR, Peterson SB. Bacterial competition: surviving and thriving in the microbial jungle. Nat Rev Microbiol 2010;8(1):15–25.

36. Fernandez M. An analytical workflow for metagenomic data and its application to the study of chronic obstructive pulmonary disease [Undergraduate Thesis]. Florida International University Undergraduate Honors Thesis; 2013.

37. Wei X, Kuhn DN, Narasimhan G. Degenerate primer design via clustering. CSB. IEEE Computer Society; 2003. p 75–83.

38. Weisburg WG, Barns SM, Pelletier DA, Lane DJ. 16s ribosomal DNA amplification for phylogenetic study. J Bacteriol 1991;173(2):697–703.

39. Turner S, Pryer KM, Miao VPW, Palmer JD. Investigating deep phylogenetic relationships among cyanobacteria and plastids by small subunit rRNA sequence analysis1. J Eukaryot Microbiol 1999;46(4):327–338.

40. Jaric M, Segal J, Silva-Herzog E, Schneper L, Mathee K, Narasimhan G. Better primer design for metagenomics applications by increasing taxonomic distinguishability. BMC Proc 2013;7 Suppl 7:S4.

41. Crosby LD, Criddle CS. Understanding bias in microbial community analysis techniques due to rrn operon copy number heterogeneity. BioTechniques 2003;34(4):790–794.

42. McHardy AC, Rigoutsos I. What's in the mix: phylogenetic classification of metagenome sequence samples. Curr Opin Microbiol 2007;10(5):499–503.

43. Meyer F et al. The metagenomics rast server–a public resource for the automatic phylogenetic and functional analysis of metagenomes. BMC Bioinformatics 2008;9(1):386.

44. Schloss PD et al Introducing mothur: open-source, platform-independent, community-supported software for describing and comparing microbial communities. Appl Environ Microbiol 2009;75(23):7537–7541.

45. Caporaso JGregory et al. QIIME allows analysis of high-throughput community sequencing data. Nat Methods 2010;7(5):335–336.

46. Arumugam M, Harrington ED, Foerstner KU, Raes J, Bork P. SmashCommunity: a metagenomic annotation and analysis tool. Bioinformatics 2010;26(23):2977–2978.

47. Angiuoli SV, Matalka M, Gussman A, Galens K, Vangala M, Riley DR, Arze C, White JR, White O, Fricke WF. CloVR: a virtual machine for automated and portable sequence analysis from the desktop using cloud computing. BMC Bioinformatics 2011;12(1):356.

48. Huse SM, Welch DBM, Voorhis A, Shipunova A, Morrison HG, Eren AM, Sogin ML. VAMPS: a website for visualization and analysis of microbial population structures. BMC Bioinformatics 2014;15(1):41.

49. Paulson JN, Stine OC, Bravo HC, Pop M. Differential abundance analysis for microbial marker-gene surveys. Nat Methods 2013;10(12):1200–1202.

50. Costea PI, Zeller G, Sunagawa S, Bork P. A fair comparison. Nat Methods 2014;11(4):359.

51. Doud M, Zeng E, Schneper L, Narasimhan G, Mathee K. Approaches to analyse dynamic microbial communities such as those seen in cystic fibrosis lung. Hum Genomics 2009;3(3):246–256.

52. Agrawal R, Srikant R. Fast algorithms for mining association rules in large databases. In: *Proceedings of the 20th International Conference on Very Large Data Bases, VLDB '94.* San Francisco (CA): Morgan Kaufmann Publishers Inc.; 1994. p 487–499.

53. Lenca P, Meyer P, Vaillant B, Lallich S. On selecting interestingness measures for association rules: user oriented description and multiple criteria decision aid. Eur J Oper Res 2008;184(2):610–626.

54. Brin S, Motwani R, Silverstein C. Beyond market baskets: generalizing association rules to correlations. In: *Proceedings of the 1997 ACM SIGMOD International Conference on Management of Data, SIGMOD '97.* New York: ACM; 1997. p 265–276.

55. Faust K, Raes J. Microbial interactions: from networks to models. Nat Rev Microbiol 2012;10(8):538–550.

56. Burke C, Steinberg P, Rusch D, Kjelleberg S, Thomas T. Bacterial community assembly based on functional genes rather than species. Proc Natl Acad Sci U S A 2011;108(34):14288–14293.

57. Knights D, Costello EK, Knight R. Supervised classification of human microbiota. Microbiol Rev 2010;35(2):343–359.

58. Langille MG et al. Predictive functional profiling of microbial communities using 16S rRNA marker gene sequences. Nat Biotechnol 2013;31(9):814–821.

59. Tamames J, Moya A. Estimating the extent of horizontal gene transfer in metagenomic sequences. BMC Genomics 2008;9:136.

60. Zhang W, Emrich SJ, Zeng E. A two-stage machine learning approach for pathway analysis. Proceedings of IEEE International Conference on Bioinformatics and Biomedicine, BIBM 2010; 2010. p 274–279.

61. Zhang W, Zeng E, Liu D, Jones S, Emrich SJ. A machine learning framework for trait based genomics. Proceedings of IEEE 2nd International Conference on Computational Advances in Bio and Medical Sciences, ICCABS 2012; 2012. p 1–6.

62. Zhang W, Zeng E, Liu D, Jones S, Emrich SJ. Mapping genomic features to functional traits through microbial whole genome sequences. Int J Bioinf Res Appl 2014;10(4-5):461–478.

63. Zhang W, Zeng E, Emrich SJ, Livermore J, Liu D, Jones SE. Predicting bacterial functional traits from whole genome sequences using random forest. Computational Advances in Bio and Medical Sciences (ICCABS), 2013 IEEE 3rd International Conference on; 2013. p 1–2.

64. Yamada T, Letunic I, Okuda S, Kanehisa M, Bork P. iPath2.0: interactive pathway explorer. Nucleic Acids Res 2011;39(Web Server issue):W412–W415.

65. Li L, Stoeckert CJ Jr., Roos DS. OrthoMCL: identification of ortholog groups for eukaryotic genomes. Genome Res 2003;13:2178–2189.

66. Tatusov RL, Koonin EV, Lipman DJ. A genomic perspective on protein families. Science 1997;278(5338):631–637.

67. Allen EE, Banfield JF. Community genomics in microbial ecology and evolution. Nat Rev Microbiol 2005;3(6):489–498.

68. West SA, Diggle SP, Buckling A, Gardner A, Griffin AS. The social lives of microbes. Annu Rev Ecol Evol Syst 2007;38(1):53–77.

69. Fruchterman TMJ, Reingold EM. Graph drawing by force?directed placement. Softw Pract Exp 1991;21(11):1129–1164.

70. Haythornthwaite C. Social network analysis: an approach and technique for the study of information exchange. Libr Inf Sci Res 1996;18(4):323–342.

71. Zaki MJ, Meira W Jr. *Data Mining and Analysis: Fundamental Concepts and Algorithms*. New York: Cambridge University Press; 2014.

72. Fernandez M, Riveros JD, Campos M, Mathee K, Narasimhan G. Microbial "social networks". BMC Genomics 2015;16 Suppl 11:S6.

73. Koller D, Friedman N. *Probabilistic Graphical Models: Principles and Techniques - Adaptive Computation and Machine Learning*. Cambridge (MA): The MIT Press; 2009.

INDEX

Computational Methods for Next Generation Sequencing Data Analysis, First Edition.
Edited by Ion I. Măndoiu and Alexander Zelikovsky.
© 2016 John Wiley & Sons, Inc. Published 2016 by John Wiley & Sons, Inc.
Companion website: www.wiley.com/go/Mandoiu/NextGenerationSequencing

Wiley Series on

Bioinformatics: Computational Techniques and Engineering

Bioinformatics and computational biology involve the comprehensive application of mathematics, statistics, science, and computer science to the understanding of living systems. Research and development in these areas require cooperation among specialists from the fields of biology, computer science, mathematics, statistics, physics, and related sciences. The objective of this book series is to provide timely treatments of the different aspects of bioinformatics spanning theory, new and established techniques, technologies and tools, and application domains. This series emphasizes algorithmic, mathematical, statistical, and computational methods that are central in bioinformatics and computational biology.

Series Editors: **Professor Yi Pan** and **Professor Albert Y. Zomaya**
pan@cs.gsu.edu   albert.zomaya@sydney.edu.au